Advances in Spatial Science

Editorial Board

Manfred M. Fischer
Geoffrey J.D. Hewings
Anna Nagurney
Peter Nijkamp
Folke Snickars (Coordinating Editor)

For further volumes:
http://www.springer.com/series/3302

Francesca Pagliara • Michiel de Bok •
David Simmonds • Alan Wilson
Editors

Employment Location in Cities and Regions

Models and Applications

Editors
Francesca Pagliara
Dept. of Transportation Engineering
University of Naples Federico II
Naples, Italy

David Simmonds
David Simmonds Consultancy
Cambridge, United Kingdom

Michiel de Bok
Significance
The Hague, Zuid-Holland
Netherlands

Alan Wilson
Centre for Advanced Spatial Analysis
University College London
London, United Kingdom

ISSN 1430-9602
ISBN 978-3-642-31778-1   ISBN 978-3-642-31779-8 (eBook)
DOI 10.1007/978-3-642-31779-8
Springer Heidelberg New York Dordrecht London

Library of Congress Control Number: 2012950599

© Springer-Verlag Berlin Heidelberg 2013
This work is subject to copyright. All rights are reserved by the Publisher, whether the whole or part of the material is concerned, specifically the rights of translation, reprinting, reuse of illustrations, recitation, broadcasting, reproduction on microfilms or in any other physical way, and transmission or information storage and retrieval, electronic adaptation, computer software, or by similar or dissimilar methodology now known or hereafter developed. Exempted from this legal reservation are brief excerpts in connection with reviews or scholarly analysis or material supplied specifically for the purpose of being entered and executed on a computer system, for exclusive use by the purchaser of the work. Duplication of this publication or parts thereof is permitted only under the provisions of the Copyright Law of the Publisher's location, in its current version, and permission for use must always be obtained from Springer. Permissions for use may be obtained through RightsLink at the Copyright Clearance Center. Violations are liable to prosecution under the respective Copyright Law.
The use of general descriptive names, registered names, trademarks, service marks, etc. in this publication does not imply, even in the absence of a specific statement, that such names are exempt from the relevant protective laws and regulations and therefore free for general use.
While the advice and information in this book are believed to be true and accurate at the date of publication, neither the authors nor the editors nor the publisher can accept any legal responsibility for any errors or omissions that may be made. The publisher makes no warranty, express or implied, with respect to the material contained herein.

Printed on acid-free paper

Springer is part of Springer Science+Business Media (www.springer.com)

Contents

1 **Employment Location Models: An Overview** 1
   Alan Wilson and Francesca Pagliara

Part I Macro-scale Approaches

2 **Employment and Labour in Urban Markets in the IRPUD Model** 11
   Michael Wegener

3 **Modelling the Economic Impacts of Transport Changes: Experience and Issues** 33
   David Simmonds and Olga Feldman

4 **A Population-Employment Interaction Model as Labour Module in TIGRIS XL** 57
   Thomas de Graaff and Barry Zondag

5 **Simulating the Spatial Distribution of Employment in Large Cities: With Applications to Greater London** 79
   Duncan A. Smith, Camilo Vargas-Ruiz, and Michael Batty

6 **Complex Urban Systems Integration: The LEAM Experiences in Coupling Economic, Land Use, and Transportation Models in Chicago, IL** 107
   Brian Deal, Jae Hong Kim, Geoffrey J.D. Hewings, and Yong Wook Kim

7 **Employment Location Modelling Within an Integrated Land Use and Transport Framework: Taking Cue from Policy Perspectives** 133
   Ying Jin and Marcial Echenique

8	**Integrating SCGE and I-O in Multiregional Modelling**	159
	Christer Anderstig and Marcus Sundberg	
9	**Interjurisdictional Competition and Land Development: A Micro-Level Analysis**	181
	Jae Hong Kim and Geoffrey J.D. Hewings	

Part II Micro-scale Approaches

10	**Occupation, Education and Social Inequalities: A Case Study Linking Survey Data Sources to an Urban Microsimulation Analysis**	203
	Paul Lambert and Mark Birkin	
11	**Firm Location Choice Versus Job Location Choice in Microscopic Simulation Models**	223
	Rolf Moeckel	
12	**Modelling Firm Failure: Towards Building a Firmographic Microsimulation Model**	243
	H. Maoh and P. Kanaroglou	
13	**Choice Set Formation in Microscopic Firm Location Models**	263
	M. de Bok and F. Pagliara	
14	**Employment Location Models: Conclusions**	283
	D. Simmonds and M. de Bok	

About the Editors 293

Contributors

Christer Anderstig WSP Analysis & Strategy, Stockholm-Globen, Sweden

Michael Batty Centre for Advanced Spatial Analysis (CASA), University College London (UCL), London, UK

Mark Birkin School of Geography, University of Leeds, Leeds, UK

M. de Bok Significance, The Hague, The Netherlands; Significance, Den Haag, The Netherlands

Brian Deal Department of Urban and Regional Planning, University of Illinois, Champaign, IL, USA

Marcial Echenique The Martin Centre for Architectural and Urban Studies, University of Cambridge, Cambridge, UK

Olga Feldman Transport for London, Strategic Analysis, London, UK

Thomas de Graaff PBL Netherlands Environmental Assessment Agency, The Hague, The Netherlands; VU University Amsterdam, The Netherlands

Geoffrey J.D. Hewings Regional Economics Applications Laboratory, University of Illinois, Urbana-Champaign, IL, USA

Ying Jin The Martin Centre for Architectural and Urban Studies, University of Cambridge, Cambridge, UK

P. Kanaroglou School of Geography and Earth Sciences, McMaster University, Hamilton, ON, Canada

Jae Hong Kim Department of Planning, Policy, and Design, University of California, Irvine, CA, USA

Paul Lambert School of Applied Social Science, University of Stirling, Stirling, UK

H. Maoh Department of Civil and Environmental Engineering, University of Windsor, Windsor, ON, Canada

Rolf Moeckel Parsons Brinckerhoff, Albuquerque, NM, USA

Francesca Pagliara Department of Transportation Engineering, University of Naples Federico II, Naples, Italy

David Simmonds David Simmonds Consultancy Ltd, Cambridge, UK; Honorary Professor, Heriot-Watt University, Edinburgh, Scotland

Duncan A Smith Centre for Advanced Spatial Analysis (CASA), University College London (UCL), London, UK

Marcus Sundberg Division of Transport and Location Analysis (TLA), KTH Royal Institute of Technology, Stocholm, Sweden

Camilo Vargas-Ruiz Centre for Advanced Spatial Analysis (CASA), University College London (UCL), London, UK

Michael Wegener Spiekermann & Wegener Urban and Regional Research (S&W), Dortmund, Germany

Alan Wilson Centre for Advanced Spatial Analysis, UCL, London, UK

Barry Zondag PBL Netherlands Environmental Assessment Agency, The Hague, The Netherlands

Chapter 1
Employment Location Models: An Overview

Alan Wilson and Francesca Pagliara

Abstract In this chapter the state-of-the-art of the modelling of the spatial distribution of employment in cities and regions is presented. Different economic approaches are described such as input–output analysis; Computable General Equilibrium models and utility maximising behaviour, translated into logit functions as probabilities. The models presented here are rich in content and build on decades of evolution.

1.1 Introduction

The modelling of the spatial distribution of employment in cities and regions is both an important subject in its own right and a crucial element of the task of building comprehensive models. The chapters of the book represent the state of the art. The fact that so many and diverse approaches are represented is a measure of the relative difficulty of the challenge. Inevitably, approximations have to be made in both theory building and model building and these different approaches in part represent alternative ways of doing this; and in part, they represent different approaches to the underlying theory and model building methods. The book is divided into two parts: the first demonstrates employment location models that have been developed within comprehensive models; the second, a range of special topics, each focussing on one element of the jigsaw puzzle that has to be assembled.

A. Wilson (✉)
Centre for Advanced Spatial Analysis, UCL, Tottenham Court Road 90,
W1T 4TJ London, UK
e-mail: a.g.wilson@ucl.ac.uk

F. Pagliara
Department of Transportation Engineering, University of Naples Federico II, Via Claudio 21,
80125 Naples, Italy
e-mail: fpagliar@unina.it

Employment is a measure of economic activity and there is a core from economic theory in all the chapters that follow. However, different economic approaches predominate in different cases. Broadly speaking, we can distinguish

- Roots in input–output analysis;
- CGE (computable general equilibrium) models;
- Roots in utility and profit maximising behaviour, usually translated into logit functions as probabilities;
- These relate to markets the outcomes of which are often expressed as spatial interaction models in either logit or entropy-maximising terms;
- The deployment of these kinds of probabilities in microsimulation models;
- Some other shifts to micro-analysis in terms of cellular and/or agent-based models;
- A key challenge is the identification of the drivers of economic development, some of these being take as exogenous in a scenario building context (often related to national policy), some from time series analysis, and some largely endogenous within an input–output framework.

These categories are not mutually exclusive, of course. For example, input–output models can be used to provide totals that constrain spatial allocation models. In the rest of this chapter, we first sketch the background in relation to the history of this model building challenge and then we outline the nature of the contributions of individual chapters against both this history and the framework sketched above.

1.2 The History as Context

Employment location choice has played a central role ever since the start of applied urban modelling. The Lowry model (1964) forecasts the location of working residents based on that of employment required by basic industries, and then predicts the location of the service industry employment based on the consumer demand of the residents. Echenique et al. (1969) extends this model to cover physical constraints upon location arising from buildings, town planning and the transport network, and to predict the distribution of employment and population within a city according to the functional relationships among urban activities subject to those constraints.

The early models used analogies to Newton's gravity law to describe the location behaviour of businesses as well as households. The modelling of location choice then adopted spatial interaction models based on entropy maximisation (Wilson 1970). Over the years the modelling approach became progressively grounded in microeconomic theory in their representation of production, consumption, trade, location choice and supply of land, buildings and transport services, whilst retaining its strength in incorporating the planning constraints. By 1975 an integrated land use and transport modelling framework has taken shape. The MEPLAN software package was developed in the early 1980s at ME&P (Echenique 2004).

Close integration of the MEPLAN modelling with Input–output Tables (Leontief 1986), random utility theory for location choice (McFadden 1974) and transport demand modelling (Ben-Akiva and Lerman 1985; Cascetta 2009) have made them effective and practical tools for exploring and assessing urban land use and transport development options at the geographic scales suitable for cost-benefit and wider impact analyses of policy initiatives.

The three models TRANUS (de la Barra 1989), MUSSA (Martínez 1996) and PECAS (Hunt and Abraham 2003) follow the MEPLAN methodology, with each further developing certain aspects of employment allocation.

The IRPUD model has been developed by Wegener (1982) and is described in Chap. 2. This was one of the very first employment location models that uses logit models based on Domencich's and McFadden's work (1975). The DELTA model developed by Simmonds (2001) is described in more detail in Chap. 3, and TIGRIS XL (Zondag et al. 2005) is described in Chap. 4.

Finding the relation between sectoral employment and labour force population and subsequently modelling the (re)distribution across regions is part of a large and international regional and urban economic literature that deals with the dynamics interaction between population and employment; hence, the name population-employment interaction models. The focus of this literature deals with the research question whether employment or population are endogeneous to each other. Note that traditional regional and urban economic model, such as that of the monocentric city (Mills 1967; Muth 1969), impose such a relation, e.g. population is endogeneous to the location of employment. Empirical studies that have tested these relations, however, found that these assumptions are hard to maintain and, if one should a pose a relation, employment is probably endogeneous to population location (Steinnes 1977; Carlino and Mills 1987; Boarnet 1994).

Transportation engineers have long known the benefits of simulation modeling, developing the traditional four-step process to produce travel forecasts as early as the 1950s. They have also recognized that land use and transportation systems are not mutually exclusive. The four-step process for example, relies on future socio-economic and land use projections as the basis for determining potential 'zonal' demand. But until recently, this relationship between land use and transportation systems has been described in static terms – i.e., a static long term land use forecast is used to drive a dynamic transportation model, or a static transportation system is used to drive a dynamic land use model. The linking of these dynamic systems has been relatively cumbersome, costly and difficult to achieve – although we will argue, critical in understanding the dynamics of urban systems. When the mutual influence of land use and travel demand are effectively captured and conveyed, the resultant information is more reliable, believable, and defensible. These coupled simulations can better inform choices made with regard to both transportation and land use policies and investments.

Waddell (2002) and Waddell et al. (2003) describe how UrbanSim, a land-use model, was integrated with a four-step travel demand model. It explicitly models individual agents in the land development process, including both the supply and demand side of the real estate market as well as the subsequent prices. Unlike a static equilibrium model, demand, supply, and prices are adjusted dynamically, UrbanSim is connected to the four-step model by projecting land use for 5 years; the

5 year projection spatially allocates new households and jobs: the new household and job information initiates the travel demand model; and the new state of the transportation network is now input back into UrbanSim for the next 5 year run.

In the last decade, several approaches modelling businesses or jobs by microsimulation have been developed. Based on the understanding that microsimulation is better suited to simulate location behaviour and, in particular, the interactions of different agents, the number of highly disaggregate approaches is growing continuously.

The first spatially explicit microsimulation of jobs was developed for the California Urban Futures (CUF) model (Landis and Zhang 1998). All potential uses, such as single-family, apartments, commercial, or industrial uses, bid for developable land.

The first complete microsimulation model for modelling business relocation and firmography instead of jobs was developed by van Wissen (2000). The model SIMFIRMS simulates businesses in the Netherlands from 1991 to 1998. Every firm is described by business type, age, size and location. Khan et al. (2002) developed a microsimulation model for small and medium-sized businesses in a synthetic study area of ten by ten raster cells. Discrete choice models are used to simulate location decisions determined by the price of floorspace. Businesses buy and sell commodities from other business location depending on good's prices. Maoh and Kanaroglou (2009) developed a model simulating small and medium-sized businesses with less than 200 employees for Hamilton, Canada. Decisions to stay, move within the study area, or out-migrate are simulated by multinomial logit models aiming at maximizing the utility. A second multinomial logit model selects a new location within a raster system of 200 by 200 m.

The Spatial Firm demographic Microsimulation (SFM) was developed by de Bok (2009) and is one of the few models that simulate firms instead of jobs.

1.3 The State of the Art: An Overview

The authors of the following chapters provide between them an excellent account of the state of the art in employment location modelling. The comprehensive models include IRPUD, DELTA, TIGRIS, LEAM, STRAGO/rAps and RELU-TRAN – an impressive collection of mnemonics that will be explained chapter by chapter in Part 1. The chapters in Part 2 are in part concerned directly with the dynamics of the location of firms and, in the main, are more micro-scale approaches, a number introducing microsimulation.

Wegener's chapter is offered first, in part because it presents a good outline of the task. He identifies five sub-markets of an urban development model, each of which connects to employment by location: the labour market, the market for non-residential buildings, reflecting changing patterns of economic activity (and hence employment), the housing market (in part reflecting access to employment), the land and construction market and the transport market (of which the journey to work will be a substantial component). He models transport, private construction and the labour and housing markets directly, but he has two interesting underpinning mechanisms as separate submodels. The first of these is an 'ageing' model which updates all stocks, with separate models for employment, population and housing – and these become critical inputs to the market models; and second, a public programmes model which allows for significant (exogenous) government investment.

Simmonds and Feldman in Chap. 3 operate with a full set of logit-based spatial interaction models at the zonal scale but, very importantly, anchor economic activity in an input–output model. This activity is then driven by an investment/disinvestment submodel – thus exposing the issue of what drives change in economic activity. They then tackle this through the exploration of various scenarios in a variety of UK applications.

In Chap. 4 De Graaff and Zondag study the impact of employment dynamics on urban development with applications in a Dutch context and look at the integration of a population-employment interaction model in the TIGRIS XL framework. The population module responses to lagged changes in employment and the employment module responses to contemporaneous changes in population. Thus, the change in population drives the change in employment. However, they argue that this assumption is too harsh, and that sectoral employment change might be more dependent upon employment change in other sectors than on population change. The test and estimation of such relations are carried out and the main outcome is that for some sectors the employment dynamics in other sectors are more important than population changes. This is achieved by an adapted version of the labour module in TIGRIS XL.

In Chap. 5 Smith, Vargas-Ruiz and Batty, in presenting a detailed transport model within a comprehensive model framework reverse the usual approach – of using the employment distribution simply as an input to the transport model – by using the transport model to explore the impact on employment distributions in London and the South-East of England. The structure they develop links input output analysis to the allocation of employment and population using traditional land use transportation interaction models. The framework then down scales these activities which are allocated to small zones to the physical level of the city using GIS-related models functioning at an even finer spatial scale. Detailed employment types generated from the input–output model are distributed to small zones through an explicit employment forecasting model in which employment types are related to functions of floor space that they use. The chapter is particularly interesting on the challenges of model calibration.

Deal, Kim and Hewings in Chap. 6, in an application to Chicago, connect economic, land-use and travel-demand simulation models through a qualitative assessment of the change in land-use and travel demand outcomes as a result of loosely coupling a macroeconomic model (CREIM), a fine scale regional dynamic spatial land use model (LEAM), and a zone based travel demand model (CMAP TDM). The comparison between land use and travel demand simulations with and without coupling is carried out in order to isolate the effects of the coupling process.

This is another very important example of how an 'out' input–output model can be used to constrain the small scale allocations in a comprehensive model.

In Chap. 7 Jin and Echenique discuss one continuously running work stream that originated as a spatial model of urban stock and activity and later became encapsulated in the MEPLAN land use and transport modelling package. One central feature of this approach is its emphasis upon simultaneously solving the employment location model with the production, trade, residential location and transport demand models for any specific year.

The remaining authors in Part 1 then represent a shift to a different kind of economic focus – translating the ideas of the computable general equilibrium model (CGE) into a spatial context – and hence SCGE.

Anderstig and Sundberg in Chap. 8 provide another example of a finer scale input–output model (rAps) being connected to an upper scale SCGE model (STRAGO). The SCGE model is developed for the nine NUTS2 regions in Sweden and is constrained in this case to national sector projections, so that these become the model drivers. There is an interesting feedback between the two layers because it is the rAps model that provides the employment constraints for the upper layer.

Kim and Hewings in Chap. 9 discuss the question concerning how metropolitan areas having a more fragmented governance structure tend to show a sprawling pattern of development. This may suggest that a fragmented institutional setting can generate a higher level of interjurisdictional competition that often hinders systematic management of the development process, thus offsetting the benefits from disaggregated local governance, such as welfare and fiscal efficiency gains. They examine how the institutional setting influences land development at a microscale. More specifically, they quantify the institutional conditions in each section, taking the jurisdictional boundaries into account and they measure its effect on land use conversion rate by employing a quasi-likelihood estimation method.

In Part 2, we consider a number of more special topics, usually at a micro scale. Lambert and Birkin in Chap. 10, for example, focus on occupations and educational attainment within a more comprehensive microsimulation population model. This forms an important element of labour market models.

Three of the remaining papers focus on the behaviour of firms, all of them using microsimulation methods. Moekel in Chap. 11 calculates his probabilities from a log model as do Maoh and Kanaroglou in their model of the survival and failure of firms in Chap. 12. These two cases focus on SMEs.

Specifically in his chapter, Moeckel describes the ILUMASS approach, which microsimulates firms, in more detail. A synthetic city is described and two models are implemented and contrasted: one simulating businesses and one simulatings employees as the decision-making unit.

Maoh and Kanaroglou analyze the survival and failure of small and medium (SME) size business establishments in the city of Hamilton, Canada. The objective is to develop a business failure module that will be used in the construction of an agent-based firmographic simulation model. Such model will be utilized within a microsimulation land use and transportation planning support system (PSS).

De Bok and Pagliara in Chap. 13 present a spatial firm demographic model (SFM) simulating changes in states of individual firms and their location choice behaviour. They then focus their contribution on the choice set composition issue in a disaggregate spatial choice context. A choice model is presented with probabilistic choice sets assuming that choice alternatives that are dominated by others, are not taken into consideration in the location decision. The estimated models have significant parameters for dominance, and they are implemented in the SFM model, to test to what extent the simulation results are improved.

The models and approaches presented here are rich in content and build on decades of evolution. However, substantial challenges remain and these are addressed in the concluding chapter by Simmonds and de Bok.

References

Ben-Akiva M, Lerman S (1985) Discrete choice discrete choice analysis: theory and application to travel demand. MIT Press, Cambridge

Boarnet MC (1994) An empirical model of intrametropolitan population and employment growth. Pap in Reg Sci 73:135–152

Carlino GA, Mills ES (1987) The determinants of country growth. J Reg Sci 27:39–54

Cascetta E (2009) Transportation systems analysis: models and applications. Springer, New York

de Bok M (2009) Estimation and validation of a microscopic model for spatial economic effects of transport infrastructure. Transp Res Part A: Policy Pract 43:44–59

de la Barra T, de la Barra T (1989) Integrated land use and transport modelling: decision chains and hierarchies, vol 12, Cambridge urban and architectural studies. Cambridge University Press, Cambridge

Domencich TA, McFadden D (1975) Urban travel demand: a behavioural analysis, vol 93, Contributions to economic analysis. North-Holland, Amsterdam/Oxford

Echenique MH (2004) Cambridge futures 2: what transport for Cambridge. The Martin Centre, University of Cambridge, Cambridge

Echenique MH, Crowther D, Lindsay W (1969) A spatial model for urban stock and activity. Reg Stud 3:281–312

Hunt JD, Abraham JE (2003) Design and application of the PECAS land use modelling system. Paper presented at the 8th international conference on computers in urban planning and urban management, Sendai, Japan, 27–29 May 2003

Khan AS, Abraham JE, Hunt JD (2002) Agent-based micro-simulation of business establishments. In: Congress of the European Regional Science Association ERSA, Dortmund

Landis J, Zhang M (1998) The second generation of the California urban futures model. Part 1: model logic and theory. Environ Plann B: Plann Des 25:657–666

Leontief W (1986) Input–output economics, 2nd edn. Oxford University Press, New York

Lowry I (1964) Model of metropolis, Memorandum RM-4035-RC. Rand Corporation, Santa Monica

Maoh H, Kanaroglou P (2009) Intrametropolitan location of business establishments. Microanalytical model for Hamilton, Ontario, Canada. Transp Res Rec: J Transp Res Board 2133:33–45

Martínez FJ (1996) MUSSA: land use model for Santiago city. Transp Res Rec 1552:126–134

McFadden D (1974) Conditional logit analysis of qualitative choice behavior. In: Zarembka P (ed) Frontiers in econometrics. Academic, New York, pp 105–142

Mills ES (1967) An aggregative model of resource allocation in a metropolitan area. Am Econ Rev 57:197–210

Muth RF (1969) Cities and housing, the University of Chicago press, Chicago. RAND Europe and Bureau Louter: 2006, System documentatie tigris xl 1.0. prepared for the Transport Research Centre, Leiden

Simmonds DC (2001) The objectives and design of a new land-use modelling package: DELTA. In: Clarke G, Madden M (eds) Regional science in business. Advances in spatial sciences. Springer, Berlin, pp 159–188

Steinnes DN (1977) Do people follow jobs' or'do jobs follow people'? A causality issue in urban economics. J Urban Econ 4:69–79

van Wissen LJG (2000) A micro-simulation model of firms: applications of concepts of the demography of the firm. Pap Reg Sci 79:111–134

Waddell P (2002) UrbanSim: modeling urban development for land use, transportation, and environmental planning. J Am Plann Assoc 68(3):297–314

Waddell P, Borning A, Noth M, Freier N, Becke M, Ulfarsson GF (2003) Microsimulation of urban development and location choice: design and implementation of UrbanSim. Netw Spat Econ 3:43–67

Wegener M (1982) Modeling urban decline: a multilevel economic-demographic model for the Dortmund region. Int Reg Sci Rev 7(2):217–241

Wilson (1970) Entropy in urban and regional modelling, Pion, London; re-issued by Routledge, London, 2011

Zondag B, Schoemakers A, Pieters M (2005) Structuring impacts of transport on the spatial distribution of residents and jobs. In: 8th Nectar conference, Las Palmas, Gran Canaria, Spain, 2–4 June 2005

Part I
Macro-scale Approaches

Chapter 2
Employment and Labour in Urban Markets in the IRPUD Model

Michael Wegener

Abstract The IRPUD model is a simulation model of intraregional location and mobility decisions in a metropolitan area. Employment and labour in the model are affected by developments in other submodels: The supply of jobs results from growth and decline, location and relocation of firms and their demand for labour by skill. The demand for jobs results from population and immigration and outmigration by age, education and skill. Intraregional labour mobility results from decisions of firms to hire or release workers and decisions of workers to start or end a job. These decisions affect commuting patterns and residential and firm location and the associated construction and real estate markets. This chapter presents the parts of the model affecting regional employment and labour and the remaining submodels as far as necessary for understanding these. The paper closes by summarising calibration results and an example application of the model.

2.1 The IRPUD Model

The IRPUD model is a simulation model of intraregional location and mobility decisions in a metropolitan area (Wegener 1982, 1983, 1985, 1986, 1994, 1996, 1998, 2011). It receives its spatial dimension by the subdivision of the study area into *zones* connected by transport networks containing the most important links of the public transport and road networks coded as an integrated, multimodal network including all past and future network changes. It receives its temporal dimension by the subdivision of time into *periods* of one or more years' duration.

M. Wegener (✉)
Spiekermann & Wegener Urban and Regional Research (S&W), Dortmund, Germany
e-mail: mw@spiekermann-wegener.de

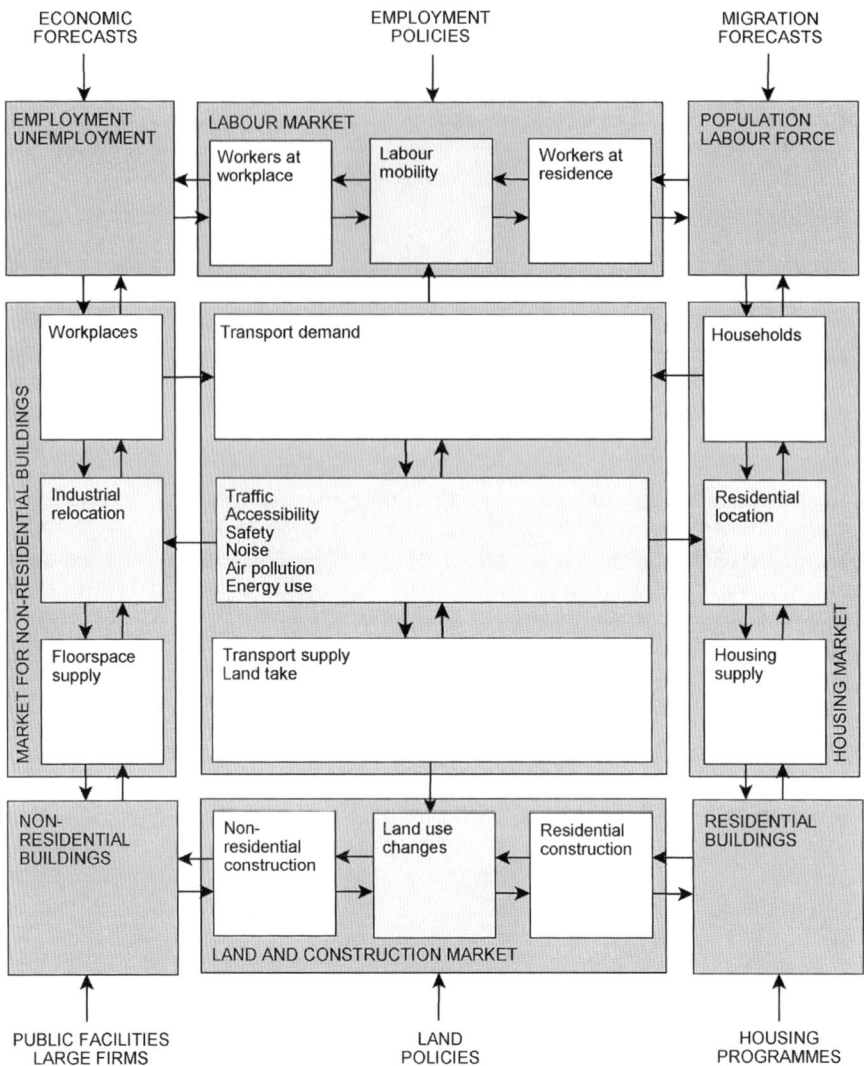

Fig. 2.1 The IRPUD model

2.1.1 Model Structure

The model predicts for each simulation period intraregional *location decisions* of industry, residential developers and households, the resulting *migration* and *travel* patterns, *construction* activity and *land-use* development and the impacts of *public policies* in the fields of industrial development, housing, public facilities and transport.

Figure 2.1 is a schematic diagram of the major subsystems considered in the model and their interactions and of the most important policy instruments.

The four square boxes in the corners of the diagram show the major stock variables of the model: *population, employment, residential buildings* (housing) and *non-residential buildings* (industrial and commercial workplaces and public facilities). The actors representing these stocks are *individuals or households, workers, housing investors* and *firms*.

These actors interact on five *submarkets* of urban development. The five submarkets treated in the model and the market transactions occurring on them are:

– *Labour market*: new jobs and redundancies,
– The *market for non-residential buildings*: new firms and firm relocations,
– The *housing market*: immigration, outmigration, new households and moves,
– The *land and construction market*: changes of land use through new construction, modernisation or demolition,
– The *transport market*: trips.

For each submarket, the diagram shows *supply* and *demand* and the resulting *market transactions*. Choice in the submarkets is constrained by supply (jobs, vacant housing, vacant land, vacant industrial or commercial floorspace) and guided by attractiveness, which in general terms is an actor-specific aggregate of *neighbourhood quality, accessibility* and *price*.

The large arrows in the diagram indicate exogenous inputs: these are either *forecasts* of regional employment and population subject to long-term economic and demographic trends or *policies* in the fields of industrial development, housing, public facilities and transport.

2.1.2 Submodels

The IRPUD model has a modular structure and consists of six interlinked submodels operating in a recursive fashion on a common spatio-temporal database:

1. The *Transport Submodel* calculates work, shopping, services/social and education trips for four socioeconomic groups and three modes, walking/cycling, public transport and car. It determines a user-optimum set of flows where car ownership, trip rates and destination, mode and route choice are in equilibrium subject to congestion in the road network.
2. The *Ageing Submodel* computes all changes of the stock variables of the model which are assumed to result from biological, technological or long-term socio-economic trends originating outside the model (i.e. which are not treated as decision-based). These changes are effected in the model by probabilistic ageing or updating models of the Markov type with dynamic transition rates. There are three such models, for employment, population and households/housing.

3. The *Public Programmes Submodel* processes a large variety of public programmes specified by the model user in the fields of employment, housing, health, welfare, education, recreation and transport.
4. The *Private Construction Submodel* considers investment and location decisions of private developers, i.e. of enterprises erecting new industrial or commercial buildings, and of residential developers who build flats or houses for sale or rent or for their own use. Thus the submodel is a model of the regional land and construction market.
5. The *Labour Market Submodel* models intraregional labour mobility as decisions of workers to change their job location in the regional labour market.
6. The *Housing Market Submodel* simulates intraregional migration decisions of households as search processes in the regional housing market. Housing search is modelled in a stochastic microsimulation framework. The results of the Housing Market Submodel are intraregional migration flows by household category between housing by category in the zones.

Figure 2.2 visualises the recursive processing of the six submodels. The Transport Submodel is an equilibrium model referring to a *point in time*. All other submodels are incremental and refer to a *period of time*. Submodels (2) to (6) are executed once in each simulation period, while the Transport Submodel (1) is processed at the beginning and the end of each simulation period. Each submodel passes information to the next submodel in the same period and to its own next iteration in the following period.

2.1.3 Employment and Labour in the Submodels

In the IRPUD model employment and labour are not modelled in one integrated model but in several interacting submodels. Change of zonal employment occurs in three different submodels:

1. Decline of zonal employment due to sectoral decline, lack of building space and intraregional relocation of firms is modelled in the *Ageing Submodel*.
2. Changes of zonal employment due to the location or removal of large plants exogenously specified by the user are executed in the *Public Programmes Submodel*.
3. Changes of zonal employment due to new jobs in vacant industrial or commercial buildings, in newly built industrial or commercial buildings or in converted residential buildings are modelled in the *Private Construction Submodel*.

Changes of zonal labour force are modelled in five different submodels:

1. Ageing of households and housing and other demographic changes of household status are modelled in the *Ageing Submodel*.
2. Public housing programmes specified by the model user are executed in the *Public Programmes Submodel*.

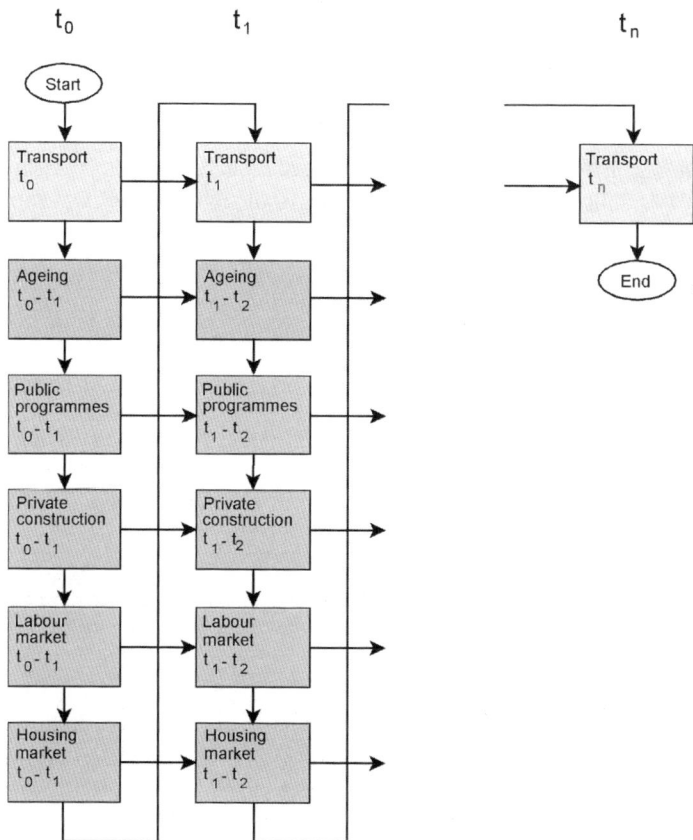

Fig. 2.2 Recursive processing of submodels

3. Private housing maintenance/upgrading and new construction investments and the resulting changes in housing and land prices are modelled in the *Private Construction Submodel*.
4. Labour mobility and the resulting changes of household income are modelled in the *Labour Market Submodel*.
5. Changes of the association of households with housing are modelled in the *Housing Market Submodel*.

2.2 Ageing Submodel

In the *Ageing Submodel* all changes of zonal stock variables are modelled which are assumed to result from biological, technological or long-term socio-economic trends or originating outside of the model, i.e. which are not treated as *decision-based*. These changes are effected by probabilistic ageing or updating, or semi-Markov, models with dynamic transition rates. There are three such models for employment, population and households and housing. Here only changes of employment are discussed.

In the Ageing Submodel decline of zonal employment due to sectoral decline, lack of building space and intraregional relocation of firms is modelled. Each of the 40 industries of the model is considered a separate submarket. The model starts from existing employment $E_{slj}(t)$ of industry s situated on land use category l in zone j at time t. There are three different ways for E_{slj} to change in this submodel:

2.2.1 Sectoral Decline

Declining industries make workers redundant. This occurs not necessarily at the same rate all over the region, but is more likely where locational conditions are less favourable:

$$R_{slj}(t,t+1) = \frac{E_{slj}(t)\exp[-\alpha_s\, u_{slj}(t)]}{\sum_{jl} E_{slj}(t)\exp[-\alpha_s\, u_{slj}(t)]}\, [E_s(t+1) - E_s(t)] \qquad (2.1)$$

is the number of workers in industry s made redundant on land-use category l in zone j between t and $t+1$. $E_s(t)$ indicates total employment of industry s in the region and $E_s(t+1)$ is the exogenous projection of total regional employment for time $t+1$. The utility $u_{slj}(t)$ expresses the attractiveness of land-use category l in zone j for industry s (see Eq. 2.7 below). R_{slj} is set to zero for growing industries.

2.2.2 Relocation

Some industries are very stationary, while others easily move from one location to another. If $r_s(t,t+1)$ is a sectoral mobility rate, then

$$M_{slj}(t,t+1) = \frac{E_{slj}(t)\exp[-\alpha_s\, u_{slj}(t)]}{\sum_{jl} E_{slj}(t)\exp[-\alpha_s\, u_{slj}(t)]}\, r_s(t,t+1)\, E_s(t) \qquad (2.2)$$

is the number of workplaces relocated from land-use category l in zone j during the period. The mobility rate $r_s(t,t+1)$ is exogenous.

2.2.3 Lack of Building Space

In most industries, mechanisation and automation tend to increase the building floorspace per workplace. Accordingly, in each period, a number of jobs S_{slj} have to be relocated because of lack of space:

$$S_{slj}(t,t+1) = E_{slj}(t)\left[1 - \frac{b_{sj}(t)}{b_{sj}(t+1)}\right] - R_{slj}(t,t+1) \qquad (2.3)$$

where $b_{sj}(t+1)$ is the projected floor space per workplace in industry s in zone j at time $t+1$, which will be always greater than or equal to its value at time t. Where redundancies exceed relocations due to lack of space, S_{slj} is set to zero. For the workers made redundant, later new buildings will be provided in the *Public Programmes* or *Private Construction Submodels*. Where decline of employment is large, buildings remain vacant, but may be reused by other industries later.

2.3 Public Programmes

In the *Public Programmes Submodel* public programmes assumed to be implemented during the simulation are processed.

Public programmes in the IRPUD model are events entered exogenously by the model user. They represent primarily public policy measures such as infrastructure investments or public housing programmes, where local, state or national governments directly intervene into the process of spatial urban development. In addition they represent "singular historical events" that are caused by private market decisions but are too unique and too large to be predicted by a model, such as the location or closure of large industrial plants, large commercial developments, such shopping malls, or large private housing developments.

Regulatory or monetary policies such as land-use plans, building regulations, taxes, public transport fares or parking fees are not part of the Public Programmes Submodel but are entered exogenously by the model user. Transport infrastructure changes are entered as dynamic transport network scenarios.

In the Public Programmes Submodel three types of events are executed: changes of *employment*, changes of *housing* and changes of *infrastructure*. Here only changes of employment are discussed.

Exogenously specified changes of employment such as the location or closure of large industrial plants or large shopping malls are entered as a code specifying the kind of change (removal or location), the number and category (industry) of jobs to be removed or located and the year and zone in which this is to take place. In the case of removal, a distinction is made whether the associated buildings are to be torn down or are to remain vacant for later use by other industries. In the case of a new location also the industrial or commercial buildings required for the new facility are constructed. If there is not enough vacant industrial land in the zone, a limited percentage of housing of low quality, if available, may be torn down. If also this is not sufficient, the number of jobs to be located is reduced.

In addition, the impact of the removal or addition of employment on household incomes in the region is calculated using the method described in the *Labour Market Submodel* below.

2.4 Private Construction

The *Private Construction* submodel considers investment and location decisions by private developers, i.e. by enterprises which erect new industrial or commercial buildings and by residential developers who build flats or houses for sale or rent or for their own use. Thus the submodel is a model of the regional land and construction market. Here only industrial or commercial construction is discussed.

The industrial location submodel makes no distinction between basic and non-basic industries, i.e. all industries are located or relocated endogenously subject to sectoral employment projections for the whole region. The model locates industrial or commercial floor space suitable as workplaces in the 40 industries considered in the model. However, as the amount of floor space occupied per worker is not constant over time and certain types of floor space can be used by several industries, the model actually locates *workplaces* or employment, which are subsequently converted to floor space. The location of workplaces of all industries may also be controlled exogenously by the user in the *Public Programmes Submodel* in order to reflect major events such as the location or closure of large plants in particular zones. Changes of zonal employment due to new jobs in vacant industrial or commercial buildings, in newly built industrial or commercial buildings or converted residential buildings are modelled in this submodel.

New workplaces are either located in existing vacant industrial or commercial buildings, in newly constructed industrial or commercial buildings or in converted residential buildings.

Before starting the location process, industries are sorted by decreasing floorspace productivity, or rent paying ability, and processed in that order. The total demand for new workplaces of industry s in the region is

$$N_s(t, t+1) = E_s(t+1) - E_s(t) + \sum_{jl} \Delta E_{slj}(t, t+1) \qquad (2.4)$$

where $\Delta E_{slj}(t, t+1)$ are net changes in employment of industry *s* on land use category *l* in zone *j* modelled in previous submodels resulting from sectoral decline, lack of building space and intraregional relocation of firms as well as from exogenously specified public programmes.

2.4.1 New Jobs in Vacant Buildings

Declining industries or relocating firms leave buildings vacant that may be used by other industries. For this purpose, the 40 industries have been divided into groups with similar space requirements.

If this demand is less than the total supply of suitable floor space, it is allocated to vacant floor space with the following allocation function:

$$V_{slj}(t, t+1) = \frac{K_{slj} \exp[\gamma_s \, u_{slj}(t)]}{\sum_{jl} K_{slj} \exp[\gamma_s \, u_{slj}(t)]} \, N_s(t, t+1) \qquad (2.5)$$

where K_{slj} is the capacity of existing buildings on land-use category l in zone j for workplaces in industry s. V_{slj} is the number of jobs accommodated.

2.4.2 New Jobs in New Buildings

For any remaining demand, new industrial or commercial buildings have to be provided. This demand is allocated to vacant industrial or commercial land with the allocation function

$$C_{slj}(t, t+1) = \frac{L_{slj} \exp[\gamma_s \, u_{slj}(t)]}{\sum_{jl} L_{slj} \exp[\gamma_s \, u_{slj}(t)]} \left[N_s(t, t+1) - \sum_{jl} V_{slj}(t, t+1) \right] \qquad (2.6)$$

where C_{slj} are new workplaces in industry s built on land-use category l in zone j between t and $t+1$. L_{slj} is the current capacity of land of land-use category l for such workplaces in zone j; since it is continuously reduced during the simulation period, it bears no time label.

The utility $u_{slj}(t)$ used in Eqs. 2.5 and 2.6 is the attractiveness of land-use category l in zone j for industry s and has three components:

$$u_{slj}(t) = [u_{sj}(t)]^{v_s} \, [u_{sl}(t)]^{w_s} \, [u_s(c_{lj})(t)]^{1-v_s-w_s} \qquad (2.7)$$

where $u_{sj}(t)$ is the attractiveness of zone j as a location for industry s, $u_{sl}(t)$ is the attractiveness of land-use category l for industry s, and $u_s(c_{lj})(t)$ is the attractiveness of the land price of land-use category l in zone j in relation to the expected profit of economic activity s. The v_s, w_s, and $1 - v_s - w_s$ are multiplicative importance weights adding up to unity. The three component utilities are constructed similarly to the components of the housing utility $u_{hki}(t)$ (see *Housing Market Submodel*). Like all utilities in the model, the $u_{slj}(t)$ remain unchanged during the simulation period as calculated at time t. The price or rent of industrial or commercial buildings is presently not represented in the model.

The land capacity L_{slj} is normally taken as being fixed as specified in the zoning plan. If a piece of land was formerly in a built-up area, its development implies the demolition of existing buildings. In addition, under certain restrictions in zones of high demand, the capacity L_{slj} may be extended by demolition of existing buildings with less profitable building uses to represent displacement processes going on within existing neighbourhoods. As L_{slj} is updated after the location of each industry, it bears no time label. All workplaces or dwellings displaced by demolition during a simulation period are replaced in the same period by iterating the industrial and residential submodels several times.

Retailing is treated like any other industry in the model except that the zonal attractiveness $u_{sj}(t)$ (see Eq. 2.7) for retailing includes a fourth attribute

$$u_{sjn}(t) = v_n \left[\frac{\sum_{qim} t_{q2ijm}(t) \; y_{qi}(t) \; / \; E_{rj}(t)}{\sum_{qijm} t_{q2ijm}(t) \; y_{qi}(t) \; / \; E_r(t)} \right] \quad (2.8)$$

where $t_{q2ijm}(t)$ are shopping trips (trip purpose $g = 2$) of households of income group q from residential zone i to shopping zone j using mode m, $y_{qi}(t)$ are retail expenses of households of income group q in zone i at time t, $E_{rj}(t)$ is retail employment in zone j at time t, $E_r(t)$ total regional retail employment, and $v_n(.)$ the value function mapping attribute n to utility. This attribute indicates retail sales per retail employee in zone j expressed in units of average turnover per retail employee in the whole region.

2.4.3 Conversion of Existing Dwellings

In the case of service workplaces, the capacity of a zone may also be extended by conversion of existing dwellings to offices where the demand for office space is high in relation to supply in order to represent the displacement of dwellings by offices observed within or near the CBD. All dwellings converted to offices during a simulation period are replaced in the same period by iterating the industrial and residential location submodels several times.

2.5 Labour Market

The urban economy is represented in the model by employment (workers at place of work) and industrial and commercial buildings classified by 40 industries. Labour force is represented by workers and unemployed persons at place of residence classified by four skill levels and by sex and nationality (native or foreign). The four skill levels correspond to the four income levels of the classification of households (see *Housing Market Submodel*). There is no distinction between employed and self-employed persons. The distribution of skill levels of workers by industry is assumed to change over time according to exogenous assumptions.

The labour market is assumed to be demand-driven; firms employ and release workers according to their needs, and this influences the distribution of employment, labour, unemployment and household incomes in the region.

2.5.1 Labour Mobility

For a variety of reasons workers change their workplace each year. If both workplace and residence are in the region, this does not normally imply a change of residence so that neither the distribution of workplaces nor the distribution of residences is changed. However, the pattern of work trips in the region is changed.

As nothing is known about work-related reasons of intraregional labour mobility, only reasons related to the work trip are modelled. It is assumed that, everything else being equal, a job nearer to home is more preferable than one farther away, so that the propensity to change job is inversely related to the trip utility of the work trip:

$$M_{qj}(t,t+1) = \sum_i \frac{t_{q1ij}(t) \, \exp\left[-\alpha \, u_{qij}(t)\right]}{\sum_{ij} t_{q1ij}(t) \, \exp\left[-\alpha \, u_{qij}(t)\right]} \, a_q(t,t+1) \, E_q(t) \qquad (2.9)$$

where $M_{qj}(t,t+1)$ are workers of skill level q working in zone j considering a change of job between time t and time $t+1$, $t_{q1ij}(t)$ are work trips (trip purpose $g=1$) of workers of skill level (income group) q between residences in zones i and workplaces in zones j at time t and $u_{qij}(t)$ is the trip utility of work trips between zones i and j for workers of skill level q aggregated over modes m:

$$u_{qij}(t) = \frac{1}{\lambda} \sum_{m \in M_h} \exp\left[\lambda \, u_{qijm}(t)\right] \qquad (2.10)$$

The mobility rate $a_q(t,t+1)$ indicating how many workers of total workers $E_q(t)$ of skill level q are likely to change their jobs between t and $t+1$ is exogenous. It is assumed that when selecting a new job, again everything else being equal, also the distance between the old and the new job plays a role (see Fig. 2.3).

Therefore a *change-of-job utility* similar to the migration utility used in the *Housing Market Submodel* (see Eq. 2.24) was defined:

$$\underline{u}_{qjj'}(t) = u'_{qij'}(t)^{w_q} \, u_{qjj}(t)^{1-w_q} \qquad (2.11)$$

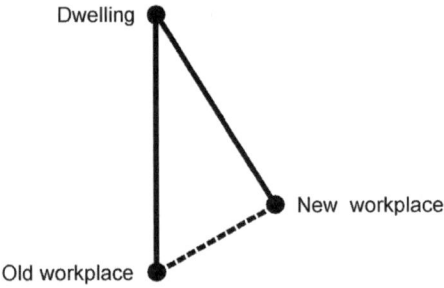

Fig. 2.3 Change of job

with

$$u'_{qij'}(t) = \frac{1}{\beta} \ln \sum_i \frac{t_{q1ij}(t) \exp[\alpha \, u_{qij}(t)]}{\sum_i t_{q1ij}(t) \exp[\alpha \, u_{qij}(t)]} \exp[\beta \, u_{qij'}(t)] \quad (2.12)$$

The first part of the change-of-job utility Eq. 2.11 is the expected utility of a work trip from the old housing zone i to the new work zone j' after the move weighted by the probability that the worker lives in zone i (Eq. 2.12), the second part evaluates the utility of a trip between the old and the new workplace. The w_q and $1 - w_q$ are multiplicative weights adding up to unity.

With the above components a doubly constrained spatial interaction model is used to model job changes between work zones j:

$$M_{qjj'}(t, t+1) = M_{qj}(t, t+1) \, A_{qj} \, M_{qj'}(t, t+1) \, B_{qj'} \exp[\gamma \, \underline{u}_{qjj'}(t)] \quad (2.13)$$

where A_{qj} and $B_{qj'}$ are balancing factors.

2.5.2 Change of Income

Household incomes are classified into four household income groups (low, medium, high, very high) corresponding to the four skill levels of workers. It is assumed that all households within one of the four income groups have the same income irrespective of their location in the region. Household incomes also determine housing budgets (used in the *Housing Market Submodel*) and travel budgets (used in the *Transport Submodel*) as well as disposable income for shopping (used in the retail location part of the *Private Construction Submodel*). Household incomes and housing, travel and shopping budgets of the four income groups are updated according to exogenously specified projections.

Changes of the income distribution of households are induced by changes of employment (see above). It is assumed that in the case of unemployment a household drops from one income group to the next lower one. Conversely, in the case of new employment the household is promoted by one income group. Changes of employment are calculated as redundancies or new jobs at places of work. It is assumed that workers are released or hired without regard of their place of residence. Therefore, using the work trip matrix (trip purpose $g = 1$) calculated for each period in the *Transport Submodel*, changes of employment at places of residence can be inferred:

$$R'_{qi}(t, t+1) = \sum_{j} \frac{t_{q1ij}(t)}{\sum_{ij} t_{q1ij}(t)} \sum_{s} R_{sqj}(t, t+1) \qquad (2.14)$$

$$N'_{qi}(t, t+1) = \sum_{j} \frac{t_{q1ij}(t)}{\sum_{ij} t_{q1ij}(t)} \sum_{s} N_{sqj}(t, t+1) \qquad (2.15)$$

where $R'_{qi}(t, t+1)$ are workers made redundant and $N'_{qi}(t, t+1)$ newly employed workers of skill level q at places of residence i between time t and time $t+1$, $R_{sqj}(t, t+1)$ are workers of skill level q made redundant in industry s in work zones j in that period due to sectoral decline, lack of building space, intraregional relocation or exogenous user specification, and $N_{sqj}(t, t+1)$ are new jobs of skill level q created in industry s in work zone j in vacant or new buildings.

As the four skill levels correspond to the four household income groups, it is assumed that workers of skill level q belong to a household of the set of household types belonging to income group q. With this assumption, for each zone a 4 × 4 matrix of transition rates between household income groups is calculated and used for updating all household distributions of the model.

2.6 Housing Market

As the supply of labour and the demand for jobs on the regional labour market are determined by the location and relocation of households, also the *Housing Market Submodel* of the IRPUD model is briefly presented here. In the Housing Market Submodel all changes of association of households with housing and the resulting changes in housing prices are modelled.

2.6.1 Household Moves

The Housing Market Submodel simulates intraregional migration decisions of households as search processes in the regional housing market. Thus it is at the same time an intraregional migration model. Housing search is modelled in a stochastic microsimulation framework. The results of the Housing Market Submodel are intraregional migration flows by household category between housing by category in the zones.

Households are represented in the model as a four-dimensional distribution classified by nationality (native, foreign), age of head (16–29, 30–59, 60+), income/skill (low, medium, high, very high) and size (1, 2, 3, 4, 5+ persons). Similarly, housing of each zone is represented as a four-dimensional distribution of dwellings classified by type of building (single-family, multi-family), tenure (owner-occupied, rented, public), quality (very low, low, medium, high) and size (1, 2, 3, 4, 5+ rooms). All changes of households and housing during the simulation are computed for these 120 household types and 120 housing types. However, where households and housing are cross-classified together, these households and housing types are aggregated to H household and K housing types, with H and K not exceeding 30.

Technically, the migration submodel is a Monte Carlo micro simulation of a sample of representative housing market *transactions*. However, it differs from other, 'list-oriented', micro simulations in that (a) sampling and aggregation are part of the simulation and (b) stocks (households and dwellings) are classified, i.e. aggregate, data. A market transaction is any successfully completed operation by which a migration occurs, i.e. a household moves into or out of a dwelling or both.

A market transaction has a *sampling phase*, a *search phase*, a *choice phase* and an *aggregation phase* (Wegener 1985; Wegener and Spiekermann 1996):

– In the *sampling phase* a household looking for a dwelling or a landlord looking for a tenant is sampled for being simulated.
– In the *search phase* the household looks for a suitable dwelling, or the landlord looks for a tenant.
– In the *choice phase* the household decides whether to accept the dwelling or not.
– In the *aggregation phase* all changes of households and dwellings resulting from the transaction, multiplied by the sampling factor, are performed.

The sampling phase and the search phase are controlled by multinomial logit choice functions. For instance, for a household looking for a dwelling,

$$p_{k|hi} = \frac{R_{hki} \exp[-\alpha_h \, u_{hki}(t)]}{\sum_k R_{hki} \exp[-\alpha_h \, u_{hki}(t)]} \tag{2.16}$$

is the probability that of all households of type h living in zone i, one occupying a dwelling of type k will be sampled for simulation,

$$p_{i'|hki} = \frac{\sum_{k'} D_{k'i'} \exp[\beta_h \, \underline{u}_{hii'}(t)]}{\sum_{i'k'} D_{k'i'} \exp[\beta_h \, \underline{u}_{hii'}(t)]} \tag{2.17}$$

is the probability that the household searches in zone i' for a new dwelling and

$$p_{k'|hkii'} = \frac{D_{k'i'} \exp[\gamma_h \, u_{hk'i'}(t)]}{\sum_{k'} D_{k'i'} \exp[\gamma_h \, u_{hk'i'}(t)]} \quad (2.18)$$

is the probability that it inspects a dwelling of type k' there before making a choice. In these equations R_{hki} is the number of households of type h living in a dwelling of type k in zone i, and $D_{k'i'}$ is the number of vacant dwellings of type k' in zone i'. The $u_{hki}(t)$ and the $\underline{u}_{hii'}(t)$ are two different kinds of utility measures expressing the attractiveness of a dwelling or a zone for a household considering a move. They are discussed in Eqs. 2.19 and 2.24. The two utilities carry the time label t, i.e. are unchanged since the beginning of the simulation period, while R_{hki} and $D_{k'i'}$ carry no time label as they are continuously updated during the microsimulation.

In the choice phase, the household decides whether to accept the inspected dwelling or not. It is assumed that it behaves as a satisficer, i.e. that it accepts the dwelling if this will improve its housing situation by a certain margin. Otherwise, it enters another search phase to find a dwelling, but after a number of unsuccessful attempts it abandons the idea of a move. The amount of improvement necessary to make a household move is assumed to depend on its prior search experience, i.e. go up with each successful and down with each unsuccessful search. In other words, households are assumed to adapt their aspiration levels to supply conditions on the market.

The results of the migration submodel are intraregional migration flows of households (including starter households and inmigrant and outmigrant households) by household type between dwellings by type in the zone.

The attractiveness of a dwelling of type k in zone i for a household of type h, $u_{hki}(t)$, is a weighted aggregate of housing attributes:

$$u_{hki}(t) = [u_{hi}(t)]^{v_h} \, [u_{hk}(t)]^{w_h} \, [u_{q(h)h}(c_{ki})(t)]^{1-v_h-w_h} \quad (2.19)$$

where $u_{hi}(t)$ is the attractiveness of zone i as a housing location for household type h, $u_{hk}(t)$ is the attractiveness of housing type k for household type h, and $u_{q(h)}(c_{ki})$ is the attractiveness of the rent or price of the dwelling in relation to the household's housing budget, which is a function of its income group $q(h)$. The v_h, w_h and $1 - v_h - w_h$ are multiplicative importance weights adding up to unity.

Both $u_{hi}(t)$ and $u_{hk}(t)$ are themselves multiattribute encompassing relevant attributes of the neighbourhood:

$$u_{hi}(t) = \sum_n a_n v_n \, f_n[\mathbf{X}_i(t), \mathbf{U}_{q(h)i}(t)] \quad (2.20)$$

or of the dwelling:

$$u_{hk}(t) = \sum_n b_n \, w_n \, g_n[\mathbf{X}_k(t)] \qquad (2.21)$$

where subscript n indicates attribute n. The a_n and b_n are importance weights adding up to unity, the $v_n(.)$ and $w_n(.)$ are value functions mapping attributes to utility, and the $f_n(.)$ and $g_n(.)$ are generation functions specifying how to calculate attributes from one or more elements of vectors $\mathbf{X}_i(t)$ or $\mathbf{X}_k(t)$ of raw attributes of zone i or dwelling type k or vectors of accessibility indices $\mathbf{U}_{q(h)i}(t)$ of zone i (see below). The housing price attractiveness $u_{q(h)}(c_{ki})$ is calculated as

$$u_{q(h)}(c_{ki}) = u_{q(h)}(c_{ki}/y_{q(h)k}) \qquad (2.22)$$

where c_{ki} is rent, or imputed rent, of dwelling type k in zone i, and $y_{q(h)k}$ is the monthly housing budget of household type h belonging to income group q for this dwelling type. The housing budgets include housing allowances and other public subsidies and are therefore different for rented and owner-occupied dwellings.

The $\mathbf{U}_{q(h)i}(t)$ are household income-group specific vectors of accessibility indices describing the location of zone i in the region with respect to activities $W_{nj}(t)$ in zones j:

$$u_{qni}(t) = \sum_{j,m \in \mathbf{M}_q} \frac{W_{nj}(t) \exp[\beta_n u_{qijm}(t)]}{\sum_{j,m \in \mathbf{M}_q} W_{nj}(t) \exp[\beta_n u_{qijm}(t)]} \, u_{qijm}(t) \qquad (2.23)$$

The accessibility is expressed in terms of mean trip utility, i.e. as a weighted average of potential trips from zone i to activities or facilities $W_{nj}(t)$ of type n in zone j using mode m with trip utility $u_{qijm}(t)$ for household income group q. The set $\mathbf{M}_q$ includes all transport modes accessible to households of income group q depending on its car ownership level.

The attractiveness measure $\underline{u}_{hii'}(t)$ used in Eq. 2.17 is a relational utility describing the attractiveness of a zone i' as a new housing location for a household of type q now living in zone i and working in any of the zones near i (see Fig. 2.4).

Corresponding to the *change-of-job* utility in Eq. 2.11, it is called *migration utility*:

$$\underline{u}_{hii'}(t) = u'_{hi'j}(t)^{w_q} \, u_{q(h)hii'}(t)^{1-w_q} \qquad (2.24)$$

with

$$u'_{hi'j}(t) = \frac{1}{\beta} \ln \sum_j \frac{t_{q(h)1ij}(t) \, \exp[\alpha \, u_{q(h)ij}(t)]}{\sum_j t_{q(h)1ij}(t) \, \exp[\alpha \, u_{q(h)ij}(t)]} \, \exp[\beta \, u_{q(h)i'j}(t)] \qquad (2.25)$$

and

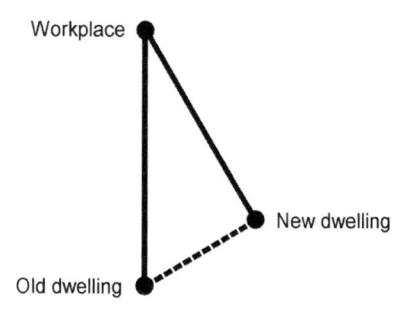

Fig. 2.4 Change of residence

$$u_{q(h)ij}(t) = \frac{1}{\lambda} \sum_{m \in \mathbf{M}_h} \exp\left[\lambda \; u_{q(h)ijm}(t)\right] \qquad (2.26)$$

The first part of the migration utility Eq. 2.24 is the expected utility of a work trip from the new housing zone i' to work zone j after the move weighted by the probability that the household works in j (Eq. 2.25) aggregated across modes (Eq. 2.26), the second part evaluates the utility of a trip between the old and the new housing zone. The w_q and $1 - w_q$ are multiplicative weights adding up to unity.

2.6.2 Price Adjustment

At the end of each simulation period housing prices and rents are adjusted to reflect changes in housing demand in the previous housing market simulation. Changes of housing prices and rents due to changes in the composition of the housing stock are dealt with in the *Private Construction Submodel*, price increases through inflation in the *Ageing Submodel*.

Housing prices and rents by housing type and zone are adjusted as a function of the demand for housing in that submarket in the period expressed by the proportion of vacant units:

$$p_{ki}(t+1) = p_{ki}(t) \left[1 + f\left(\frac{V_{ki}(t+1)}{D_{ki}(t+1)}\right)\right] \qquad (2.27)$$

where $p_{ki}(t)$ is monthly rent or imputed rent per square metre of housing floorspace of dwelling type k in zone i at time t, $V_{ki}(t+1)$ is the number of vacant dwellings of housing type k in zone i at time $t+1$, and $D_{ki}(t)$ is the total number of dwellings of that type in the zone at time $t+1$. The function f(.) is an inverted S-shaped elasticity curve entered exogenously resulting in a reduction of housing prices and rents if there is a large percentage of vacant dwellings of that kind not bought or rented in the previous housing market simulation, and in a price or rent increase if there are no or only few vacant dwellings left. No attempt is made to determine equilibrium housing prices or rents. The price adjustment model reflects price adjustment behaviour by landlords. If they reduce or increase prices or rents too much, this will be corrected in the subsequent simulation period.

2.7 Calibration and Validation

The calibration and validation of the IRPUD model cannot be presented in detail here. Due to the large number of its submodels not all of them could be estimated statistically using maximum likelihood techniques because of lack of data.

In addition, the intended use of the model to assess scenarios quite different from past experiences made estimation of the model based on past behaviour less meaningful. Instead most attention was paid to capturing the constraints on mobility and location behaviour imposed by household disposable incomes, in particular travel and housing budgets. Accordingly the utility functions of firms, households and developers were largely determined by expert judgment based on theory and the available literature.

In the absence of calibration data, validation of the model becomes all important. This is why the model simulations always start in the year 1970 in order to iteratively adjust model parameters until the major spatial processes observed in the past are adequately reproduced.

As an example for the successful reproduction of characteristics of the labour market of the region, observed and modelled travel times and travel distances of commuter trips are compared in Fig. 2.5. The Figure shows cumulated frequencies of travel times (left) and travel distances (right) of work trips in the year 2000 aggregated for municipalities. The grey line shows travel times and distances between internal municipalities and to and from external municipalities (i.e. without internal travel times and work trips between external municipalities) forecast by the model, whereas the black lines show the empirically observed work trips of the North-Rhine Westphalia State Statistical Office multiplied by the travel times and travel distances of the *Transport Submodel*.

Figure 2.6 shows a comparison of population and employment of the municipalities of the study area as forecast by the model with observed population and employment data. The two diagrams show the quality of the forecasts based only on 1970 base year data. In particular the population data of the large cities Dortmund and Bochum were matched almost exactly, whereas Hagen is somewhat over- and Hamm and Herten somewhat underestimated.

The employment numbers of Dortmund were reproduced well, whereas those of Bochum, Hagen and Iserlohn were slightly overestimated. The examples demonstrate that the model is able to reproduce the essential spatial processes in the region. The small deviations from reality have no influence on the relevance of the results, as in their interpretation not absolute values but differences between scenarios are analysed.

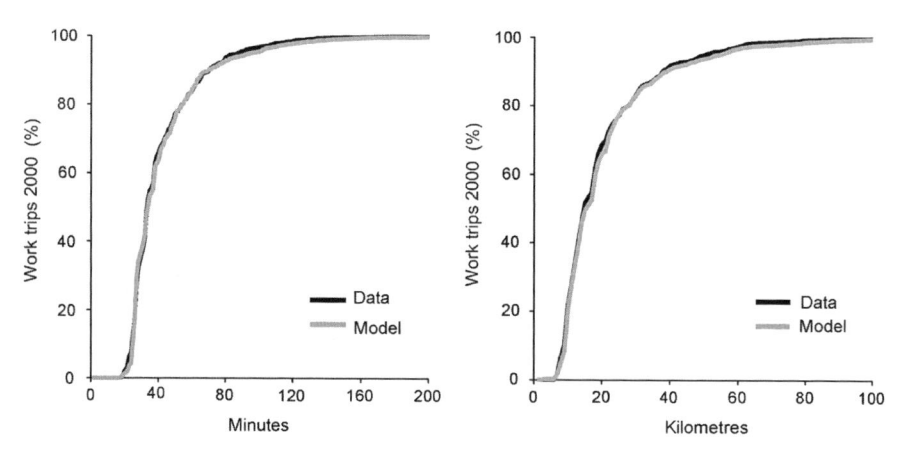

Fig. 2.5 Observed and modelled cumulated travel times and travel distances of work trips 2000

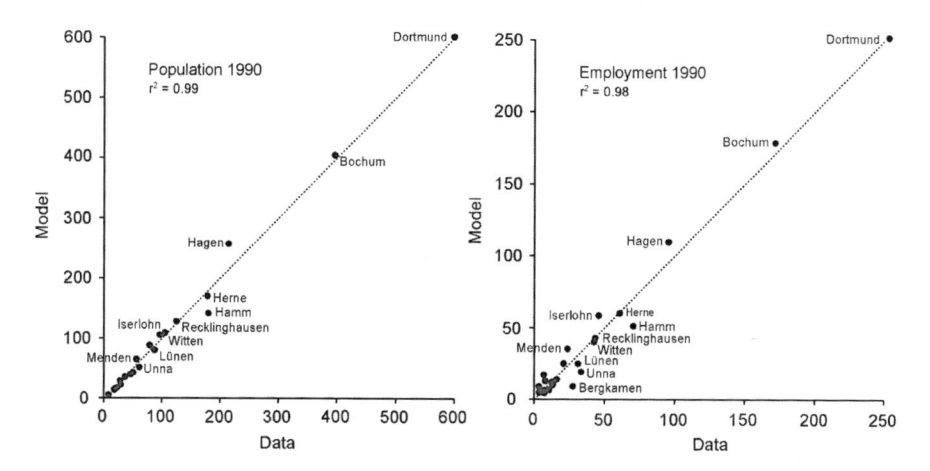

Fig. 2.6 Observed and modelled population and employment 1990

2.8 Model Applications

The IRPUD model has been applied in projects for the European Commission and regional authorities, such as the EU projects PROPOLIS: Planning and Research of Policies for Land Use and Transport for Increasing Urban Sustainability (Lautso et al. 2004) and STEPs: Scenarios for the Transport System and Energy Supply and their Potential Effects (Fiorello et al. 2006) and a project for the State government of North-Rhine Westphalia (Spiekermann and Wegener 2005) and is presently extended to the whole Ruhr area (Huber et al. 2011).

It is impossible to show in this chapter the full range of policy issues investigated with the model in these projects. Here only one example will be given.

In the STEPs project it was explored what impact on travel behaviour would result from a combination of travel demand management and land use policies on the location of households and firms (Fiorello et al. 2006).

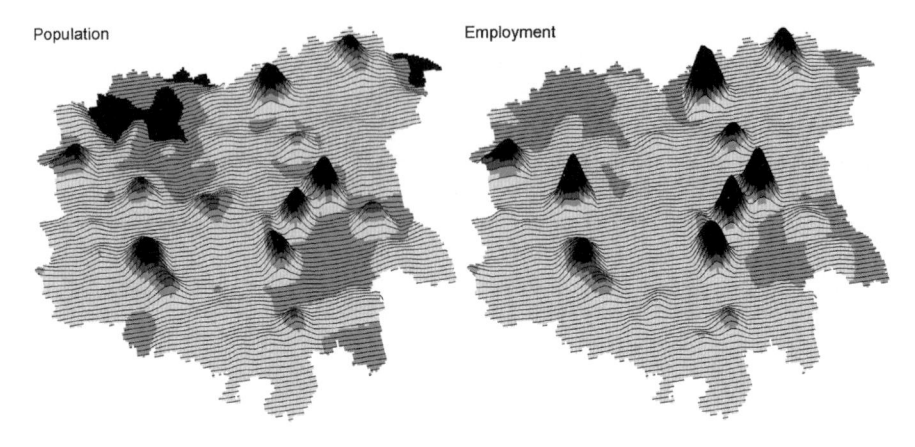

Fig. 2.7 Scenarios of concentration of population and employment at public transport stations

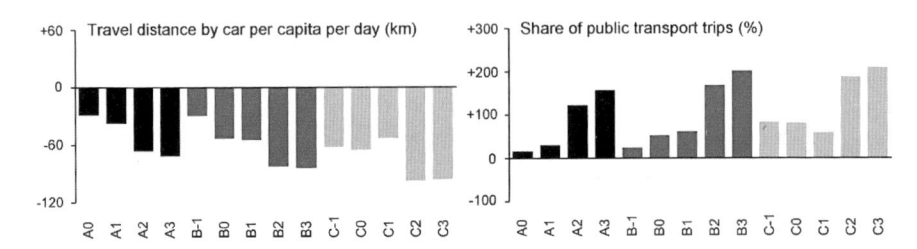

Fig. 2.8 The impacts of ant-sprawl land-use policies and other transport policies on travel distances by car and share of public transport trips

Figure 2.7 shows the effects of a scenario in which transport-related policies, such as higher fuel taxes and parking restrictions are combined with a land use plan restricting development to locations near rail stations. The two surfaces in the upper part of the figure show the effects on population and employment location as differences to the reference scenario.

Figure 2.8 shows the impacts of a range of scenarios on total distances travelled by car and modal shares. It can be seen that the scenarios with anti-sprawl land use controls (A2, A3, B2, B3 and C2 and C3) perform better in terms of environmental quality, energy conservation and reduction of greenhouse gas emissions.

2.9 Conclusions

This chapter presented a model of the urban labour market which is distinct from other models by its close integration between demand for labour by firms and the supply of labour by households.

Employment and labour in the model are affected by developments in other submodels: The supply of jobs results from growth and decline, location and relocation of firms and their demand for labour by skill. The demand for jobs results from population and immigration and outmigration by age, education and skill. Intraregional labour mobility results from decisions of firms to hire or release workers and decisions of workers to start or end a job. These decisions affect commuting patterns and residential and firm location and the associated construction and real estate markets.

It was demonstrated that the model could be calibrated with a minimum of data and produces relevant answers to urgent current urban problems.

References

Fiorello D, Huismans G, López E, Marques C, Steenberghen T, Wegener M, Zografos G (2006) Transport strategies under the scarcity of energy supply. In: Monzon A, Nuijten A (eds) STEPs final report. Buck Consultants International. The Hague. http://www.steps-eu.com/reports.htm. Accessed on 26 August 2012

Huber F, Spiekermann K, Wegener M (2011) Cities and climate change: a simulation model for the Ruhr Area 2050. In: Schrenk M, Popovich VV, Zeile P (eds) Tagungsband der Tagung REAL CORP 2011, Essen, 18–20 May 2011. http://www.corp.at/archive/CORP2011_94.pdf. Accessed on 26 August 2012

Lautso K, Spiekermann K, Wegener M, Sheppard I, Steadman P, Martino A, Domingo R, Gayda S (2004) PROPOLIS: planning and research of policies for land use and transport for increasing urban sustainability. PROPOLIS final report. LT Consultants, Helsinki. http://www.ltcon.fi/propolis/. Accessed on 26 August 2012

Spiekermann K, Wegener M (2005) Räumliche Szenarien für das östliche Ruhrgebiet. Schlussbericht. Institut für Landes- und Stadtentwicklungsforschung und Bauwesen des Landes Nordrhein-Westfalen (ILS NRW), Dortmund. http://www.ils-forschung.de/down/raum-szenarien.pdf. Accessed on 26 August 2012

Wegener M (1982) Modelling urban decline – a multilevel economic-demographic model for the Dortmund region. Int Reg Sci Rev 7:217–241

Wegener M (1983) Description of the Dortmund region model. Working Paper 8. Institute of Spatial Planning, University of Dortmund, Dortmund

Wegener M (1985) The Dortmund housing market model: a Monte Carlo simulation of a regional housing market. In: Stahl K (ed) Microeconomic models of housing markets, vol 239, Lecture notes in economics and mathematical systems. Springer, Berlin/Heidelberg/New York, pp 144–191

Wegener M (1986) Transport network equilibrium and regional deconcentration. Environ Plann A 18:437–456

Wegener M (1994) Die Stadt der kurzen Wege – müssen wir unsere Städte umbauen? Working Paper 136. Institute of Spatial Planning, University of Dortmund, Dortmund

Wegener M (1996) Reduction of CO_2 emissions of transport by reorganisation of urban activities. In: Hayashi Y, Roy J (eds) Land use, transport and the environment. Kluwer Academic, Dordrecht, pp 103–124

Wegener M (1998) The IRPUD model: overview. http://irpud.raumplanung.tu-dortmund.de/irpud/pro/mod/mod_e.htm. Accessed on 26 August 2012

Wegener M (2011) The IRPUD model. Working Paper 11/01. Spiekermann & Wegener Urban and Regional Research, Dortmund. http://www.spiekermann-wegener.de/mod/pdf/AP_1101_IRPUD_Model.pdf. Accessed on 26 August 2012

Wegener M, Spiekermann K (1996) The potential of microsimulation for urban models. In: Clarke GP (ed) Microsimulation for urban and regional policy analysis, vol 6, European research in regional science. Pion, London, pp 149–163

Chapter 3
Modelling the Economic Impacts of Transport Changes: Experience and Issues

David Simmonds and Olga Feldman

Abstract This paper describes and illustrates the approach to land-use/transport/economic interaction modeling that the authors and colleagues have implemented as the DELTA software package and have used to forecast and appraise the employment and economic impacts of a range of proposed or possible transport strategies and schemes under consideration in different parts of the United Kingdom. It first describes the background to the development of the approach. Secondly it outlines the models developed within this approach, focusing in particular on the two-level representation of space and on the links between transport and the economy. The workings of the model in practice are then illustrated with some example forecasts from one major study. We then outline the current approach to appraisal of wider economic impacts, and discuss the appraisal results corresponding with the example results. The final section discusses some current developments.

3.1 Introduction and Structure

This paper describes the family of models which have been implemented using full applications of the DELTA software package, with an illustration of the model's use and discussion of some current areas of development in relation to employment forecasting. The first section of the paper describes the background to the development of this particular modeling approach. The second section presents some of the

D. Simmonds (✉)
David Simmonds Consultancy Ltd, Suite 14, Millers Yard, Mill Lane, Cambridge CB2 1RQ, UK

Honorary Professor, Heriot-Watt University, Edinburgh, Scotland
e-mail: david.simmonds@davidsimmonds.com

O. Feldman
Transport for London, Transport Strategy and Planning, 9th floor Windsor house, 42–50 Victoria street, London SW1H 0TL, UK
e-mail: OlgaFeldman@tfl.gov.uk

detail of the approach, in particular focusing on the ways in which the transport system affects employment and the economy. The working of the model is illustrated with results from one particular application. The fourth section then outlines the approach to appraisal of economic impacts which is currently mandated by the UK Department for Transport (DfT), and discusses the appraisal results corresponding with the example forecasts previously presented. The final section considers issues arising from the assessment of wider economic impacts and some of the ways in which the modeling approach is being further developed in response to these issues.

In UK transport planning practice, the *ex ante* estimation of the costs and benefits which are expected to accrue from a proposed intervention is referred to as **appraisal**, whilst *ex post* estimation of the costs and benefits flowing from a completed intervention is referred to as **evaluation**. That convention is observed in this paper.

3.2 Background

The DELTA package was originally developed as a single-level urban model intended to model a medium-sized city as a single economic unit (the design and prototype applications of this version were described in Simmonds and Still 1998 and Simmonds 1999). The design concentrated on representing incremental change through inter-related processes operating over time, the main processes of change being the development of building stocks, changes in housing quality; household change (reflecting demographic and social change); household and employment location; and the employment status (working/not working) of household members.

Initial applications of this purely "urban" model in Edinburgh and Manchester led to an invitation to apply the same approach to the much larger area of the Trans-Pennine Corridor, extending some 250Km across the full width of Northern England, covering four metropolitan conurbations, a number of other major urban areas and some more rural areas. In the early phases of this project we had to consider the widespread agreement in the literature of spatial economics and location analysis that the variables affecting households' or businesses' choices of where to locate within a city are to some extent different from those affecting their choices of location between cities. In the demographic field, for example, Gordon and Molho (1998) have shown that different variables drive household moves over short, medium and long distances (even within the limited distances within the UK). The equivalent point about economic phenomena is emphasized by McCann (2001, p. 3).

The differences between "intra-urban" and "inter-urban" processes of change and location meant that we could not develop the Trans-Pennine model simply by scaling up the earlier Manchester model. We therefore extended the DELTA design to add higher-level economic and migration models, representing choices or processes working between rather than within urban areas. The urban (zonal) level models remained largely unchanged in the expansion to the two-level system, but now take their controlling inputs for each modelled year from the higher level models rather than from exogenous inputs to the system, and return values to the

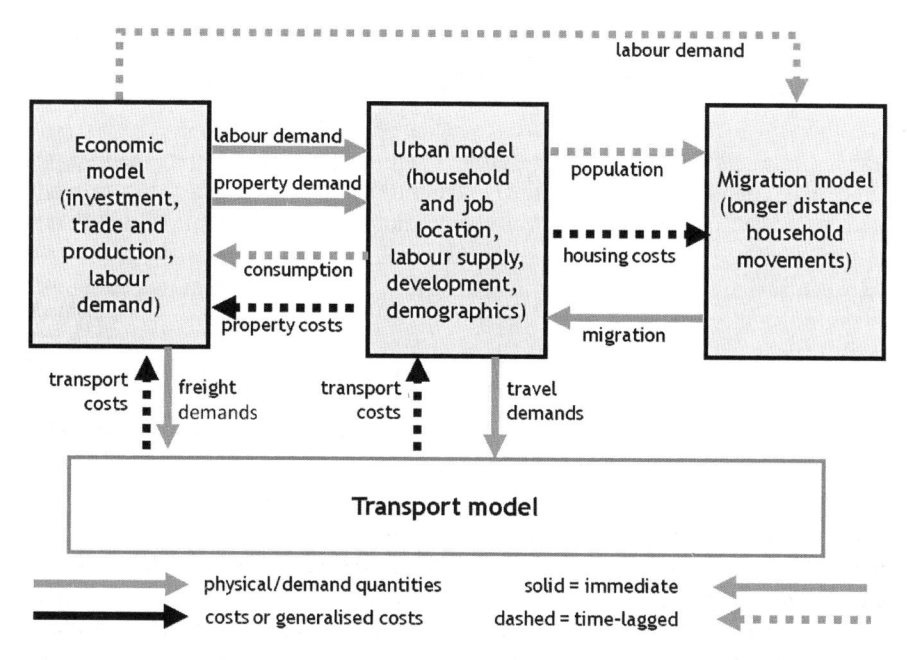

Fig. 3.1 Structure of the DELTA model: main components

higher level models which influence subsequent migration and economic change. The resulting design, summarized in Fig. 3.1, has provided the basis for the DELTA package over the past decade.

The fact that the DELTA package allows for two distinct levels of spatial choice, representing local choices (usually within urban areas) and sub-regional or higher choices (usually between urban areas) is an important feature and one which we believe is unusual if not unique in land-use/transport/economic interaction modeling.

3.3 The Regional Economic Model in DELTA

Within the limits of this paper we cannot attempt to describe the whole DELTA design in detail. Instead, we concentrate on the higher-level representation of economic change, ie the regional economic model, and its interactions both with the lower, zonal level and with the transport model.

The regional economic model itself consists of two components: a cross-sectional spatial input–output model, and an incremental model of investment or disinvestment over time in each sector in each sub-region. The spatial input–output framework is widely used in land-use/transport interaction modelling. Unlike some model systems where the spatial input–output framework provides most or all of the model mechanism, its use within DELTA is limited to representing the short-term economic outcome for economic sectors at the higher spatial level (unlike some other applications, it is not for example used to forecast household numbers or locations). The final demand consists (quite conventionally) of exports, government expenditure and fixed capital formation, which are exogenous inputs defining the modelled scenario, and of consumer expenditure which is endogenous to the model and calculated from the number, mix and employment status of households. The pattern of trade is forecast within the model and is sensitive to the costs of transport between sub-regions; these include both freight haulage and business travel costs, and are calculated from the associated transport model outputs so that they vary with changes in transport policy, transport infrastructure or forecast levels of congestion.

The spatial input–output model can be summarized in three main equations. First, the demand for commodity s in area j at time $(t + 1)$ is, as usual, defined as being the sum of final and intermediate demands:

$$Y^s_{(t+1)j} = Y(F)^s_{(t+1)j} + \sum_r a^{sr}_p . P^r_{(t+1)j} \qquad (3.1)$$

where

$Y^s_{(t+1)j}$	is the total demand for commodity s in area j at time $(t + 1)$
$Y(F)^s_{(t+1)j}$	is the final demand for commodity s in area j at time $(t + 1)$ (inclusive of consumer final demand, which is endogenous to the model)
a^{sr}_p	is the technical coefficients of the input–output matrix, ie the quantity of s input per unit of r output during period p
$P^r_{(t+1)j}$	is the total production of r in j at time $(t + 1)$.

Secondly, the demand for commodity s in area j is allocated to production areas by a logit trade model which takes account of the production capacity of each area and the costs of delivery from production area to consumption area:

$$T^s_{(t+1)ij} = Y^s_{(t+1)j} \left[1 - m^s_p\right] \frac{K^s_{(t+1)i} . \exp\left[-\lambda^s_t (p^s_{(t)i} + c^s_{Tij} + b^s_{(t+1)ij} + r^s_{ij})\right]}{\sum_i K^s_{(t+1)i} . \exp\left[-\lambda^s_t (p^s_{(t)i} + c^s_{Tij} + b^s_{(t+1)ij} + r^s_{ij})\right]} \qquad (3.2)$$

where

$T^s_{(t+1)ij}$	is the trade in s from i to j at time $t + 1$

(continued)

m_p^s	is the proportion of demand for s which is met by separately modelled imports in period p
$K_{(t+1)i}^s$	is the capacity of zone i to produce s at time $t + 1$
λ_t^s	is the distribution coefficient for s in period p
$p_{(t)i}^s$	is the production price of s at area i and time t
c_{Tij}^s	is the cost of transporting one unit of s from i to j in the previous transport model year T
$b_{(t+1)ij}^s$	is the non-transport cost (border effects, etc.) per unit of trade in s between i and j at time $t + 1$ (if used)
r_{ij}^s	is the residual disutility per unit of trade in s between i and j at time t (if used).

Thirdly, the production of each commodity in each area is defined as the sum of the trades it supplies:

$$P_{(t+1)i}^s = \sum_j T_{(t+1)ij}^s \qquad (3.3)$$

The model is solved iteratively, with the levels of production calculated in the third equation being substituted back into the Eq. 3.1 until the system converges having accounted for the total economy of the modeled region.

The sub-regional capacities by sector which constitute the "size" terms in the logit models for the trade distribution model are forecast by the model of investment distribution, which represents longer-term business decisions, and a simple model of disinvestment, based on a fixed depreciation rate. The investment model is an incremental model which is sensitive both to the costs of producing each sector's output in each one sub-region and to the sub-region's accessibility to the market for that sector. The key equation is

$$K(N)_{pa}^s = K(N)_{p*}^s \frac{K_{ta}^s \cdot \left[\frac{A_{t_A a}^s}{A_{t_B a}^s}\right]^{\alpha_p^s} \cdot \left[\frac{c_{t_A a}^s}{c_{t_B a}^s}\right]^{\gamma_p^s}}{\sum_a K_{ta}^s \cdot \left[\frac{A_{t_A a}^s}{A_{t_B a}^s}\right]^{\alpha_p^s} \cdot \left[\frac{c_{t_A a}^s}{c_{t_B a}^s}\right]^{\gamma_p^s}} \qquad (3.4)$$

where

$K(N)_{pa}^s$	is the quantity of new investment in sector s allocated to area a during period p
$K(N)_{p*}^s$	is the total new investment in sector s to be allocated during p
K_{ta}^s	is the pre-existing capacity of sector s in area a (before disinvestment or new investment)

(continued)

$c_{t_Aa}^s, c_{t_Ba}^s$	are recent and time-lagged measures of cost of producing s in area a
α_p^s, γ_p^s	are coefficients defining the sensitivity of investment location to accessibility and to cost; and
$A_{t_Aa}^s, A_{t_Ba}^s$	are more and less recent time-lagged Hansen measures of accessibility to consumption, using the same variables and distribution coefficient as the trade distribution model (Eq. 3.2)

$$A_{ta}^s = \sum_z \left\{ Y_{tj}^s \cdot \exp\left(\lambda_t^s \left[c_{taz}^s + b_{(t+1)ij}^s + r_{ij}^s \right]\right) \right\} \quad (3.5)$$

(Note that only the main equations for the modelled areas are shown here, excluding various complications relating mainly to imports and exports.)

3.4 Interaction Between Economy, Urban Activities and Transport

DELTA's economic model provides the "drivers" for its modeling of changes in employment at the zonal level. Employment is assumed to be determined partly by capacity (typically "management" jobs) and partly by the level of production. These employment outputs determine both the the total demands for floorspace by each sector in each area, and the demand for labour by sector and socio-economic group in each area. These linkages are represented by the downwards arrows from the area level (the regional model) to the zonal level Fig. 3.2. The modelled employment sectors are often rather more disaggregate at the zonal level, mainly to take account of different types of employment within one sector using different types of floorspace (e.g. office rather than industrial).

The zonal-level location models allocate employment changes to zones, taking account of the availability of space (a very strong effect) and accessibility to labour, customers, etc. (generally a weaker effect at the zonal level, but varying markedly between sectors). The form of employment location model is typically an incremental logit model of the form

$$E(LM)_{p(i \in a)}^s = \left[\sum_{i \in a} E(M)_{pi}^s \right] \cdot \left\{ \frac{E(M)_{pi}^s \cdot \left(\frac{F(A)_{pi}^u}{F(M)_{pi}^u} \right) \cdot \exp(\Delta V_{pi}^s)}{\sum_{i \in a} \left[E(M)_{pi}^s \cdot \left(\frac{F(A)_{pi}^u}{F(M)_{pi}^u} \right) \cdot \exp(\Delta V_{pi}^s) \right]} \right\}$$

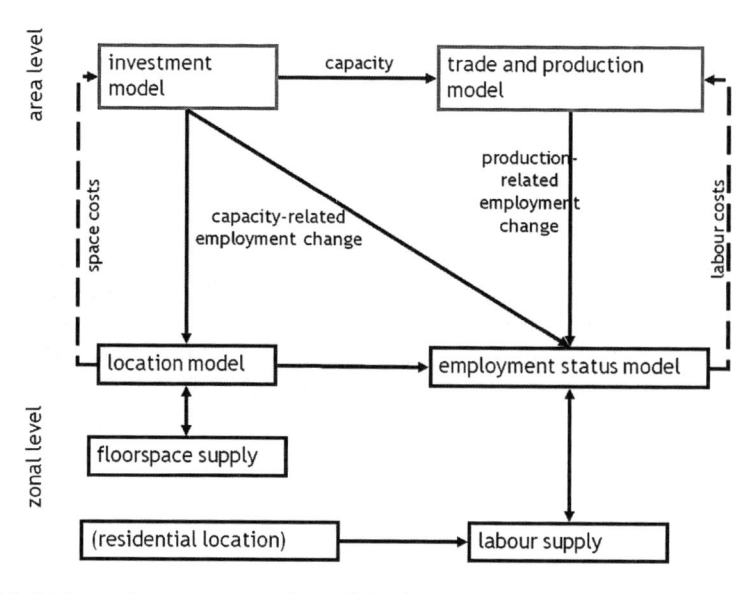

3 Modelling the Economic Impacts of Transport Changes: Experience and Issues

Fig. 3.2 Linkages between area and zonal level: economic/employment. (NB other inputs affecting floorspace supply and residential location not shown)

where:

$E(LM)^s_{pi}$	is mobile employment in sector s located to zone i;
$E(M)^s_{pi}$	is mobile employment in sector s initially located in zone i;
$F(A)^u_{pi}$	is available floorspace type u within which s can locate;
$F(O)^u_{ti}$	is previous occupied floorspace of type u; and
$F(M)^u_{pi}$	is space of type u previously occupied by employment (of any sector) now classified as "mobile"
ΔV^s_{pi}	is the change in utility of location over a defined period of years:

$$\Delta V^s_{pi} = \theta^{sC}_p \left(c^s_{pi} - c^s_{(tB(U,s))i}\right) + \theta^{sA}_p \left(A^s_{(tA(A,s))i} - A^s_{(tB(A,s))i}\right)$$

$\left(c^s_{pi} - c^s_{(tB(U,s))i}\right)$	is the change in cost of location over the relevant period
$\left(A^s_{(tA(A,s))i} - A^s_{(tB(A,s))i}\right)$	is the change in accessibility over the same period (the accessibility variable itself being a weighted composite of accessibility to labour, to customers, to suppliers)

(continued)

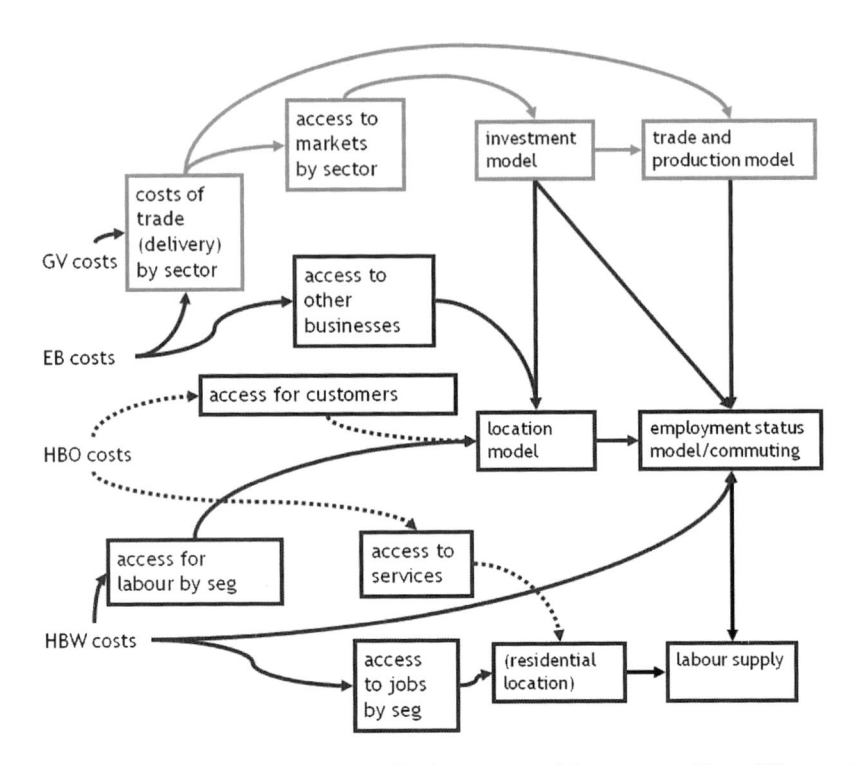

Fig. 3.3 Linkages from transport generalised costs to model processes. (Note: GV = goods vehicle, EB = employer's business, HBW = home-based work, HBO = home-based other (i.e. non-working travel, including commute). "Costs" = generalized costs)

θ_p^{sC}	is the coefficient on cost of location
θ_p^{sA}	is the coefficient on accessibility.

The employment status model distributes the resulting located jobs to residents, given the supply of workers at each point in time, determining both the mixture of working/non-working residents in each zone and the matrices of commuting. Both the supply of floorspace for employment activities, and the number and mix of residents (the supply of labour) are the results of other sub-models responding to change over time (and not necessarily in equilibrium with the economic model).

From the results of the zonal sub-models, the levels of consumer demand by area (calculated from households and their incomes), and the average floorspace costs are calculated and passed back up (with time-lags) to the area level.

Transport costs feed into the modeling of economic and employment change in a variety of ways, illustrated in Fig. 3.3, where the key components of Fig. 3.2 are in the upper-right part of the diagram. In a typical application,

- Goods vehicle costs (GV) and costs of travel on employer's business (EB) are converted into typical costs per unit of trade which enter directly into both the production/trade submodel and into the accessibility-to-market calculations for the investment model (they also enter indirectly from the production/trade submodel to the accessibility-to-market calculations as part of the costs of inputs)
- The generalized costs of home-based travel (to work and to other destinations, including shopping and services), and of employer's business travel, enter into the zonal-level calculations of accessibility which affect the changes in employment location within each area (and of course within the available floorspace supply); they also enter into the accessibility calculations which affects households' choices of location (again of course within the available housing supply).

In relation to Fig. 3.3, note that not all the linkages are shown: in particular, the location of employment (by socio-economic level) is an essential variable in the accessibility to jobs calculation, particular types of jobs or particular types of floorspace are used in the accessibility to services, and numbers of workers by place of residence in accessibility to labour.

Different components of change in the transport system therefore affect different processes of change – and different actors within those processes – in the land-use/economic system. This can in some cases result in complex influences of transport upon land-use, especially if the nature of a transport intervention under consideration is such as to have differential impacts on different segments of transport demand.

For example, a congestion charging scheme which increases the generalized costs of commuting but reduces the generalized costs of business travel and goods distribution will tend to make some locations more attractive to certain activities whilst at the same time making them less attractive to others. Moreover the different activities and processes affected have different rates of response within the dynamic structure of the model, sometimes leading to complex effects over time (especially if the interventions are substantial enough that the indirect effects are also significant).

Some general patterns of impacts can be discerned, generalizing across different applications of the model. Major transport changes affecting connectivity between urban areas tend

- To have their main direct impacts on the distribution of economic activity, and hence of employment, across the areas modeled;
- If the transport changes are important enough, significant multiplier effects will arise both through supplier and income effects, potentially affecting sectors which may have been largely unaffected by the initial change;
- Impacts on the location of population come about first through the impact of changing employment levels on incomes (which affects locational preferences both by modifying the affordability of housing, and by affecting car ownership and hence the perception of accessibility), and then more gradually as the change in the pattern of employment opportunities modifies the pattern of migration between areas.

In contrast, transport changes mainly affecting local travel tend to show that

- Their main direct impacts are on residential location (and hence on commuting patterns) within urban areas;
- Their employment impacts arise mainly in terms of changes in the quantity and distribution of service employment resulting from the changes in residential location and incomes, with further small-scale multiplier effects following from these;
- Most of the impacts (measured as percentage changes in zonal variables from the equivalent year of the reference case) occurs within 5–7 years of the initial intervention.

3.5 Applications

The two-level form of DELTA has been used, in conjunction with various four-stage or more elaborate transport models, in a number of major studies and ongoing projects in different parts of the UK, as listed in the following table.

In all of the cases listed the usual practice is that the LUTI modelling (DELTA + transport model) represents a substantial region, whilst the transport and land-use planning proposals is carried out on the basis of fixed economic and demographic scenarios for a large region; the additional models of wider economic impacts which have been developed by the UK Department for Transport are used to assess the net economic change for that region as a whole. Those models of wider economic impacts and their use in conjunction with a DELTA-based LUTI model were described in detail in Feldman et al. (2008) and are summarized below. It should also be noted that whilst there is an increasing emphasis on forecasting the impact of transport improvements (or transport management) on total economic activity, there is also an increasing requirement for spatial detail in order to represent land-use planning policies in more detail and to take account of network detail in the transport model's representation of highway and transit networks.

3.6 Example Results

We illustrate the workings and outputs of a typical DELTA model using a set of comparative results from the SWYSM model for South and West Yorkshire. The model was extensively used in 2005-6 to forecast and appraise the impacts of a wide range of largely hypothetical schemes, as input to the Eddington Review (ETSR 2006); the results quoted are taken from one of these tests. The model itself is a full DELTA model application covering the South and West Yorkshire subregions and a substantial adjoining area. This fully modelled area, shown in the maps below, had a base year population of just under 3.5 million persons; this is surrounded by

Table 3.1 DELTA applications using full regional/zonal approach

Core area/study	Transport modelling package used	Comments	Reference(s)
Trans-Pennine Corridor/ Strategic Environmental Assessment in the Trans-Pennine Corridor	START	Urban/regional model	Simmonds and Skinner (2001)
South & West Yorkshire Strategic Model (SWYSM)/South and West Yorkshire Multi-Modal Study, Eddington Study and other projects and scheme appraisals	START	Urban/regional model	Simmonds and Skinner (2002, 2004), Feldman et al. (2007)
Glasgow & Clyde Valley/ Central Scotland Transport Corridors Study, Clyde Corridor Study, appraisals of major motorway schemes	TRIPS (CSTM3A application)	Urban model focused on West Central Scotland; regional model covers whole of Scotland	
Edinburgh & Lothian (3)/ New Transport Initiative	TRAM	Urban model focused on Edinburgh and Lothian; regional model covers whole of Scotland	Simmods et al. (2005)
Strathclyde(SITLUM)/for use by transport and land-use planning agencies	STM	Urban model focused on Strathclyde; regional model covers whole of Scotland	Aramu et al. (2006)
TELMoS (Transport/ Economic/Land-use Model of Scotland)/ Scotland (various successive versions)	TRIPS, then CUBE (Transport Model for Scotland applications)	Urban/regional model	Nicoll et al. (2006), Bosredon et al. (2009)
GMSPM2 (Greater Manchester Strategy Planning Model 2)	TRAM	Urban/regional model, focused on Manchester City Region	Dobson et al. (2009)
LonLUTI (London, East and South-East England)	LTS	Urban/regional model, focused on London	Feldman et al. (2010)

a very substantial Buffer Area, modelled in less detail, occupied by a further seven million persons. Further detail of the model is given in the references for SWYSM in Table 3.1.

These results are from a test looking at the possible improvement of the highway links between Leeds and Sheffield, ie between the two main centres of the region under consideration, compared with a Reference Case without that improvement. The scheme tested involved widening the existing M1 motorway by one lane (i.e. from dual three to dual four lane), and widening the connecting roads linking the M1 to central Leeds and to central Sheffield. The focus is on the impacts of this improvement (referred to for simplicity as the M1 improvement test) relative to the Reference Case, rather than upon the absolute forecasts.

The following paragraphs describe the results in the order transport-economy-population, because this is the dominant logic of the model in relation to this particular test (at least in looking at the regional rather than the local pattern of impacts). It should however be kept in mind that there are numerous interactions between these three areas of the model in the course of the forecast, e.g. there is feedback from the land-use economic modelling to the transport model as the forecasts run forward over time.

3.6.1 Transport Impacts

The overall travel demand effects of this test across the Fully Modelled Area (FMA) are relatively small in proportional terms, relative to the Reference Case. This is quite typical at this level of modelling; although the improvement of the highway links would be a major investment, it is still a relatively modest change in the overall transport supply of the region.

Over time, car trip making rises slightly faster than in the Reference Case, as one would expect given the increase in highway capacity. The impact on car trip kilometres is slightly higher, with the increase beginning at 2005 (the year in which this hypothetical scheme was assumed to open). The overall results for goods vehicles traffic are similar. Bus passenger trips and trip kilometres show an increase as a result of this intervention (2.5 % for trips and 1.7 % for trip kilometres in 2020). This is because of the benefits to bus operations from the highway capacity increase. By contrast, as would be expected, rail shows a slight decrease in 2005 (where only transport impacts are present). In the later years, the impact of economic activity stimulated by the intervention leads to a small increase (less than 1 %) in rail travel, an effect that is also occurring for bus.

Changes in vehicle kilometres, as well as being small in overall terms, do not exhibit any great variation between districts or by time of day. In 2020, Wakefield, Barnsley, Sheffield and Rotherham exhibit increases in traffic flows consistently around the 1 % level across the day.

The location-specific transport results for this test can be summarised as follows:

- By 2020 there are significant numbers of extra commute and business trips, particularly by car. These are mainly in Leeds and Sheffield, but growth occurs in many other districts with connections to this improved motorway alignment;
- Sheffield experiences significant growth in goods vehicle trips;
- As would be expected given the provision of extra road capacity, Sheffield to Leeds and Rotherham to Leeds experience the largest reduction in congestion; and

Employment: Test HP-HO

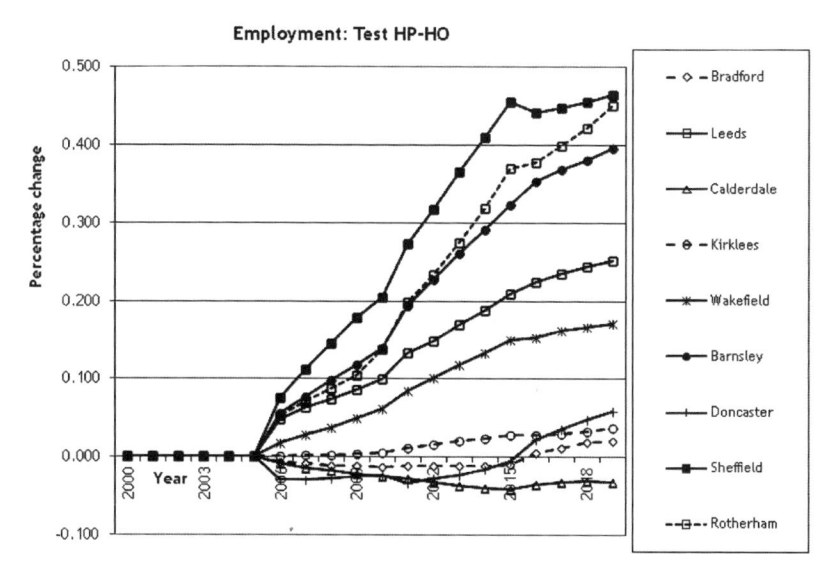

Fig. 3.4 Employment differences between M1 test and reference case

- Some increases in congestion occur where trip making has increased (note that for Sheffield the test includes some urban highway capacity increases to connect the city centre to the M1 motorway northwards).

3.6.2 Land-use/Economic Impacts

Employment: The employment effect of the test is concentrated on the Leeds to Sheffield corridor. The effect on the total employment by local authority for the highway test (HP) compared with the Reference Case (HO) is shown in Fig. 3.4. As can be seen, 2021 employment with the highway improvements is slightly higher than the in the Reference Case for all Districts within South and West Yorkshire except Calderdale. The more significant positive impacts are for the districts along the Leeds-Sheffield highway route, i.e. Leeds, Wakefield, Barnsley, Rotherham and Sheffield.

These results are a good illustration of a common feature in forecasting regional economic impacts: small percentage changes representing rather large absolute numbers. The impacts of the test on Sheffield and Rotherham are increases of less than 0.5 %, but this represents a gain of nearly 2000 jobs compared with the Reference Case.

The impacts come about in the first instance through the workings of the regional economic model. The main effect is in the investment model. The improvement in transport along the M1 corridor slightly improves the accessibility of locations in that corridor to some of their markets and suppliers. As a result, a slightly higher proportion of the investment or reinvestment that takes place in each sector, each year, is attracted to the areas in the corridor. This higher rate of investment, and the higher levels of production that result from it, require slightly more labour than was employed in the Reference Case, resulting in the increases shown in Fig. 3.4.[1]

The transport improvement, introduced as a single step change in 2005, has a direct effect on investment for the following 10 years. This period is related to the rates of reinvestment: the model assumes that one-tenth of all investment has to be replaced (and is influenced by changed circumstances) each year and hence that all investment is influenced within 10 years of a transport change. Changes in area employment **after** 2015 (beyond 10 years from the transport change) are therefore due to a combination of:

- Continuing multiplier effects between industrial sectors;
- Multiplier effects as a result of population and consumer expenditure effects (see below), including in particular those from continuing migration in response to earlier employment changes;
- The transport improvement itself having increasing effects over time: the time savings due to the improvement (compared with the Reference Case situation) tend to increase in significance in later years as a result of increasing values of time and increasing levels of congestion in the Reference Case; and
- Changes in the supply of commercial floorspace, themselves induced by the changes in demand resulting from the transport improvement.

The graph indicates that whilst the three districts most remote from the M1 Corridor – Bradford, Calderdale and Doncaster – all suffer slightly negative effects in the early years after 2005 as a proportion of investment is diverted into the areas that become more accessible, there are slight upturns in their results in the late years of the forecast, indicating that these longer-term multiplier and similar effects are more widely distributed, through spillovers (or "leakage") between areas, rather than being concentrated close to the initial transport improvement.

The impacts of the road improvements on employment by zone in the final year of the forecast (2020) are shown in Figs. 3.5 and 3.6. Figure 3.5 shows the absolute changes in numbers of jobs; the area of the symbol for each zone is proportional to the difference in the number of jobs between the two tests. Increases (more jobs as a result of the M1 investment) are shown as dark circles, and decreases (fewer jobs) as light circles. Figure 3.6 shows the same changes as percentages of the numbers of jobs in the Reference Case forecast for 2020: darker shading shows gains, and lighter shading losses. The maps confirm that the impacts are fairly broadly distributed around the highways affected and not limited to immediately adjacent zones.

---

[1] Note that unless otherwise stated, "employment" results are always employment by workplace. The numbers of persons in work could also be tabulated by their place of residence if required, for example in order to examine impacts on areas of social deprivation.

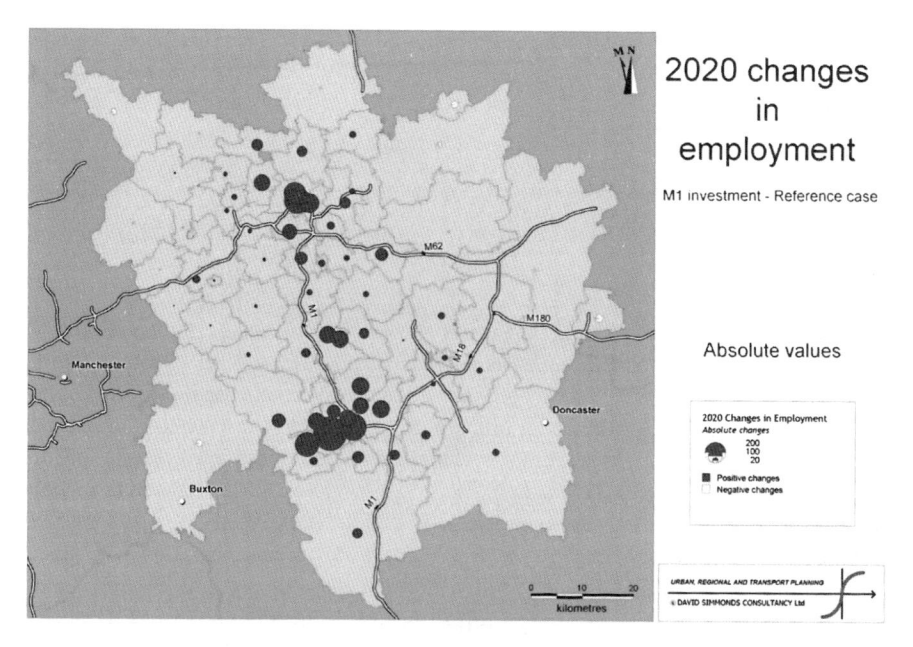

Fig. 3.5 2020 changes in employment, M1 investment vs. reference case

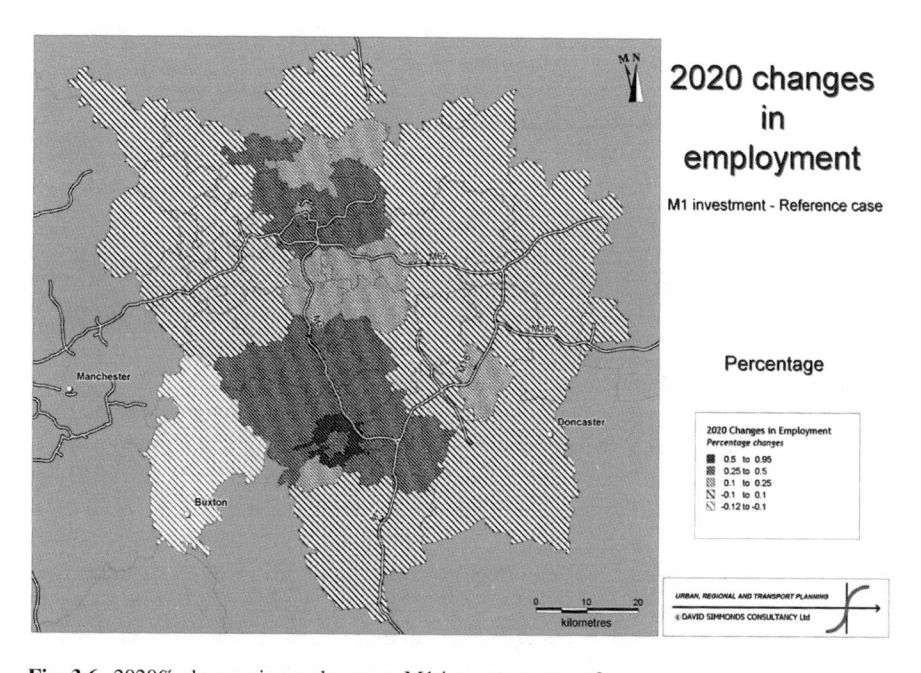

Fig. 3.6 2020% changes in employment, M1 investment vs. reference case

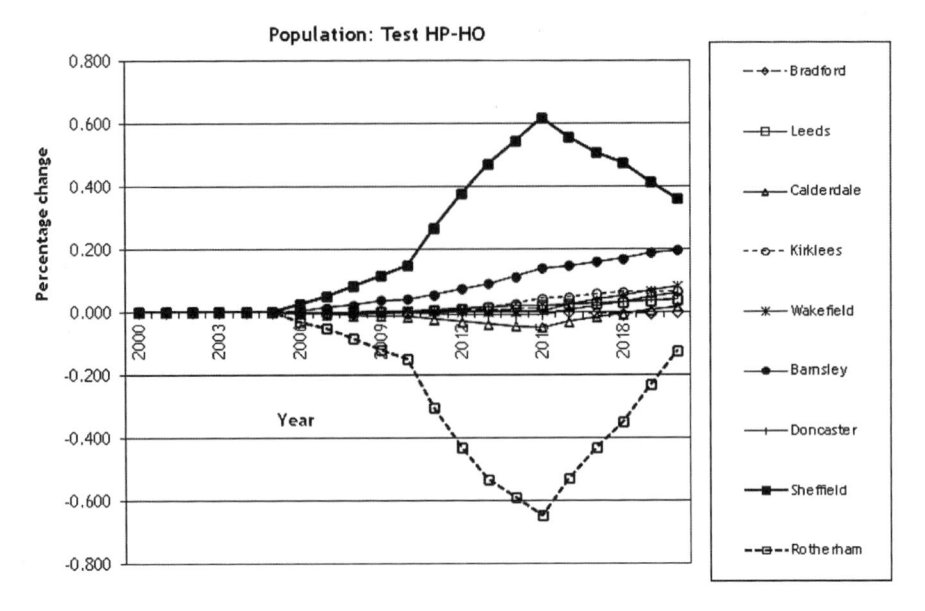

Fig. 3.7 Population differences between M1 test and the reference case

Population: Figure 3.7 shows that the main population effects of the test relative to the Reference Case are gains in Sheffield-Rotherham and in Barnsley; by 2020 there are smaller gains or no net change in all the other districts of South and West Yorkshire. It can be seen that Calderdale has marginally the most negative population result of these districts, as it did for employment. It is also apparent that the area-level population impacts are smaller than the employment impacts for the four districts where the employment impacts are most significant. (Note that Sheffield and Rotherham are within one area for regional economic and migration modeling, and the gain in Sheffield at the expense of Rotherham is due to local rather than regional factors).

These impacts are the results of changes in the pattern of migration, plus some slight redistribution of natural demographic changes as a result of those migration changes (e.g. that children are born in different places because their parents have migrated to a different place). The impacts of the transport test on the pattern of migration come about because the modelled migration is influenced by employment opportunities (a strong positive influence) and by housing costs (a weak negative influence). Younger households are more likely to move than the older or the retired. Areas where employment opportunities improve relative to the Reference Case therefore attract or retain more people than in the Reference Case, and to experience some further population growth in later years as a result of this.

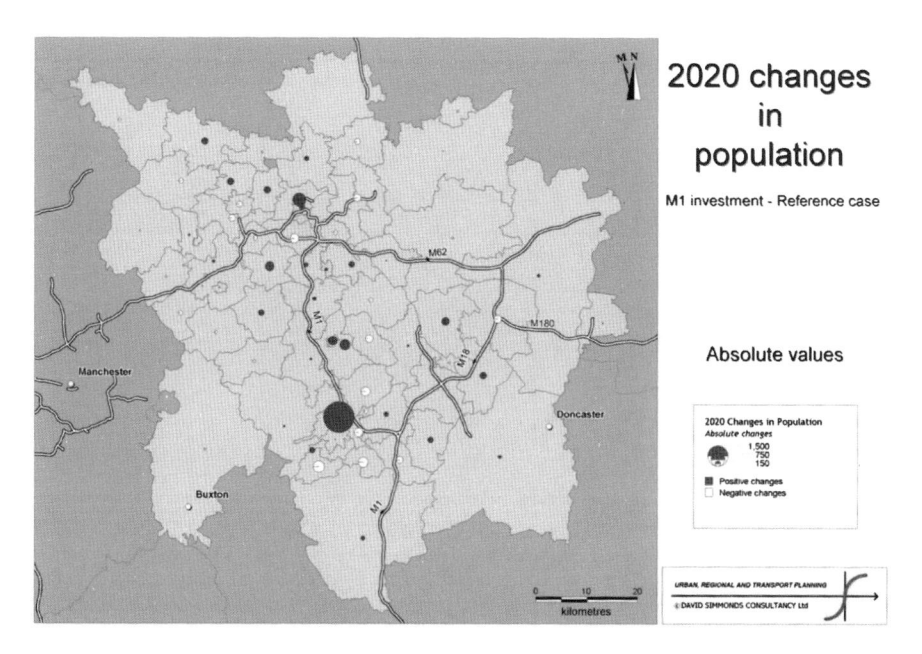

Fig. 3.8 2020 changes in population, M1 investment vs. reference case

All the migration effects come about as changes in the balance between in-flows and out-flows for each area. Migration tends to increase towards areas which have higher proportion of residents in work. More jobs in Leeds as a result of the highway scheme will therefore result in more in-migrants to, and fewer out-migrants from, the areas which supply labour to Leeds, not just to Leeds itself.

There is generally migration in both directions between any two areas. Longer-distance migration flows, though important in this analysis, are generally small in number compared with the volume of local moves, which are mainly driven by lifecycle and housing factors. These local moves can also result in households relocating across area (generally district) boundaries, so the population impacts shown in the graphs include some element of these short-distance moves.

Figures 3.8 and 3.9 show the absolute and percentage impacts of the highway investments in 2020, in the same formats as the employment maps but with blue rather than green tones for positive impacts.

The maps show that the impacts of the inter-urban highway investment on population are more widely dispersed relative to the improved road links than the employment impacts. The population impacts are essentially the effects of improved accessibility to work and service opportunities, but also take account of the (relatively small, but absolutely rather significant) employment effects shown above. (There is also of course an element of employment relocation in response to the population effects; we are here presenting the results of a recursive process in a linear form.)

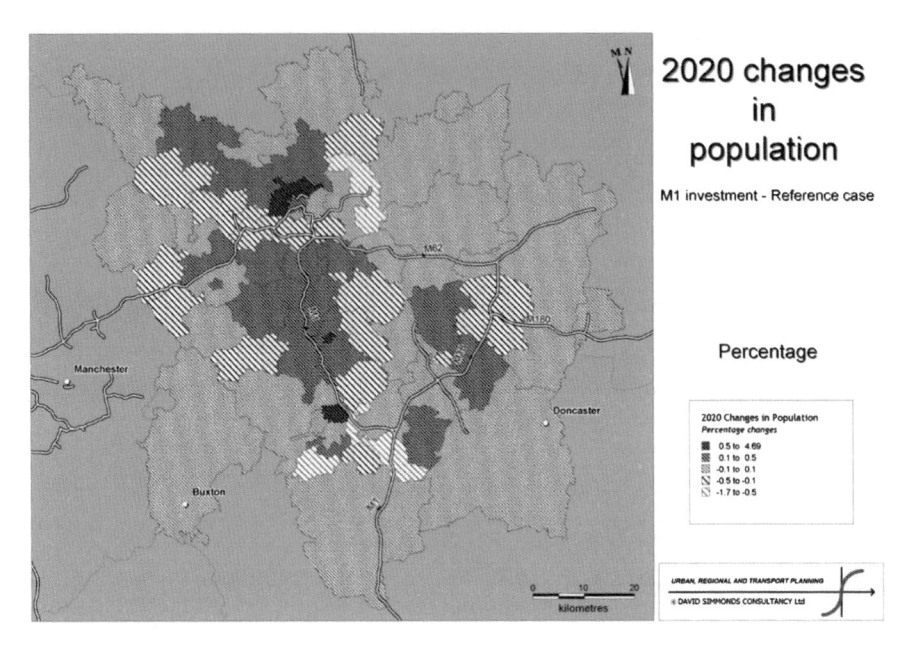

Fig. 3.9 2020% changes in population, M1 investment vs. reference case

3.7 Wider Economic Impacts of the Example Test

Once the land-use and transport consequences of proposed interventions have been forecast, their effect on agglomeration economies can be calculated using the methods developed for the Department for Transport's "Wider Impacts" calculations.

The DfT methodology identifies a number of types of potential welfare benefits which are either missing or only partly counted in the standard appraisal of transport economic efficiency. These include the following (for full list and discussion see Feldman et al. 2008):

- Agglomeration economies: these describe the productivity benefits that some firms derive from being located in higher density areas, due to better functioning labour and supply markets and to spillovers between firms;
- Benefits from improved labour supply, due to improved travel for commuters leading some people to work who would otherwise choose not to work
- Moves to more productive jobs: the DfT approach argues that certain areas display higher productivity than other areas, and hence that, if the effect of a transport intervention is to move jobs from a lower-productivity area to a higher-productivity one, there is an additional economic gain.

These benefits (or disbenefits) are shown in Table 3.2 for each of the 10 local authority areas in South and West Yorkshire, plus the fringe area which makes up the rest of the Fully Modelled Area shown in the maps above, and the surrounding

3 Modelling the Economic Impacts of Transport Changes: Experience and Issues 51

Table 3.2 Wider economic benefits of the M1 investment test, £M, 2002

	Agglomeration	More people working	Move to more productive jobs
Bradford	10	−0.2	63
Leeds	83	−1.3	1338
Calderdale	2	0.1	−35
Kirklees	9	0.3	82
Wakefield	17	0.0	286
Barnsley	15	0.0	407
Doncaster	8	0.1	98
Sheffield	61	5.2	1643
Rotherham	23	1.9	535
Fringe areas	*27*	*1.5*	*−62*
Buffer zones	*32*	*6.6*	*−4465*
Total (NPV)	**287**	**14.4**	**−111**

Buffer Area (not shown on the maps). (These are extracted from Tables 3.13, 3.14 and 3.15 of the Appendix to the original report, see ETSR 2006) Note that these are all converted to present value (£M, 2002) terms, i.e. they represent the total value of these benefits over the entire appraisal period. Note also that the move to more productive jobs is the present value of the change in value added due to more (or fewer) jobs being located in the area; the figures shown for the Fully Modelled Area are offset by negative values in the surrounding Buffer Zones giving an overall negative result of the test in this particular effect.

The Wider Economic Benefits of the M1 investment test are chiefly due to the positive effect on agglomeration. As illustrated in Feldman et al. (2008), this is typical of the relative scale of the different benefits from a variety of transport interventions. The pattern of agglomeration benefits is one of positive results throughout South and West Yorkshire, the largest contributions being from the gains in Leeds and Sheffield. Even areas which are relatively disadvantaged by the improvement of transport between Leeds and Sheffield, and where the employment effects are neutral or very marginally negative, gain in agglomeration due to a combination of more jobs relatively close by (in Leeds) and better access to a range of employment centres further afield. (Note that more recent research (Graham et al. 2009) suggests both more concentrated effects and, for some sectors, lower elasticities. The effect of this would probably be to reduce agglomeration benefits, particularly in the case of benefits which appear to be arising at a significant distance from the transport changes or their major economic impacts). The benefits from improvement in labour supply – improved commuting conditions drawing more people into work – are very small compared with the agglomeration benefits.

The positive benefits in terms of increased productivity from improved agglomeration are significantly offset by the negative impact of jobs relocating to less productive areas. This is primarily because the transport change induces a net relocation of jobs to South Yorkshire, where (in the base data) productivity per job is distinctly lower than in any other part of the modeled region. (Changes in productivity due to increased agglomeration are all attributed to the agglomeration effect, and it is assumed that no other changes in differential productivity will occur.)

3.8 Further Developments and Conclusion

DELTA models have contributed to a number of major studies and to decisions about major schemes over the past decade, by providing a reasoned and quantified basis for assessing the impacts of proposals particularly in terms of the impacts of transport interventions on land-use and economic change. As ever, there is ample scope for further work to extend the formal calibration of the model. In addition, we are continuing to develop the modeling approach further both to refine the existing design and to extent its scope in order to meet emerging requirements. (This is of course in addition to the equivalent refinements and extensions that are being pursued in the associated transport models).

As an example of potentially important current work in refinement and attention we return to the methods currently used in the UK for analysis of wider economic impacts, summarized and illustrated in the preceding section. These are not just a means of appraisal but a set of models of additional processes of change within the economy. One area of current work is to consider the possibilities and implications of bringing these wider economic impacts within the dynamic LUTI framework itself.

At present these calculations are done (in DELTA and elsewhere) as part of the appraisal process, after the forecasting process has been completed. The appraisal calculations assume that the agglomeration changes result in changes to incomes. Ideally we should calculate and use these income changes within the forecast, rather than calculating them only after the forecasts have been completed. A range of consequences will follow from the increases in incomes brought about by improvements in agglomeration. These can be examined without any change to the existing models. These would include

- Increases in employment (by sector) as a result of the increased incomes being spent (at least partly) in the regional economy;
- Increases in car ownership as a result of the increased incomes (perhaps more markedly amongst people who become employed, rather than those whose productivity is initially enhanced);
- Increases in housing demand (likewise, possibly more amongst people who go from unemployed to employed);
- Increased employment will lead to more travel to work and higher car ownership will generally mean that more of that travel is by car – so unless other policy interventions are put in place, the success in enhancing agglomeration and expanding the economy will have negative impacts on congestion and pollution. These negative consequences may be exacerbated if greater demand for housing leads to more suburban or ex-urban development in locations where private car is the only viable mode of transport.

All of these effects will tend to have spatial consequences, in the sense that increases in housing demand or traffic will not be uniform and will not necessarily arise in exactly the places where agglomeration has been increased.

The component of wider economic impacts which is closest to being implemented in DELTA is the "more people in work" effect – the extra benefits (outside the conventional consumer surplus calculations) from increased labour supply[2] due to better transport effectively increasing real wages (net of commuting time and cost). The issue that has to be addressed in bringing the labour supply response into the LUTI modeling is that the existing DELTA design, like most models in the LUTI tradition, constrains the labour supplied – residents in work plus net in-commuting – to equal the labour demand determined by the workings of the regional economic model. From a modelling point of view, it is obviously unreasonable to represent an increase in the number of residents in work without considering where they work (or indeed whether there are further workers available); moreover, to record a benefit from "more people in work" without accounting for the additional congestion and environmental impact that will result from their travel to work is difficult if not impossible to defend.

To implement a working model therefore requires that the response should be treated as one of labour **demand** rather than labour **supply**. The **demand** for labour has to be made elastic with respect to the changes in real wages resulting from the transport scheme under appraisal; the model's normal mechanisms for matching supply to demand will then ensure that the additional jobs are filled (or the reduction in jobs is matched), and impacts on commuting patterns, congestion and environment will follow automatically. In addition, the model will also take account of multiplier effects arising from the additional employment, including not only the increase in consumer demand resulting from the net increase in incomes but also the impact of increased income (and of increased employment itself) on car-ownership. Increases in car ownership will of course further modify the travel, congestion and environmental impacts – and may quite possibly arise from public transport (transit) improvements as well as from highway investment.

Another area for improvement is the possibility of bringing the productivity changes due to changes in agglomeration within the model itself – again for consistency, to ensure that the impacts of higher productivity are taken into account. There is a need for better understanding of the different components of agglomeration economies, because the three conventional components of agglomeration –deeper labour markets, better matching of suppliers to customers, and knowledge spillovers between firms – must be influenced by different aspects of transport supply. The labour market must be very strongly influenced by the conditions of transport for commuter travel; in contrast relations between suppliers

---

[2] More precisely, the DfT methodology identifies the extra benefit as the additional direct taxes paid by people who would be brought into work as a result of the transport improvement; it assumes that the net benefit to those people is captured as part of the conventional consumer surplus measure in the transport economic efficiency calculations. This will not always be the case: in many models the constraints within the transport model will not allow the labour supply response that would produce that element of consumer surplus. In that case the whole increase in GDP resulting from the additional employment should be counted, rather than just the tax component.

of goods and services and business consumers must be influenced by the conditions affecting business travel and the delivery of goods. There seems to be the beginnings of a consensus that the labour market element of agglomeration economies is generally the most important; in contrast there is some evidence that distance and transport now play very little part in some forms of knowledge spillover. For example, Gallié (2009) concludes that within collaborative technological networks, "the distance that separates partners does not constitute a barrier to knowledge flows" when the partners are located within France or within the European Union, even though it is true that face-to-face meetings are required within such partnerships. She finds that the situation is different between European and non-European partners, though to what extent distance is the barrier, to what extent institutional and cultural factors slow knowledge flows, requires further study. This suggests that in some sectors, some forms of knowledge spillover may take place at a level such that the agglomeration of activities is at a continental rather than a regional or urban level – though of course in other sectors or for other aspects of knowledge it may still be that daily face-to-face contact is essential for such spillovers to occur.

Agglomeration is not a condition which can be directly created by planning policies, though policies can be devised which deliberately seek to achieve greater agglomeration in pursuit of higher economic growth. It is also the case that "agglomeration" implies (at some scale) higher densities and hence carries some inherent dangers such as increased congestion and associated environmental problems. It is therefore important to consider whether interventions that are intended to increase agglomeration – whether they are specific projects such as major transport schemes, or policies intended to be applied across a large number of smaller decisions – will actually have that effect, taking into account feedback effects. Policies aimed at achieving growth through greater agglomeration need to be assessed within a framework which can address three distinct rounds of questions, namely:

- Will the interventions that are being considered actually succeed in increasing the critical measures of agglomeration – and what other consequences (beneficial or otherwise) will these interventions have?
- What impact will the increase in measures of agglomeration have on productivity and economic output?
- What will be the further (feedback?) consequences of that increase in productivity?

The form of land-use/transport/economic model implemented in DELTA-based models such as GMSPM2, TELMoS and LonLUTI provides a powerful structure which already deals extensively with the first and third questions, drawing on a wide variety of research in urban geography and economics, and in particular into household and firm behaviour. The central question – the increase in productivity itself – can readily be incorporated into this structure.

A congestion charging scheme which successfully prices commuters of the congested roads in order to improve conditions for business travel and goods movement would therefore contribute to greater agglomeration economies in supplier-consumer relations but would create diseconomies in the labour market. The present DfT methodology is a compromise, measuring agglomeration as accessibility to jobs based on a weighted average of commuting, business travel and goods generalized costs. Further refinement of this calculation is highly desirable, and clearly possible making use of the different types of transport costs and different measures of accessibility already modelled (as shown in Fig. 3.3).

The modelling framework we have described has proved effective and valuable in practical planning applications. It is very much open to further refinement in the light of further research and calibration, but we believe the process-oriented, dynamic framework will continue to provide an effective context for such work and for the use of models to inform both strategic policy-making and the decisions on major transport and planning projects.

References

Aramu A, Ash A, Dunlop J, Simmonds DC (2006) SITLUM – the strathclyde integrated transport/land-use model, Proceedings of the EWGT 2006 international joint conferences, Politecnico di Bari

Bosredon M, Dobson AC, Simmonds DC, Minta P, Simpson T, Andrade K, Gillies H, Lumsden K (2009) Transport/economic/land-use model of Scotland: land-use modelling with DELTA. Paper presented to the CUPUM conference, Hong Kong

Dobson AC, Richmond EC, Simmonds DC, Palmer I, Benbow N (2009) Design and use of the new Greater Manchester land-use/transport interaction model (GMSPM2). Paper presented to the European Transport Conference 2009. Available at www.etcproceedings.org

ETSR (2006) Wider economic impacts of transport interventions, Eddington transport study report, research annex, vol 3. Available at http://www.dft.gov.uk/about/strategy/eddingtonstudy/researchannexes/researchannexesvolume3/widereconomicimpacts http://www.dft.gov.uk/about/strategy/eddingtonstudy/researchannexes/researchannexesvolume3/widereconomicimpactsappendices

Feldman O, Simmonds DC, Nicoll J (2007) Modelling the regional economic and employment impacts of transport policies. Paper presented at the 20th Pacific Regional Science Conference, Vancouver, Available at www.davidsimmonds.com. May 2007

Feldman O, Nicoll J, Simmonds DC, Sinclair C, Skinner A (2008) Integrated transport land use models in the wider economic benefits calculations of transport schemes. Transport Res Rec 2076:161–170

Feldman O, Simmonds DC, Simpson T, Raha N, Mills C (2010) Land-use/transport interaction modeling of London. In: Proceedings of the 12th World Conference on Transport Research, Lisbon, 11–15 Jul 2010

Gallié E-P (2009) Is geographical proximity necessary for knowledge spillovers within a cooperative technological network? – the case of the French biotechnology sector. Reg Stud 43 (1):33–42

Gordon IR, Molho I (1998) A multi-stream analysis of the changing pattern of interregional migration in Great Britain 1960-1991. Reg Stud 32(4):309–324

Graham DJ, Gibbons S, Martin R (2009) Transport investment and the distance decay of agglomeration benefits. Centre for Transport Studies, Imperial College, Mimeo. Available at http://personal.lse.ac.uk/gibbons/Papers/Agglomeration%20and%20Distance%20Decay%20Jan%202009.pdf

McCann P (2001) Urban and regional economics. Oxford University Press, Oxford

Nicoll EJ, Aramu A, Simmonds DC (2006) Land-use/transport interaction modelling of the Bathgate-Airdrie railway reopening. In: Proceedings of the European Transport Conference 2006, Available at www.etcproceedings.org

Simmonds DC (1999) The design of the DELTA land-use modelling package. Environ Plann B 25 (5):665–684

Simmonds DC, Skinner A (2001) Modelling land-use/transport interaction on a regional scale: the trans-pennine corridor model. In: Proceedings of the European Transport Conference 2001, Available at www.etcproceedings.org

Simmonds DC, Skinner A (2002) The South and West Yorkshire strategic model, Yorkshire and Humber regional review, 12

Simmonds DC, Skinner A (2004) The South and West Yorkshire strategic land-use/transport model. In: Clarke G, Stillwell J (eds) Applied GIS and spatial analysis. Wiley, Chichester

Simmonds DC, Still BG (1998) DELTA/START: adding land-use analysis to integrated transport models. In: Meersman H, van de Voorde E, Winkelmans W (eds) Transport policy. Elsevier, Amsterdam

Simmonds DC, Leitham S, White V, Williamson S (2005) Using a land-use/transport/economic model in Edinburgh. In: Proceedings of the European Transport Conference 2005, Available at www.etcproceedings.org

Chapter 4
A Population-Employment Interaction Model as Labour Module in TIGRIS XL

Thomas de Graaff and Barry Zondag

Abstract This chapter looks at the integration of a population-employment interaction model in the TIGRIS XL framework. TIGRIS XL model is an integrated land-use and transport model and is actually a system of sub-models (or modules) that allows for dynamic interaction between them. Currently, the population module responses to lagged changes in employment and the employment module responses to contemporaneous changes in population. Thus, the change in population mainly drives the change in employment—an assumption which, given the strict Dutch restrictions on population location, is not entirely unrealistic. We, however, argue that this assumption is too harsh, and that sectoral employment change might be more dependent upon employment change in other sectors than on population change. We, therefore, test and estimate such relations and conclude indeed that for some sectors the employment dynamics in other sectors are more important than population changes and propose an adapted version of the labour module in TIGRIS XL. The new methodology within TIGRIS XL is assessed by looking at a (stylized) case study. This study concerned the doubling of the size of the new town Almere located 20 km east of Amsterdam. About 60,000 dwellings are to be built and 100,000 jobs are to be added in the period up to 2030. The accessibility benefits of a particular land use planning variant, with a tailored public transport investment alternative, are examined for the new labour module in TIGRIS XL. Both models predict, in case of the construction of 60,000 dwellings, a number of additional jobs much lower than 100,000. The model results of the new

T. de Graaff (✉)
PBL Netherlands Environmental Assessment Agency, Oranjebuitensingel 6, 2511 VE The Hague, The Netherlands

VU University Amsterdam, De Boelelaan 1105, 1081 HV, The Netherlands
e-mail: Thomas.deGraaff@pbl.nl; t.de.graaff@vu.nl

B. Zondag
PBL Netherlands Environmental Assessment Agency, Oranjebuitensingel 6, 2511 VE The Hague, The Netherlands
e-mail: Barry.Zondag@pbl.nl

module shows that the population-employment interaction model reacts slower and on shorter distances on population changes than the old model. Thus, employment does not seem to mold that easily to population changes as earlier Dutch employment models have predicted.

4.1 Introduction

Traditionally, integrated Land-Use and Transport Interaction (LUTI) models have difficulties with modeling sectoral and regional employment. This is partly due to the unpredictable nature of employment in itself. However, the need in these models to postulate a priori an endogeneous relation between especially population and employment is cumbersome as well. E.g., for the Netherlands it is usually assumed that the change in population drives primarily the change in employment–an assumption which, given the strict Dutch restrictions on population location, is not entirely unrealistic (Vermeulen and van Ommeren 2009).

The assumption that employment is endogeneous to population is usually justified by the empirical findings of a large (international) empirical literature, such as Boarnet (1994a,b), Carruthers and Vias (2005), Carruthers and Mulligan (2007) and Carruthers and Mulligan (2008) and – specifically for The Netherlands – Hoogstra et al. (2003) and Vermeulen and van Ommeren (2009). However, as, e.g. Hoogstra et al. (2005, 2011) clearly show, the empirical evidence is far from conclusive and subject to national or regional characteristics, such as the institutional organization of the housing and labour market. Moreover, PBL (2008) shows that the dynamics between economic sectors can be significantly higher than the dynamics between employment and population.[1]

This chapter uses the insights of this literature above by implementing a population-interaction model (which we coin Weber[2]) into an already well studied and described Dutch land-use and transport interaction model: namely, the TIGRIS XL model (RAND Europe and Bureau Louter 2006).We specifically focus on (*i*) avoiding postulating a priori an endogeneous relation between population and employment and (*ii*) on decomposing employment into various interrelated economic sectors. Moreover, our estimation procedure allows us to control for spatial spillovers in the endogenous variables and incorporate spatially correlated errors. Our model is estimated using the GS3SLS estimator of Kelejian and Prucha (2004), and incorporates endogenous spatial spillovers as well as intrasectoral and intersectoral linkages. Although, in a sense, the implementation of this new labour market module is still preliminary, it is constructed such that future implementations are general enough to model, e.g., a further liberalisation of the Dutch housing market.

---

[1] They label such endogeneity between employment in different sectors as: "Jobs follow jobs".

[2] An acronym for the Dutch phrase "Werkgelegenheid En BEroepsbevolking Ramingen".

We test the performance of our new labour market module, Weber, by evaluating an already well studied scenario analysis: namely, the expansion of the city of Almere with 60,000 newly to be built houses in the next 20 years. Although our results by no means constitute a full blown scenario analysis – let alone, a cost-benefit analysis –, the results suggest that our model performance is comparable with that of the previous labour market module with two exceptions: the new employment dynamics are (1) more local (so where one municipality 'gains', adjacent municipalities are likely to 'lose') and they are (2) less pronounced. Typically, our model culminates in less 'new' jobs for the city of Almere (due to the growth of population) than more traditional models.

The organization of the remainder of this chapter is as follows. In Sect. 4.2 we provide more detail on the TIGRIS XL model with a specific focus on the labour market module and its relation with other land use and transport models. Section 4.3 summarizes concisely the empirics and the theoretical framework of population employment interaction models. Section 4.4 implements such a model in TIGRIS XL, provides a description of the data as well as the overall results for an analysis of municipalities in The Netherlands using data for the period 1996–2008. In addition, Sect. 4.5 looks at the behavior of the new labour market module Weber within the TIGRIS XL framework during the forecast period 2008–2028 for the case of Almere. Finally, Sect. 4.6 concludes.

4.2 Structure of TIGRIS XL

The TIGRIS XL model is part of the family of Land-Use and Transport Interaction (LUTI) models and has substantial similarities to other LUTI models following a dynamic system approach, such as the DELTA modeling package in the UK (Simmonds 1999), the Urbansim model (Waddell 2001) and the IRPUD model for the Dortmund region (Wegener 1998b). The TIGRIS XL model is an integrated system of sub-models addressing specific sectors. The model uses time steps of 1 year for most of its modules, and the modules influence each other outcomes either within a year or in the next year. The underlying assumption is that the system is not in equilibrium at a certain moment in time, therefore no general equilibrium is simulated within one time step, but that the system moves towards an equilibrium. For example, a high demand for houses at a certain location can result in additional housing construction at that location in the following years. The land-use model is fully integrated with the National Transport Model (LMS) of the Netherlands and the land-use modules and transport model interact, for reasons of computation time, every 5 years.

The TIGRIS XL model consists of five modules addressing specific markets. Core modules in TIGRIS XL are the housing market and labour market module; these modules include the mutual interaction between the population and jobs and the effect of changes in transport on residential or firm location behaviour. The land and real estate module simulates supply constraints arising from the amount of

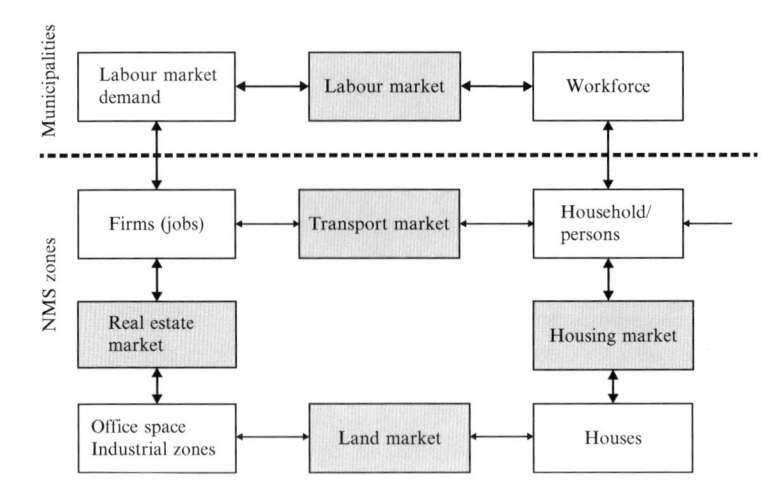

Fig. 4.1 Functional design of the TIGRIS XL model

available land, land-use restrictions and construction. In this module the user can specify different assumptions on the levels of government influence on the residential land market, ranging from a fully regulated towards a free land market. In case of a less regulated land market feedback loops exist within the model and demand preferences of households influence the amount and location of new housing. A demographic module is included in the framework to simulate demographic developments at the local level. At the national level the model output is consistent with existing external socio-economic forecasts or scenarios.

The model has a multi-scale set-up and two spatial scale levels are distinguished, namely the regional level (COROP, 40 regions in the Netherlands) and local transport zones of the National Model System (LMS sub-zones, 1308 sub-zones covering the Netherlands). Figure 4.1 presents an overview of the Tigris XL (TXL) model and the main relationships between the modules, for a more detailed description we refer to RAND Europe and Bureau Louter (2006) and Zondag (2007).

4.2.1 Demography Module

The demographic module addresses the transition processes of the population and households. It deals with persons by category (gender, age) as well as households by category (size, income, etc.). The demographic module operates at the local zone level and processes, besides the transitions (e.g. ageing), the migration flows calculated in the housing market module. The demography module ensures consistency between households and persons categories for all the zones.

4.2.2 Housing Market Module

The aim of the housing market module is to simulate the annual moves (if any) of households. The module interacts with the demographic, land and real estate, labour and transport market module to account, for example, for demographic changes and changes in the supply of houses. The housing market module simulates two choices, namely the choice to move or stay and the residential location choice, conditional on a move. The residential location choice has a nested logit structure and contains a regional and local scale level: there is a choice of region and a choice of location within the region. At each level a wide set of explanatory variables have been tested in the model estimation to address differences in household characteristics, local amenities, prices and accessibility. Another important variable is the distance (travel time) related impedance, that was included to reflect the geographical dimension of the moves. The parameters of the move/stay and residential location choice function, for each household type, have been estimated on a large four-annual housing market survey in the Netherlands with over 100,000 households of 2002 (see WBO 2002).

4.2.3 Transport Market

The transport module calculates the changes in transport demand and accessibility. The National Transport Model (LMS) is integrated for this purpose within the TIGRIS XL framework. The land-use market modules of the TIGRIS XL model generate the socio–economic input data for the transport module and the transport module calculates accessibility indicators, based on the changes in socio-economic data and/or transport policy measures, as input for the land-use modules. The transport module interacts with the land-use modules every 5 years. In the TIGRIS XL model so-called logsum accessibility measures are used to differentiate accessibility by transport conditions, geographical patterns and characteristics of the actors.

4.2.4 Land and Real Estate Market

The land and real estate market module processes the changes in land-use and buildings, office space and houses, and addresses both brown field and green field developments. The land and real estate market module interacts with the housing market and labour market module. The modelling of the changes in land-use depends on the user settings for the level of market regulation by the government. This can vary from a regulated residential land-use planning system to a free residential land market. In a regulated market, all supply changes are planned by

4.2.5 Labour Market Module

The labour market module in TIGRIS XL simulates the changes in number of jobs by sector at the level of municipalities and transport zones. At the municipality level specific models have been estimated for seven economic sectors to account for the differences in location behaviour between economic sectors. For each sector the influence of accessibility on the spatial distribution has been modelled in combination with a set of other explanatory variables. The parameters have been estimated on a historical data set (1986–2000) including employment figures by sector at a municipality level. The developments in employment by sector calculated at the municipality level are subdivided to the zonal level by simple allocation rules, based on population, industrial sites or office space. These allocation rules are sector specific. The model is not estimated at a zonal level and therefore it can only be used to model structuring effects of transport measures on employment at a municipality level. The labour market module interacts with the demographic-, land and real estate-, housing market- and transport modules.

The changes in the number of jobs for a sector in a region depends on the changes in the number of jobs by sector at the national level, the so-called changes in economic structure, and on regional characteristics, here addressed as location factors. The location preferences of firms, in the regional labour market module, depend on the value of location variables and the parameter values. In the model the behavior of firms is modeled at the level of jobs. This simplification is similar to other LUTI-models like IRPUD (Wegener 1998a), MEPLAN (Echenique 2004) and DELTA (Simmonds 1999), although more recent research work is available focusing on modeling firm demography and mobility (Moeckel 2005; de Bok 2007).

To address the large variety in preferences between economic sectors, the level of sectoral detail is crucial. The economic sectors differ in their land-use, interaction with the population and in their response to changes in accessibility. The seven sectors in TIGRIS XL have been defined based on expected differences in behavior and on consistency with economic sector classifications used at the national level. The economic sectors in TIGRIS XL are: agriculture, industry, logistics, retail, consumer services, business services and government and other non-commercial services.

For each of the sectors labour demand will be estimated at a municipality level; the following equation is applied:

$$\frac{E_{ge}(t)/E_{ge}(t-1)}{E_{NLe}(t)/E_{NLe}(t-1)} = \prod_x LF_{xg}^{\alpha_{xe}}(t). \qquad (4.1)$$

where $E_{ge}(t)$ is employment of municipality g, time t, and economic sector e. NL stands for the Netherlands. $LF_{xg}(t)$ is location factor x for municipality g at time t, e.g. population or accessibility. α_{xe} is the estimated model parameter for location factor x and economic sector e.

Log-linearising Eq. 4.1 yields:

$$\ln E_{ge}(t) = \ln E_{ge}(t-1) + (\ln E_{NLe}(t) - \ln E_{NLe}(t-1)) + \alpha_{0e} + \alpha_{1e} \\ \times \ln LF_{1g}(t) + \ldots \alpha_{ne} \times \ln LF_{ng}(t) \quad (4.2)$$

All parameter values α_{1e} until α_{ne} are sector specific. Some of the location factors or explanatory variables are output from other sub models of TIGRIS XL and their value will be endogenously influenced in the model. For example accessibility indicators link the transport market model and labour market model. These accessibility indicators are included as explanatory variables for different sectors; accessibility indicators in the labour market are a logsum measure for business travel, a mirrored logsum measure for commuting and an accessibility measure for freight transport. Other explanatory variables at the municipality level are population developments, the level of urbanization, size of agglomeration, European location and the location quotient for the sector in the zone.

The parameters have been estimated for the time period 1986–2000 and the main conclusions from the labour market estimation are (see for more details RAND Europe and Bureau Louter 2006):

- Accessibility indicators have a significant influence on the regional distribution of employment in most economic sectors. The logsum variable for commuting has a significant and positive influence on employment in the industry or other consumer services. The logsum variable for business has a significant and positive influence on employment in business services. The freight transport accessibility indicator, generated from travel times by truck, has a positive influence on employment in the logistic sector. It can be concluded that changes in accessibility, through transport measures, will have a significant structuring effect on employment in most sectors.
- A higher level of urbanization negatively influences economic sectors with an extensive land-use. Growth of urban land-use in a region will result in higher pressure on land and higher land prices;
- Population development in a region is an important variable for service sectors such as retail and government services. These sectors do not respond on changes in accessibility but they respond only on population developments within the region itself. The regional scale of the model, which is often larger than the market area for most services, makes the developments in surrounding regions of less importance;
- The relative share of a sector within a region has for almost all economic sectors a significant and negative parameter value. This indicator expresses for a region the share of a sector in total regional employment compared to the share of this sector in total national employment. Arguments for this characteristic can be found in the economic filtering-down theory (Thompson 1968).

Although the performance of the labour market module was in general quite adequate, a need for further development has arisen; mainly to be able to take a future liberalisation of the Dutch housing market and possible intrasectoral and intersectoral linkages into account. The next section offers a concise treatment how to do so.

4.3 Population-Employment Interaction Models

This section first treats concisely the previous literature regarding population-employment interaction models and subsequently offers a modelling framework.

4.3.1 Previous Literature

Finding the relation between sectoral employment and labour force population and subsequently modelling the (re)distribution across regions is part of a large and international regional and urban economic literature that deals with the dynamics interaction between population and employment; hence, the name population-employment interaction models. The focus of this literature deals with the research question whether employment or population depend on each other. Note that traditional regional and urban economic model, such as that of the monocentric city (Mills 1967; Muth 1969), impose such a relation, e.g. population depends on the location of employment. Empirical studies that have tested these relations, however, found that these assumptions are hard to maintain and, if one should a pose a relation, employment is probably endogenous to population location (Steinnes 1977; Carlino and Mills 1987; Boarnet 1994a).

The theoretical development of this literature starts with the simultaneous equation model by Steinnes and Fischer (1974) and later Steinnes (1977) with the now frequently used title: 'do people follow jobs' or 'do jobs follow people'? However, this model focused on levels of employment and population instead of growth. It were Carlino and Mills (1987) who introduced a model with adjustment lags toward an equilibrium.[3] Finally, Boarnet (1994a, b) introduced spatial lags to identify the model. The regions of the research of Boarnet are basically municipalities, which leaves no reason to doubt that there is considerable spatial interaction not only within but also between the municipalities.

These theoretical models have provoked a rather large empirical literature that focuses on the causality issue and which used more or less the framework suggested by Carlino and Mills (1987) and Boarnet (1994b). For a qualitative review we refer

---

[3] Hence, the commonly used name 'Carlino-Mills' studies for this literature.

to Mulligan et al. (1999) and for a quantitative meta-analysis to Hoogstra et al. (2005, 2011). All these studies produce a rather diffuse image, revealing that the empirical evidence is still inconclusive, although most results points toward employment being endogenous to population – or 'jobs follow people', especially for The Netherlands.

Usually, this literature however assumes homogeneous employment and population. There is usually no disaggregation into sectors or population segments. A notable exception is Thurston and Yezer (1994), but their study does not take into account possible spatial interaction. They found only a small number of significant effects – one of them that the service and retail sector were positively affected by population growth.

For The Netherlands, not much empirical research has been done up to now. Bruinsma et al. (2002) conducted a study for the province of South-Holland at a four-digit zip code level in which they concluded that employment growth was caused by population growth and not vice versa. These findings were corroborated by Hoogstra et al. (2003) who looked at the case of the province of Groningen. Recently, Vermeulen and van Ommeren (2009) looked into this relation on a national scale with a specific focus on local labour markets. They found that employment and population growth was primarily caused by supply side factors, such as spatial planning policy, and much less by the interaction between employment and population growth.

All studies referring to The Netherlands conclude that employment growth depends to a large extent on population growth, while there is little to no evidence that this relation works in the other direction as well. PBL (2008) however argues that a particular institutional feature causes this effect. Population-employment dynamics are an equilibrium-adjustment mechanism in a regional supply and demand model. If households are restricted in their locational choice, they will become less mobile relative to firms that are least restricted. This is particularly the case for The Netherlands, where residential locational choice is to a large extent restricted, especially in the densely populated western part of the country. This has two main causes. First, there exists a shortage of suitable housing in popular areas (for the Netherlands, the supply elasticity of housing has been estimated not to exceed 5 % (Vermeulen and Rouwendal 2007)). Second, the government still determines the residential location choice to a large extent (van den Burg 2004). Considering this, the fact that in the Netherlands previous studies unanimously find that employment is endogenous to population should largely be contributed to the limited supply of suitable housing properties. And imposing this relation in scenario analyses that include further liberalisation of the housing market might be inconsistent. Although policy makers already use housing as key planning instrument in The Netherlands, it remains unclear whether jobs follow people is the dominant influence for all economic sectors, controlling for other influences as well as spatial and temporal dependence (see as well the results in PBL, 2008). In addition, PBL (2008) finds large regional differences in population-employment interaction models. This is to a large extent related with the driving forces of housing and the underlying sectoral composition. Rural and less densely populated regions face different population-employment interactions than urbanized and densely population regions if only for differences in sector composition. The next subsection therefore focuses on a model that explicitly accounts for cross-sectoral interactions in the context of a model specifying population-employment dynamics.

4.3.2 Modelling Framework

As argued above, the dynamics between population and employment is a simultaneous process; both might influence each other at the same time. To analyse these dynamics, we develop a multisectoral simultaneous model along the lines of Boarnet (1994b), with as crucial difference that we differentiate with respect to various economic sectors.

For this model, we start by assuming the following simultaneous system of equations.

$$P^*_{r,t} = f\left(X_{r,t}, \bar{E}^{*,s}_{r,t}\right)$$

$$E^*_{r,t} = f\left(Y_{r,t}, \bar{P}^*_{r,t}\right) \tag{4.3}$$

where P denotes population, E employment, r the municapality, t the time period and s an economic sector. The asterix indicates that both employment and population are in equilibrium. The variables X_r and Y_r are exogenous factors which explain the population and employment location, but which are – ideally – not related with characteristics of the local labour market. Finally, the bar above E and P indicates that total local labour market. So, both population and employment in municipality r depend on, respectively, employment and population in the total local labour market. Note that this model implies that the spatial level should be smaller than local labour markets, because the latter is used as the equilibrium mechanism.

The model in Eq. 4.3 indicates that the level of population $P^*_{r,t}$ in a municipality depends on the employment ($\bar{E}^{*,s}_{r,t}$) in the surrounding local labour market, ceteris paribus. Analogously, the level of employment in sector s ($E^*_{r,t}$) in a municipality depends on the population ($\bar{P}^*_{r,t}$) in the surrounding local labour market, ceteris paribus. So, theoretically, there is an equilibrium state for both population and employment. Usually, however, local labour markets are not in equilibrium, because of, e.g., changes in location of households and firms due to all kinds of asymmetric shocks. Therefore, we assume that local labour markets are not in equilibrium but that they are continuously converging to such an equilibrium – by migration of the same firms and households.

We assume the following specification for such a process of convergence:

$$\Delta P_{r,t} = P_{r,t} - P_{r,t-1} = \lambda_P \left(P_{r,t}^* - P_{r,t-1} \right)$$

$$\Delta E_{r,t}^s = E_{r,t}^s - E_{r,t-1}^s = \lambda_E \left(E_{r,t}^{s,*} - E_{r,t-1}^s \right) \quad (4.4)$$

Basically, the model in Eq. 4.4 indicates that when the level population or sectoral employment deviates from their equilibrium level, they will change in the direction of this equilibrium level. The λ's denote the speed of this convergence process. In the extreme case of λ being one equilibrium states will be attained instantaneously. If λ is smaller than one then the convergence process will take longer than one period. The further population and employment are out of equilibrium the faster they will grow towards that equilibrium. Theoretically, all values of λ should be positive: if they are negative, employment and population will grow further from their equilibrium state.

Combining Eq.(4.3) and (4.4) and assuming a linear relation in Eq.(4.3) yields the following system of equations:

$$\Delta P_{r,t} = X_{r,t}\beta_P + \sum_s \gamma_P^s \Delta \bar{E}_{r,t}^s + \sum_s \delta_P^s \bar{E}_{r,t-1}^s - \lambda_P P_{r,t-1}$$

$$\Delta E_{r,t}^s = Y_{r,t}\beta_{E^s} + \gamma_{E^s} \Delta \bar{P}_{r,t} + \delta_{E^s} \bar{P}_{r,t-1} + \sum_{j \neq s} \gamma_P^j \Delta \bar{E}_{r,t}^j + \sum_{j \neq s} \delta_P^j \bar{E}_{r,t-1}^j - \lambda_{E^s} E_{r,t-1}^s$$

(4.5)

Model (5) basically shows that changes in population and employment depend on a set of exogenous variables ($X_{r,t}$) and ($Y_{r,t}$), changes in employment and population in the local labour market respectively, and the level of population and employment in the previous period. If the latter was small then growth will be large and if the latter was large then growth will be small.

Within this model, the definition of the local labour market is crucial. Following Boarnet (1994b) we use spatial econometric techniques to do so. We first define a spatial weight matrix W which denotes for all municipalities r a distance relation. Using this spatial weight matrix we can now indicate whether two municipalities belong to the same local labour market. The local labour market can now be denoted as well as (the following subsection will implement W further):

$$\bar{P}_{r,t} = (I + W) \times P_{r,t}$$

$$\bar{E}_{r,t} = (I + W) \times E_{r,t} \quad (4.6)$$

Finally, combining our definition of the local labour market Eq.(4.6) with model (5), yields:

$$\Delta P_{r,t} = X_{r,t}\beta_P + \sum_s \gamma_P^s \Delta(W \times E_{r,t}^s) + \sum_s \delta_P^s (W \times E_{r,t-1}^s) - \lambda_P P_{r,t-1}$$

$$\Delta E_{r,t}^s = Y_{r,t}\beta_{E^s} + \gamma_{E^s}\Delta(W \times P_{r,t}) + \delta_{E^s}(W \times P_{r,t-1}) + \sum_{j \neq s}\gamma_P^j\Delta\left(W \times E_{r,t}^j\right)$$
$$+ \sum_{j \neq s}\delta_P^j\left(W \times E_{r,t-1}^j\right) - \lambda_{E^s}E_{r,t-1}^s \tag{4.7}$$

which completes our empirical model specification. In model (7) the population growth depends on the employment growth in the surrounding local labour market and of values of employment growth and population growth in the previous year. Likewise, the population growth depends on the poulation growth in the surrounding labour market and of values of employment growth and population growth in the previous year.

Note that the important policy question "Do jobs follow people of people follow jobs?" can be answered by assessing the parameters γ_P^s and γ_{E^s}. If γ_P^s is larger than zero then population follow jobs in the sector s. En if $\gamma_{E^s} > 0$ then jobs in sector s follows population. In this model it is thus very well conceivable that both processes "Jobs follow people and people follow jobs" occur. Because our system of equations can even model several economic sectors, we can even have "Jobs follow jobs".

4.4 Empirical Specification & Estimation of a New Labour Module

This section first discusses how we implement the general population interaction model displayed in Eq. 4.7 in the TIGRIS XL model. Subsequently, it deals with the available data for estimation and the last subsection concisely deals with the estimation results.

4.4.1 Implementation

Because the TIGRIS XL model consists of modules – as laid out in Fig. 4.1 – that have clearly defined relations between each other, it is not possible to directly implement Eq. 4.8. Most notably, population changes come from a demography module and should be considered as exogeneous. Further, we have chosen to omit the simultaneous changes in sectoral employment as well in order to fit the TIGRIS XL model structure. The loss of loosing these endogenous relations are partly offset by the gain in model stability; something that is clearly useful for scenario analyses. We would like to stress here that we still estimate the model as if population changes are endogeneous, so that we can control for possible endogeneity.

The equation we finally implement looks as follows:

$$\Delta E^1_{r,t} = \gamma_1 \Delta(I+W) P_{r,t} + \delta^P_1 \Delta(I+W) P_{r,t-1} + \sum_{i \neq 1}^{S} \delta_{1iP}(I+W) E^i_{r,t-1} + \lambda_{E^1} E^1_{r,t-1}$$
$$+ X_1 \beta_1 + \mu_1$$

$$\Delta E^2_{r,t} = \gamma_2 \Delta(I+W) P_{r,t} + \delta^P_2 \Delta(I+W) P_{r,t-1} + \sum_{i \neq 2}^{S} \delta_{2iP}(I+W) E^i_{r,t-1} + \lambda_{E^2} E^2_{r,t-1}$$
$$+ X_2 \beta_2 + \mu_2$$

$$\Delta E^S_{r,t} = \gamma_S \Delta(I+W) P_{r,t} + \delta^P_S \Delta(I+W) P_{r,t-1} + \sum_{i \neq S}^{S} \delta_{SiP}(I+W) E^S_{r,t-1}$$
$$+ \lambda_{E^S} E^S_{r,t-1} + X_S \beta_S + \mu_S \tag{4.8}$$

where both the employment and population level are shares of the national total and in logarithms. Note that the logarithm of the change in employment shares now equals the logarithm of the left hand side of Eq. 4.1.

We implement the spatial weight matrix by estimating the following distance-decay function[4]:

$$W^m_{ij} = \frac{\alpha}{1 + \exp\left(\beta + \gamma \times \ln(c_{ij})\right)},$$

where α, β and γ are parameters to be estimated, m denotes the modality and c_{ij} are generalised transport costs. Our final spatial weight matrix is composed as $W_{ij} = 0.86 W^{car}_{ij} + 0.14 W^{public\ transport}_{ij}$.[5]

The next subsection briefly discusses the data we use for employment, population and the set of exogenous matrices $X^s_{r,t}$.

4.4.2 Data

We estimate the model with Dutch panel data from 1996 to 2008. We use all 443 municipalities in The Netherlands as our spatial unit of analysis.

---

[4] As a matter of fact, this approach fills in the complete matrix I + W and not only W, because data is available for intra-municipality traffic as well.

[5] We estimate the parameters using the Dutch Mobility Research database (MON) where α, β and γ are 0.0155, 0.1073 and 1.9038 for car mobility and 0.0002, −5.2408 and 3.1459 for public transport mobility. This results in a rather steep distance-decay function.

Table 4.1 Sector composition in TIGRIS XL and weber

ID	Code	Name
1	AG	Agriculture
2	IN	Industry
3	LO	Logistics
4	RT	Retail
5	CS	Consumer services (excl. retail)
6	FS	Financial and business services
7	GO	Government, education and healthcare

Dutch municipalities are usually smaller than local labour markets, but they are typically large enough to capture most urban forces of attraction for both firms and households. The employment data are taken from the so-called LISA database, the "National Information System for Employment" (see http://www.lisa.nl/). Table 4.1 displays the sector composition which matches that of the former labour module and is consistent with sector classifications of similar models.

Population data are taken from the population registries of the municipalities. As the main variable of interest, we chose the size of the labour force, defined as the population between 15 and 65 years of age. To measure the attractiveness of a municipality for households we have incorporated the following conditioning variables: the size of the municipality (in hectares) to account for population density differences, the percentage of land used for built area, nature, agriculture, water, recreation, et cetera, and the age structure of the population within a municipality operationalized as the percentage of the population that is younger than 25, or older than 65 years of age.

In addition to changes in the levels and growth rates of the regional population and employment shares, as given in Eq. 4.8, we have modeled several attraction factors for firms at the level of municipalities. Typically, these are concerned with accessibility characteristics, land prices and land use restrictions, and regional economic conditions. In order to capture these factors we include the size of the municipality, which accounts for density effects as well, and the percentage of specific land uses within that municipality (see also Boarnet et al. 2005). To correct for potentially rationed location space for firms we include the municipality's growth in industrial parks. Finally, we included a relative representation variable (measured as the log difference between the ratio sectoral employment and labour force within a municipality and the ratio total sectoral employment and total labour force within The Netherlands), which can alternatively be denoted as:

$$\ln\left(\frac{E^s_{r,t-1}}{P_{r,t-1}}\right) - \ln\left(\frac{E^s_{NL,t-1}}{P_{NL,t-1}}\right)$$

Table 4.2 Estimation results (statistically significant at 5 %; number of observations is 5,316)

Variabel	Δ AG	Δ IN	Δ LO	Δ CS	Δ RT	Δ FS	Δ GO
γ_1							
δ_P					0.0486		0.0523
δ_{i1}		0.0152	0.0190			−0.0243	
δ_{i2}							
δ_{i3}						0.0366	
δ_{i4}	0.1293	0.0515					
δ_{i5}	−0.0898	−0.0333					
δ_{i6}					0.0247		
δ_{i7}		−0.0310					
λ_i	−0.0356	−0.0044	−0.0034	−0.0050	−0.0081	−0.0104	0.0080
β_i							
% < 25 jr.							
% > 65 jr.							
Δ Houses				1.29E-05		1.50E-05	
Surface r							
Landuse variables							
% Water					0.0006		
% Agriculture						0.0005	
% Nature					0.0004	0.0005	
% Semi-built			0.0027				
% built							
% infrastructure						0.0031	
Δ Industrial sites		0.0006					
Rel. representation							−0.0338

4.4.3 Estimation Results

Estimation of Eq. 4.8 is performed utilizing the generalized spatial three-stage least squares (3SLS) estimator as described in Kelejian and Prucha (1998) and Kelejian and Prucha (2004). The results are denoted in Table 4.2 where we only report those results that are significant at a 5 % level (these are also the results we use for scenario analyses).[6]

The main conclusion we can drawn from Table 4.2 is that employment growth does not seem to depend (positively) on population dynamics: both γ_1 and δ_{i1} are non-significant or negative. The main driver for population-employment dynamics seem to be changes in the housing stock – and in this model to a lesser extent to changes in employment in various sectors. More specific, financial services and

---

[6] There is some debate about this approach. Ideally, one would use all parameters (regardless of their significance level) and then performance a scenario analysis with an associated confidence interval. The latter can be obtained by, e.g., bootstrap techniques. Confidence intervals for the scenario analysis can then be obtained by using a large number of draws from a multivariate normal distribution centered on the estimated parameters and taking into account the estimated variances and covariances.

retail seem to depend heavily on changes in the housing stock, and via these sectors the other sectors are affected as well, because of the dynamic relations constructed by parameters $\delta_{i2}-\delta_{i8}$. Because of this intertwined behaviour and the endogenous structure created by the spatial weight matrix, long-run marginal effects cannot be readily inferred, but should be calculated by, e.g., scenario analyses. Due to the fact that population is seen as endogenous within the TIGRIS XL module, short run marginal effects (yearly changes) can however be seen directly from Table 4.2.

The most remarkable results from Table 4.2 is that employment in consumer services and government are negatively related with population growth in the previous year. This might be caused by competition for land. So changes in housing stock and not demographic changes is the main driver for employment dynamics (compare Vermeulen and van Ommeren 2009). Indeed, the last decades in the Netherlands has shown that where housing stock increases (basically the Western most urbanised area of the Netherlands), employment increases.

When looking at the exogenous variables, we see first that employment in the industry is the only sector that is positively affected by increases in industrial sites. Moreover, employment in logistics does seem to concur with semi-built land use. Employment in retail and consumer services seems to depend on land use such as surface water, nature and recreation, respectively. Employment in financial services is predominantly positively related with infrastructure and to a lesser degree agriculture and nature. One should be careful in a direct interpretation of the impacts of these land use variables. For instance, sectoral employment is probably not affected by the presence of agricultural land, but more by the possible availability of land – which is rather scarce in the Netherlands. However, as a general observation we can note that employment growth in most sectors seem to be affected by their spatial location

The government, healthcare and education sector is a bit of an outlier in this system. If the relative representation is high, then the share of this sector decreases and vice versa. Basically, it shows that this sector is extremely homogeneously distributed over the various municipalities and is quite sticky in its changes.

Moreover, in conformity with theory, all but one λ parameters are negative, which entails that all sectoral employment shares grow faster when the system is more out of equilibrium. The only exception is the government, healthcare and education sector, which again points to dynamics of this sectors which are more deviant from other sectors.

As mentioned above, to gain more insight in the performance of this model over several years we need simulation or scenario exercises. This is something we will do in the next section.

4.5 Application: Scenario Analyses

In the coming decades there is a large need for new dwellings in the western part of The Netherlands–also known as the Randstad. Land available for construction is scarce in this part of The Netherlands, which has lead to an upward pressure on housing prices and might perhaps act as well as a restriction on (regional) economic growth. The only region which still has a relative abundance of land in this area is the province of Flevoland which has been reclaimed from the former sea 'de Zuiderzee' in the 1950s and 60s. The largest town in this province is Almere and is situated in the western part of this region, just to the east of Amsterdam.

At the moment Almere is with 190,000 inhabitants a middle-sized town, but is projected to become the fifth largest town in the Netherlands in 20 years time with 350,000 inhabitants and an additional 100,000 newly to be created jobs. Local and national policy makers envisage to do so by investing both in urban planning (60,000 new houses will be constructed) and infrastructure projects (CPB 2009). The main component of the latter is a large improvement of rail transport between Schiphol Airport, Amsterdam and Almere. Ultimately, the project should transform the city of Almere from a dormitory town of Amsterdam to a vibrant, dynamic and entrepreneurial city, with endogenous economic growth potential (PBL 2007).

In the cost-benefit analysis of the public transport investments the population and employment developments of Almere were considered to be exogenous (CPB 2009). In addition, sensitivity runs were made with the traditional TIGRIS XL model illustrating that the construction of 60,000 houses is likely to result in a much lower, more than a factor 2, number of additional jobs than the targeted number of 100,000 additional jobs. The latter is considered important because at the moment there is a large spatial mismatch between the number of jobs and workers available in the city of Almere, which causes severe traffic congestion, especially on the highway west of Almere that, via a large bridge, leads to Amsterdam (Amsterdambrief 2011).

Without performing a full scenario analysis (see, e.g., PBL 2011), we will look in more detail to our model implementation performance by looking at the 'Water town' scenario, where the majority of the new housing stock will be built to the west of Almere at the border of Lake IJssel. Apart from a spatial planning scenario this compasses as well an almost doubling of the rail and highway infrastructure capacity.

Figure 4.2a–f offer the results of this scenario analysis for the various sectors as displayed in Table 4.2. Clearly, a large increase in the housing stock causes eventually large increases in local employment for all sectors, except maybe for logistics. Most 'benefits' of this policy will thus accrue locally. However, because this primarily of model regional distribution, some other municipalities faces 'costs' as well; they lose employment. Notably, these costs can be found in the direct adjacent municipalities of Almere. Thus, Almere's employment growth causes relative employment losses in nearby regions. The dynamics of logistics, consumer services and government, education and health care are very local. This is especially important for the latter sector, because this is the only sector that is projected to grow significantly in the coming 20 years.

Compared to other models and especially the TIGRIX XL model with the old labour module as explained in Sect. 4.2, the dynamics in this model are far more local – and, in absolute numbers, less pronounced. For one part this is caused by the modelling of accessibility by the specification in Eq. 4.4 which can be seen as more

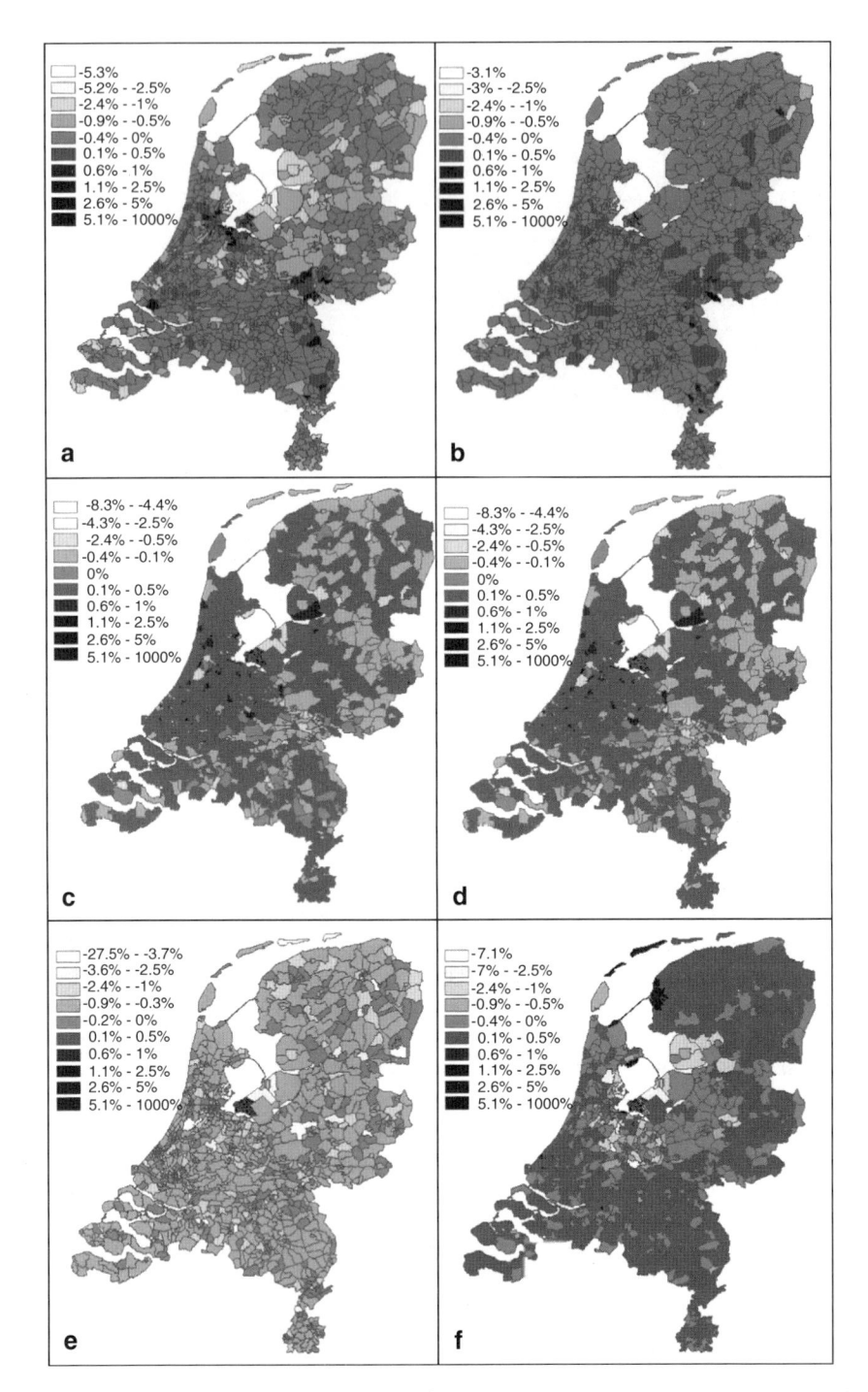

Fig. 4.2 Sectoral employment changes (in percentage point differences) between a 20 year simulation with and without the Almere Waterstad scenario for the TIGRIX XL model including Weber (**a**) Industry, (**b**) Logistics, (**c**) Consumer services, (**d**) Retail, (**e**) Producer services, (**f**) Government, education and healthcare

restrictive as the traditional log-sum approach. For another part, however, this is caused by estimating the coefficients in a simultaneous framework. Thus, by allowing for endogeneity, the effects of population and employment changes upon each other become less strong.

4.6 Conclusion

The main aim of this chapter was to research the integration a population-employment interaction model in the TIGRIS XL modelling framework. Our main reasons to do so were (*I*) to avoid postulating a priori an endogeneous relation between population and employment and (*ii*) to decompose employment into various interrelated economic sectors. An additional advantage of this approach is that we estimate first the relation between employment and population (and between employment in various sectors), thereby controlling for possible endogeneity, and that we subsequently use these estimations in a scenario model.

One of the results of this exercise is that the scenario outcomes are comparable to that of other models – including the former labour module in TIGRIS XL – with two main exceptions. First, employment dynamics are more local. And second they are less pronounced. For Dutch policies geared towards improving employment, this has some important implications. First, one has to be careful in assessing the number of jobs that accrue to a housing market policy. Jobs probably do not follow people that easily. Moreover, the gains from such a policy could be quite local, where adjacent municipalities might suffer significantly in terms of employment losses. And finally, the dynamics between sectors might be larger than between population and employment. Where some sectors gain in employment, others might even lose. Employment in logistics, e.g., does not seem react upon changes in population. Therefore, policy makers need to be aware that population increase only affects certain sectors.

Obviously, there is still ample room for further research. First of all, additional exogenous variables (such as accessibility measures to other activities than working) should be taken into account. In combination with scenario analyses, the marginal effects of all exogenous variables should be assessed, by, e.g. bootstrapping techniques. Unfortunately, although the Weber model is capable of estimating fully endogenous systems of equations (such as in Eq. 4.7), the architecture behind TIGRIS XL is not yet geared to incorporate these estimates. It might be fruitful to incorporate them in further versions of the new labour module. Note, however, that this creates a trade-off between realism and tractability, because fully endogenous systems can become unstable – they might not converge over time. Another issue for improvement is the concept of accessibility. At the moment, this is incorporated in our spatial weight matrix and it defines the local labour market. However, the used distance decay function seems rather steep. Ideally, one would like to use the logsum measure accruing from the transport market in TIGRIS XL (the National Transport Model) to have a new fully integrated labour market module.

Acknowledgements This research was done at the Netherlands Environmental Assessment Agency. We would like thank Michiel de Bok, Raymond J.G.M. Florax and Frank van Oort for useful comments and remarks. Finally, the authors would like to particularly thank Significance for calculating some of the scenario analyses. The usual disclaimer applies.

References

Amsterdambrief (2011) Economic development perspectives for the northwing of the Randstad. Ministry of Economic Affairs (in Dutch), The Hague
Boarnet MC (1994a) An empirical model of intrametropolitan population and employment growth. Pap Reg Sci 73:135–152
Boarnet MC (1994b) The monocentric model and employment location. J Urban Econ 36:79–97
Boarnet MC, Chalermpong S, Geho E (2005) Specification issues in models of population and employment growth. Pap Reg Sci 84:21–46
Bruinsma F, Florax RJGM, van Oort F, Sorber M (2002) Wonen en werken: wringen binnen rode contouren. Economische Statistische Berichten 87:384–387
RAND Europe and Bureau Louter (2006) Systeem documentatie tigris xl 1.0. prepared for the Transport Research Centre, Leiden, The Netherlands
Carlino GA, Mills ES (1987) The determinants of country growth. J Reg Sci 27:39–54
Carruthers JI, Mulligan GF (2007) Land absorption in US metropolitan areas: estimates and projections from regional adjustment models. Geogr Anal 39:78–104
Carruthers JI, Mulligan GF (2008) A locational analysis of growth and change in American metropolitan areas. Pap Reg Sci 87:155–171
Carruthers JI, Vias AC (2005) Urban, suburban, and exurban sprawl in the Rocky Mountain West: evidence from regional adjustment models. J Reg Sci 45:21–48
CPB (2009) Cost-benefit analysis of urbanisation and public transport alternatives for Almere, Netherlands bureau for economic policy analysis (CPB; in Dutch), The Hague
de Bok M (2007) Infrastructure and firm dynamics: a micro-simulation approach. T2007/5, trail thesis series, Delft, The Netherlands
Echenique MH, Hargreaves AJ (2004) Cambridge futures 2: what transport for Cambridge. The Martin Centre, University of Cambridge, Cambridge
Hoogstra GJ, Florax RJGM, van Dijk J (2003) Specification issues and sample selection in population-employment interaction models. working paper, Universiteit Groningen
Hoogstra GJ, Florax RJGM, van Dijk J (2005) Do'jobs follow people' or'people follow jobs'? a meta-analysis of carlino-mills studies. Working paper, Universiteit Groningen
Hoogstra GJ, van Dijk J, Florax RJGM (2011) Determinants of variation in population-employment interaction findings: a quasi-experimental meta-analysis. Geogr Anal 43:14–37
Kelejian HH, Prucha IR (1998) A generalized spatial two-stage least squares procedure for estimating a spatial autoregressive model with autoregressive disturbances. J Real Estate Financ Econ 17:99–121
Kelejian HH, Prucha IR (2004) Estimation of simultaneous systems of spatially interrelated cross sectional equations. J Econometrics 118:27–50
Mills ES (1967) An aggregative model of resource allocation in a metropolitan area. Am Econ Rev 57:197–210
Moeckel R (2005) Microsimulation of firm location decisions. Proceedings of the 9th CUPUM conference, London

Mulligan GF, Vias AC, Glavac SM (1999) Initial diagnostics of a regional adjustment model. Environ Plann A 31:855–876

Muth RF (1969) Cities and housing. The University of Chicago Press, Chicago

PBL (2007) Adolescent almere. how a city is made. NAi Uitgevers/Ruimtelijk Planbureau, Rotterdam/Den Haag

PBL (2011) Nederland in 2040: een land van regio's. Ruimtelijke Verkenning 2011. Planbureau voor de Leefomgeving, Den Haag

PBL (2008) Woon-werkdynamiek in Nederlandse gemeenten. NAi Uitgevers/Ruimtelijk Planbureau, Rotterdam/Den Haag

Simmonds DC (1999) The design of the DELTA land-use modelling package. Environ Plann B 26:665–684

Steinnes DN (1977) 'Do people follow jobs' or'do jobs follow people'? A causality issue in urban economics. J Urban Econ 4:69–79

Steinnes DN, Fischer WD (1974) An econometric model of intraurban location. J Reg Sci 14:65–80

Thompson WR (1968) Internal and external factors in the development of urban economies. In: Perloff HS, Wingo L Jr (eds) Issues in urban economies. John Hopkins Press, Baltimore, pp 43–62, chapter 4

Thurston L, Yezer AMJ (1994) Causality in the suburbanization of population and employment. J Urban Econ 35:105–118

van den Burg A (2004) Ruimtelijk beleid in nederland op nationaal niveau, Geordend Landschap. 3000 jaar ruimtelijke ordening in Nederland, Verloren, chapter van Heeringen RM, Cordfunke EHP, Ilsink M, Sarfatij H, p 159–176

Vermeulen W, Rouwendal J (2007) Housing supply and land use regulation in the Netherlands. tinbergen institute discussion paper, TI 2007-058/3, Amsterdam

Vermeulen W, van Ommeren J (2009) Does land use planning shape regional economies? A simultaneous analysis of housing supply, internal migration and local employment growth in the Netherlands. J Hous Econ 18:294–310

Waddell P (2001) Between politics and planning:urbansim as a decision support system for metropolitan planning, in building urban planning support systems. Center for Urban Policy Research, The State University of New Jersey, New Jersey

Wegener M (1998a) The IRPUD model: labour market submodel. Website of the Institute of Spatial Planning, University of Dortmund

Wegener M (1998b) The IRPUD model: overview. Website of the Institute of Spatial Planning, University of Dortmund

Zondag B (2007) Joint modeling of land-use, transport and the economy. T2007/4, Trail Thesis series, Delft, The Netherlands

Chapter 5
Simulating the Spatial Distribution of Employment in Large Cities: With Applications to Greater London

Duncan A. Smith, Camilo Vargas-Ruiz, and Michael Batty

Abstract In this chapter, we first review the development of employment location models as they have been developed for integrated models of land use transportation interaction (LUTI) where the focus is on the allocation of population and employment. We begin by sketching how employment models based on input–output and multiplier relationships are used to predict future employment aggregates by type and then we illustrate how these aggregates are distributed to small zones of an urban region in ways that make them consistent with the distribution of population and service employment allocated using spatial interaction-allocation models. In essence, the structure we are developing, which is part of an integrated assessment of resilience to extreme events, links input–output analysis to the allocation of employment and population using traditional land use transportation interaction models. The framework then down scales these activities which are allocated to small zones to the physical level of the city using GIS-related models functioning at an even finer spatial scale.

The crucial link in this chain is how we distribute detailed employment types generated from the input–output model to small zones consistent with the way population and services are allocated using the LUTI model. To achieve this, we introduce an explicit employment forecasting model in which employment types are related to functions of floorspace that they use. These are estimated using linear regression analysis which enables future predictions of their location to be scaled in proportion to the totals generated from the input–output framework. These future estimates of floorspace condition the supply side of the model, and combined with accessibility indicators, provide the heart of the employment location model. We have developed this model for London and its region – south east England – and

---

D.A. Smith • C. Vargas-Ruiz • M. Batty (✉)
Centre for Advanced Spatial Analysis (CASA), University College London (UCL), 90 Tottenham Court Road, London WC1E 6BT, UK
e-mail: duncan.a.smith@ucl.ac.uk; camilo.ruiz@ucl.ac.uk; m.batty@ucl.ac.uk

after presenting the results of the model, we sketch how the integrated framework is being used to generate scenarios for future employment and population change.

5.1 Introduction

Integrated land use transport interaction (LUTI) models originating from the notion that the spatial distributions of different geodemographic and economic activities define the way various processes of location operate in space. These are usually composed of activities that are coupled together in pairs with one activity determining another. These couplings form chains of activities that are sometimes entirely looped with no one activity taking precedence in the simulation, thus illustrating how urban activities determine one another simultaneously, but more usually, this chain is broken in some way, with specific activities being defined exogenously. In examining urban structure, it is not easy to determine where the causal chain of dependence can be broken but usually integrated land use transport models, define certain categories of employment, rather than population as being exogenous to the simulation, thus determining these variables in advance as initial conditions that drive the simulation. One of the most obvious distinctions is between employment which is clearly associated with export-orientated activities, and employment that is entirely dependent on population and other employment in the city. Export-orientated is usually difficult to predict with respect to the internal processes of location within the city, hence often being specified as exogenous to the simulation. This long-standing division is sometimes referred to as a distinction between 'basic' and 'non-basic' employment, which 50 years ago was the division between primary-manufacturing and service employment, but any division into activities that cannot be easily predicted and those which are easier to predict in terms of the particular city in question, are relevant to this definition.

Since these types of spatially disaggregate model were first developed following Lowry's (1964) pioneering model of Pittsburgh, there has been continued pressure to relax this distinction between exogenous and endogenous employment. Almost as soon as such model structures were generalised, the notion of building predictive models of different employment categories, traditionally considered exogenous, was broached. Putman (1983, 1991) was the first to develop explicit models of employment location which were embedded in land use transportation model structures, developing his EMPAL (EMPloyment Allocation) suite of models that essentially fused spatial interaction with land development, based on econometric estimation with the use of accessibility potential functions reflecting spatial interaction in the manner first used by Hansen (1959). As with many LUTI models which have gradually moved away from their strict comparative static origins, the EMPAL model is now set within a semi-dynamic incremental framework that involves variables lagged in time but most models of this kind are still largely calibrated using trial and error estimation rather than within any strict econometric framework. More recently, individual firm location models based on agent-based and microsimulation approaches have found favour in the context of LUTI models; but progress has been slow due to the immense data requirements required for such models and problems over providing rich enough input data for scenario testing (see Maoh and Kanaroglou 2013, this volume, and Moeckel 2013, this volume).

Here we will focus on building an employment location model that is deeply embedded in the LUTI model where it is used to generate locational distributions that are used in other sectors such as residential and services location. This is then nested within a wider process of integrated modelling that embeds the LUTI model within a sequence that begins with demographic prediction, moves to aggregate employment prediction using input–output analysis and then predicts the employment distributions using simple regression. The results of this sequence are then used in the LUTI model which in turn generates predictions of population which are further scaled down to greater spatial detail using physical land development modelling based on GIS. The integrated assessment model was first developed for a project involving the evaluation of the impacts of long term climate change in the London region on the distribution of population and employment, particularly with respect to sea level rise over the next 100 years. The various models involved in the sequence were designed and built by different groups with commensurate expertise in demographics, input–output modelling, LUTI modelling (which represents our own contribution) and fine scale GIS[1] (Walsh et al. 2011). The integrated assessment is now being simplified and continued with a stronger focus on assessing resilience not only over the long term but over the short term where the emphasis is on the fracturing of urban networks and constraints on location of which climate and energy change are significant parts[2].

Here we will focus on the link between the input–output and the LUTI models, demonstrating the way we disaggregate the aggregate employment predictions for each employment type which emanate from the input–output model. These are then input to the LUTI model and various feedback loops are necessitated to ensure consistency and balance. In the next section, we discuss the sequence of aggregate, thence spatially disaggregate prediction in terms of the sequence of models used. We formally specify these models in generic and more formal terms, and then we develop the employment forecasting model that links the input–output and LUTI models. We calibrate this model to the London region (which nests the region used for the LUTI model – Greater London and the outer metropolitan area – within

---

[1] The consortium involved in building the integrated suite of models comprised the University of Newcastle upon Tyne Earth Sciences group, responsible for the overall project, and for the transportation networks, GIS and flooding models, Cambridge Econometrics (University of Cambridge Land Economy Environmental group) responsible for the input–output model, and ourselves at CASA-UCL responsible for the LUTI model. Other groups at the Leeds University Institute of Transport Studies, Loughborough University Transport Group, and Manchester University Centre for Urban Regional Ecology were responsible for parallel pollution and energy studies. The project was organised under the auspices of the Tyndall Centre for Climate Change see http://www.tyndall.ac.uk/sites/default/files/engineeringcities.pdf

[2] ARCADIA: **A**daptation and **R**esilience in **C**ities: **A**nalysis and **D**ecision making using **I**ntegrated **A**ssessment, http://www.ukcip-arcc.org.uk/content/view/628/9/

South East England) and we present a critique of the results. We then show how this model is used to couple the input–output and LUTI models together, commenting on potential feedback loops. We finally sketch possible scenarios with the model which are under current development and we conclude with an assessment of the state of the art in such modelling and suggest how this might be improved.

5.2 The Integrated Model Framework

5.2.1 The Generic Structure

In essence, the model sequence takes aggregate estimates of employment E and population P for the region, and first disaggregates these into m employment types E_m using demand from the population H_m for employment in these types. It organises these within an input–output framework and then allocates each employment type E_{im} to zones i of the urban region using the employment location model which is the main focus of this chapter. From these predictions, E_{im}, it simulates the location of populations P_j associated with these employments using spatial interaction models which are coupled together to form the land use transportation interaction (LUTI) model. This sequence is illustrated in the block diagram shown in Fig. 5.1 which also indicates a number of key feedback loops that we will discuss in the following description. Before we do so however, we will describe the generic process that is implied in this structure.

The main drivers of the model framework are aggregate quantities of employment and population with population broken down into household demand associated with employment types m. These are illustrated by the dark boxes in the block diagram in Fig. 5.1. They are linked to each other through the usual input–output coefficients that determine how much of one employment type m is used in the creation of another n and the resulting outputs – employment types E_m – are then input to the employment location model. The supply side of the urban system is driven by floorspace and travel cost variables which determine the attraction of locations and the deterrence to travelling between them. The employment models generate employment types in small zones E_{im} and these are distributed to residential zones using the residential location model which in turn generates population in zones j, P_j, from which the demand for service employment is generated. This is a final demand, consistent with one type of employment in the input–output model. This is constrained to the total required but all the service demand model does is distribute this quantity to the zones of the system.

The population and service demand models form the heart of the LUTI model although it might be argued that the employment model is also part of this extended framework despite the fact that in Fig. 5.1, we show them as distinct from one another. There are many possible feedbacks in the system. First there are land and

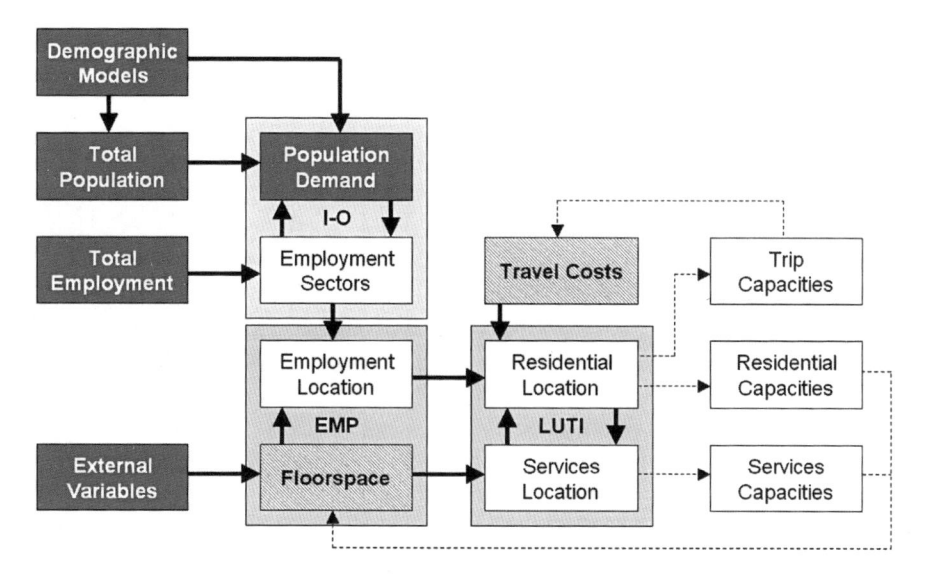

Fig. 5.1 The integrated model composed of input–output (I-O), employment location (EMP), and land use transportation Interaction (LUTI) Models

floorspace constraints on the generation of activities. Once employment and population have been predicted, it is possible to work out their associated land and floorspace requirements. If floorspace differs as it is likely to do so, it is possible to change the floorspace inputs to the employment and LUTI models and reiterate this sequence. If land supply constraints are breached which in turn are fixed by the subsequent GIS model which translates the employment and population activities to a finer spatial scale, then these models are also reiterated. Trip distributions are also central to the logic and these are assigned to the underlying networks; if capacities are breached in terms of such assignments, then travels costs are changed to reflect congestion and the model is reiterated to meet these constraints. These various loops, the exogenous (dark blocks in Fig. 5.1) and endogenous variables (light blocks in Fig. 5.1) and those that act in both ways (the stippled blocks in Fig. 5.1) indicate the flow of simulation in Fig. 5.1 with respect to various components of the integrated framework. The iterative loops that are used to balance the structure and bring the three models (I-O, Employment Location, and LUTI) into equilibrium are given by the broken lines and arrows.

5.2.2 The Input–Output Structure

Aggregate demand for population or employment is usually simulated using a linear structure in which exogenous demand X drives endogenous demand Y which is usually taken as a function of total or final demand Z. We can write this equation as $Z = X + Y$ and if we consider the function of total demand to be $Y = \lambda Z$, we are able to write the equation for total demand as

$$Z = X + \lambda Z, \tag{5.1}$$

which simplifies to

$$Z = X(1 - \lambda)^{-1}. \tag{5.2}$$

$(1 - \lambda)^{-1}$ is the multiplier effect that scales exogenous to total demand. There is nothing in this structure that tells us what is total, exogenous or endogenous demand. It depends on the formulation of the problem. In traditional LUTI models, which are embedded within a loose input–output structure, demand is usually defined in terms of employment activities with Z as total, X export-orientated or basic, and Y services or non-basic employment dependent on the level of total employment. However in a traditional input–output framework, it is often argued that the exogenous input which drives the model is the demand by the population for employment which is usually called final or household demand, with the employment being that supplied to meet this demand. This difference in orientation implies that such linear input–output structures can be oblique to one other, and their precise form depends upon the way the system and problem are articulated.

In fact, the models in this integrated structure are what we might call 'loosely-coupled' in that different groups are responsible for their design and construction. The input–output model is a conventional structure where household demand drives employment structure rather than exogenous inputs of employment driving the resulting total employment. The model is part of a suite developed by Cambridge Econometrics and the one used for this application is disaggregated by UK regions with the East, London, and South East England representing the focus for the predicted employment structure[3] (Junankar et al. 2007). It has not been possible to restructure the input–output model to reflect exogenous employment but in any case, in this application, it is preferable to consider future changes in household demand as having more significance with respect to the climate change scenarios which lie at the core of the application. Moreover it is somewhat easier to predict household demand than employment demand. Although our treatment here is generic and not quite the precise form that is used, the model is structured along the following lines.

The model is formulated in terms of predicting total demand for employment in category m, E_m, from the final household demand for employment in each category n, H_n, which is calculated as

---

[3] http://www.camecon.com/ModellingTraining/suite_economic_models/MDM-E3/MDM-E3_overview.aspx

$$H_n = \omega_n \alpha^{-1} P = \alpha^{-1} P_n, \tag{5.3}$$

where ω_n is the ratio of employment in category n to total employment E, α is an activity rate defined as P/E and P_n is the actual household demand measured in terms of population. We then define the coefficients of dependence between the employment sectors as λ_{mn} and the input–output equation can then be written as

$$E_m = H_m + \sum_n \lambda_{mn} E_n. \tag{5.4}$$

We can write the equivalent of the multiplier relationship in matrix form as $\mathbf{e} = \mathbf{h}(\mathbf{I} - \mathbf{\Lambda})^{-1}$ where $\mathbf{e}$ and $\mathbf{h}$ are appropriate row vectors for E_m and H_m and $(\mathbf{I} - \mathbf{\Lambda})^{-1}$ is a matrix inverse where $\mathbf{I}$ is the identity matrix and $\mathbf{\Lambda}$ the matrix of the technical coefficients λ_{mn}. Equation 5.4 predicts the employment by category/type which is then input to the employment location model that scales these quantities to employment in zonal locations E_{im} which we will sketch in the next section.

5.2.3 The Employment Location Model

Before we introduce the model, it is worth noting that there are several developments of LUTI models that extend the spatial interaction structure of these models to embrace spatial interactions between different employment sectors; in short, disaggregating the classic input–output model in Eq. 5.4 into zonal as well as sectoral categories. Following earlier work by Macgill (1977), one of the authors (Batty 1986) has developed a generic framework for this but the one that is being currently used in operational practice is the extended MEPLAN model which is built around a spatially disaggregate input–output structure (Echenique 2004). In fact, Echenique's model is rather more general than any of those that we have detailed here in that he defines two types of sector – production and consumption, or supply and demand, not specifying formally whether these sectors are measured in terms of population or employment. The model as it has been applied, defines production as employment and consumption as household demand, measured in units of employment and thus this structure is similar to the input–output model used here. As the model is iterative in structure, what begins the process can be either inputs or outputs and there are many possible variants. However, the key difference is that the model structure here separates aggregate predictions for employment from their spatial allocation which in turn are separated from the residential and service locations and interactions. In the Echenique and related models, all these stages are fused.

For the rapid assessment of scenarios, we need to develop a model whose independent variables are relatively easy to forecast in their own right. To this end, we first predict floorspace as a linear function of various independent variables that are easy to specify for future scenarios and then use these predictions of floorspace to produce employment by category and location using a second linear prediction model. Immediately an issue arises as to the transmission of errors in this two stage process but although error is clearly passed on from the first to the second models, both models are heavily constrained to lie within certain limits. It is not possible in these kinds of model to work with negative quantities and moreover, we need to make sure that the total employment which is predicted for each category sums to the known totals generated from the input output model. These complicate the structure and in general, compromise the strict statistical interpretations of the model fits.

To provide some sense of these two models, the first activity that is predicted – floorspace y'_i say – is estimated as a function of various independent variables x_{ig} where the coefficients $a_o...a_g$ are estimated using least squares regression. These variables are then input to a second regression which predicts an employment category z'_i as a function of another set of independent variables v_{ig} with coefficients $b_o...b_g$ and the floorspace variable from the first stage $\vartheta y'_i$ where ϑ is the appropriate weight from the floorspace model. These two models are written as

$$y'_i = a_0 + \sum_g a_g x_{ig}, \qquad (5.5)$$

$$z'_i = b_0 + \sum_g b_g v_{ig} + \vartheta y'_i. \qquad (5.6)$$

It might be remarked that these equations could be collapsed into one but at each stage we constrain the estimates to meet minimum values that are greater than or less than zero. There are two ways of doing this. First if the estimates that are below zero are small in number and value, we simply exclude the variable from the model and replace the value of the predicted floorspace or employment by its observed equivalent. That is, for Eqs. 5.5 and 5.6 respectively

$$\text{If } y'_i \leq 0 \text{ then } y'_i = y_i \qquad (5.7)$$

$$\text{If } z'_i \leq 0 \text{ then } z'_i = z_i, \qquad (5.8)$$

where the primed variable denotes the predicted value and the non-primed variable the observed value. The second method is to simply scale the value by adding the minimum value to each prediction; that is

$$\text{If any } y'_i \leq 0 \text{ then } y'_i = \min_\ell y'_\ell + y'_i \qquad (5.9)$$

$$\text{If any } z'_i \leq 0 \text{ then } z_i = \min_\ell z'_\ell + z'_i. \qquad (5.10)$$

In fact we have used the first method for the first model whereas for the second model where we predict employment by category, we scale the estimates so that the sum of these employments meet the predetermined input–output totals.

We can now state the model as follows. First we predict the floorspace F'_{im} associated with employment category m as

$$\left.\begin{array}{ll} F'_{im} = a_0 + \sum_g a_g x_{ig}, & \text{if } F'_{im} > 0 \\ F'_{im} = F_{im}, & \text{if } F_{im} \leq 0 \end{array}\right\}, \quad (5.11)$$

and then we use this floorspace to predict the employment in category m as E'_{im} from

$$\left.\begin{array}{ll} \hat{E}_{im} = b_0 + \sum_g b_g v_{ig} + \vartheta F'_{im}, & \text{if } \hat{E}_{im} > 0 \\ \hat{E}_{im} = E_{im}, & \text{if } \hat{E}_{im} \leq 0 \end{array}\right\}. \quad (5.12)$$

The final employment location is scaled to ensure that

$$E'_{im} = E_m \frac{\hat{E}_{im}}{\sum_m \hat{E}_{im}}. \quad (5.13)$$

Equations 5.12 and 5.13 that define the model are applicable to each employment category associated with the input–output model. When we detail the calibration below, we will present the actual independent variables and categories used.

5.2.4 The Land Use Transportation Interaction (LUTI) Model

We can state the land use transportation models quite succinctly. We allocate employment E_{im} to residential areas using a spatial interaction model that computes the work trips T^q_{ij} by travel mode q from workplace zone i to residential zone j as a function of the residential floorspace R_j, the modal travel cost c^q_{ij} from i to j, and the friction parameter β^q. We state this model as

$$T^q_{ij} = \left(\sum_m E_{im}\right) \frac{R_j \exp\left(-\beta^q c^q_{ij}\right)}{\sum_\ell R_\ell \sum_q \exp\left(-\beta^q c^q_{i\ell}\right)}. \quad (5.14)$$

Note that the script ℓ is used throughout the text as a floating index pertaining summation over zones, employment sectors or floorspace types to make a distinction from the actual flows or volumes. Population in zone j is computed by summing trips over modes q and employment zones i and scaling the result by the activity rate α as

$$P_j = \alpha \sum_i \sum_q T_{ij}^q. \tag{5.15}$$

We now compute the demand for employment in category n at zone c as a function of the household demand in j, thus repeating in some way the sort of structure that is represented at both the input–output and employment location models stages. Then

$$S_{jcn}^q = \xi_n P_j \frac{F'_{cn} \exp\left(-\varphi^q c_{jc}^q\right)}{\sum_\ell F'_{\ell n} \sum_q \exp\left(-\varphi^q c_{j\ell}^q\right)}. \tag{5.16}$$

where the service centre zone is notated as c, ξ_n is the employment demand coefficient for category n and φ^q the modal friction parameter. We are then able to predict the employment of each category in zone c by summing Eq. 5.16 over j and q as

$$E'_{cn} = \sum_j \sum_q S_{jcn}^q. \tag{5.17}$$

We see an immediate simultaneity in Eqs. 5.14, 5.15, 5.16, and 5.17 where an element of the employment input which we take from the prior employment allocation models is also predicted by the LUTI model. Strictly this leads to iteration over Eqs. 5.14, 5.15, 5.16, and 5.17 until balance occurs (Wilson 1970). However what we effectively do is divide the categories of employment into two sets: those that we consider cannot be predicted as a function of population which is akin to true exogenous employment in the traditional basic sense, and employment in that category that is clearly a function of population – in short, service or non-basic employment.

Essentially we apply Eqs. 5.16 and 5.17 to the retail category of service employment, that is $n = $ *retail employment* and we aggregate this to one class. Employment which is not retail employment is predicted using the employment location model in the previous section and thus the two employment models are quite separate. However within the overall structure, there exists the potential for iterating between the LUTI and employment location models if this appears appropriate. Other loops involving these various models relate to (a) capacity constraints and (b) locational attractors. First there are capacity constraints on the floorspace and land areas consumed by population and employment. We convert population and employment into residential and commercial floorspace as $R'_j = \eta P_j$ and $F'_{cn} = \gamma E'_{cn}$ where η and γ are appropriate conversion coefficients.

Then if $R'_j > R_j^{max}$ and $F'_{cn} > F_{cn}^{max}$, we scale these attractors to produce less activity in the next iteration of Eqs. 5.14, 5.15, 5.16, and 5.17. This reduces the attraction until the constraints are met. We should note that these floorspace variables are also used in the employment location model and thus another iteration can be set up which involves both the LUTI and its precursor, the employment model. Finally if the trip distributions T_{ij}^q and S_{jcn}^q exceed their link capacities, this sets up yet a further iteration which involves ensuring that these link capacities are met and this involves altering the travel cost matrices accordingly. There are many possible loops of this kind and we are exploring all of these within the wider project (Batty et al. 2011).

5.3 Applications to London and South East England

5.3.1 Defining the Regional Model and Classifying Employment

Our suite of integrated models is being developed for the urban region comprising Greater London and the Outer Metropolitan Area which is a region containing 6.82 million (m) jobs and 13.42 m population. It is broadly circular whose radius varies between 40 and 50 miles from central London. It includes all the major London airports and green belts but does not extend to include the university towns of Oxford and Cambridge, the new town of Milton Keynes, or the south coast ports and resort towns. This region however cuts across three official English regions – London, the East, and the South East – and this complicates the employment modelling for the input–output model is built at this regional level. The employment model which disaggregates these regional totals for the finer scale population zones is thus built for these three regions which we refer to as the Greater South East Region. This larger region is about twice the size of the actual LUTI model region with a population of 20.56 m and employment of some 10.09 m (which is composed of 3.84 m in London, 2.82 m in the East and 4.15 m in the South East). The LUTI model region is composed of some 1,767 zones based on wards and this is nested within the larger employment model region which contains some 3,202 zones. The zoning systems for the three models are illustrated in Fig. 5.2. In fact it is worth noting that the ratio of population to employment – the activity rate α – in the larger region to the smaller is quite close, 1.967 compared to 2.038 which indicates that the London region is only slightly more active than its wider periphery.

To achieve consistency between the input–output and the LUTI models, the employment model will be developed for the larger region but only applied to the smaller region which is a proper subset of the larger without any iterative couplings between the sectors. In short this is only possible because the employment model is composed of separable and distinct sectors. We now need to define these sectors. There are 42 sectors in the input–output model including miscellaneous but to give an idea of what the region would look like if employment were distributed

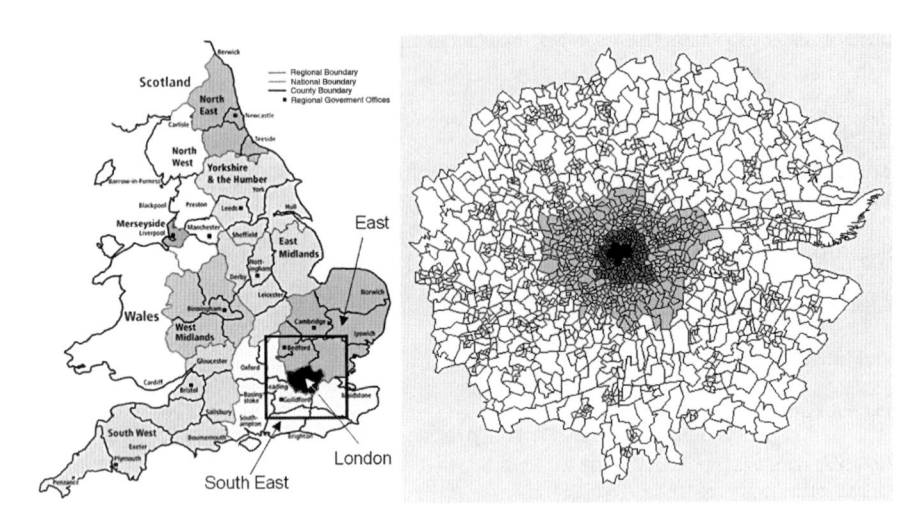

Fig. 5.2 The input–output, employment and LUTI model regions. *Left*: the three regions used in the I-O and employment models, with the inset → *Right*: the London and outer metropolitan area, with the Greater London Authority area (*mid grey*), Inner London (*dark grey*) and the Central Boroughs (*black*) nested within one another

uniformly between these sectors and over all 3,202 zones, we would only have 75 employees in each sector in each zone. In short, many of these sectors are far too small to be capable of being simulated in locational terms. 24 of these sectors or about 60% account for only a tiny fraction, some 5% of the employment, and we thus need to collapse these according to some obvious logic that takes account of their magnitude and spatial distribution. The services sector in this region is enormous and constitutes over 80% of all employment. Many of these services are highly clustered. 2% of all employment, each in different types, is in just four categories x zones which is 0.001% of the number of categories x zones while 10% of all employment is located in just 0.01% of the categories x zones. Indeed if we plot the spatial distribution of employment across the larger region, it is very clear that this is highly polarised with some 10%, 20%, 30% 40% and 50% of all employment in 0.4%, 1.5%, 3.6%, 6.9% and 11.6% of all zones. This is far more extreme than the oft-quoted skew distribution of income which suggests that 80% of all wealth is in the hands of 20% of the population (the so-called 80–20 rule which describes the long tail).

A good measure of inequality of the distribution of employment is given by the entropy function defined as

$$H = -\sum_i \frac{E_i}{E} \log \frac{E_i}{E} \tag{5.18}$$

which varies from 0 to a maximum value of $\log(n) = \log(3202) = 3.505$, where 0 denotes a situation where all employment is in one zone and $\log(n)$ a situation

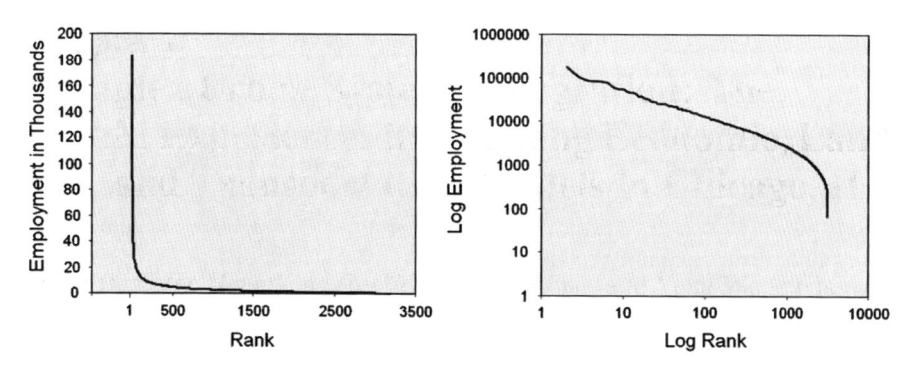

Fig. 5.3 Inequalities in the employment size distribution by ranked locations

where all the employment is evenly spread. A measure of inequality is given by the redundancy which is defined as

$$R = \frac{\log(n) - H}{\log(n)} = 1 - \frac{H}{\log(n)} \qquad (5.19)$$

which varies from 1, complete inequality to 0, complete uniformity or equality. The entropy H in the employment distribution is 3.179 and thus the redundancy R is 0.093, apparently much closer to uniformity than extreme inequality, notwithstanding the apparently highly skewed nature of the distribution. In Fig. 5.3, we show the employment distribution in terms of its rank size in both its raw and its log transformed plots and it is clear that there is substantial polarisation. But given the number of zones, this distribution is moderated by the fact that there is employment in all zones, hence the value of the redundancy statistic. Were we to aggregate the zones, the distribution would become more extreme.

Another perspective on the structure of employment in the region and one that is required in grouping the 42 sectors from the input–output model into a more manageable and representational set of classes involves examining the clustering of employment types. The services sector is enormous constituting over 80% – manufacturing and transport only 10%, and thus we need to disaggregate different kinds of services – financial and business and probably IT from retail and from public services. It is clear that the locational demands of these various services are quite different as our analysis of clustering reveals. A good measure of spatial clustering is the Getis-Ord G statistic (Getis and Ord 1992) which essentially determines how close a particular activity in a particular zone is to all other activities of that type in all other zones. This statistic has been computed for all the zones in the system with respect to each of the 42 categories of employment that are defined in the input–output model. We use an inverse distance weighting with a seven mile radius cut-off to encapsulate relevant zones near to those which we consider clusters might appear in. The observed employment at zonal (ward) level is taken from the Annual Business Index from the Office of National Statistics which is a 10% yearly sample, and we combine multiple years (2005–2007) to produce reliable estimates of the actual distribution.

The distribution of these cluster statistics for the 42 sectors is highly polarised and we use these to provide indicative measures of the extent to which sectors need to be specific. The largest cluster values which are more than twice the average are of two kinds: banking and finance, large in employment, and specific one-off activity locations such as oil, gas and water utilities which tend to be rather small in employment. Banking and finance are heavily concentrated in the core of the region. The second less intense set of clusters involve professional, miscellaneous, public, and hotel-restaurant services which form more diffuse clusters. These are still concentrated in Greater London but are much larger in scale constituting some 1.5 m jobs in the region. In short, we see two kinds of services – those which are very niche and concentrated in the core and those which are concentrated in urban areas which define the polycentric structure of towns that form the wider region but are spread more diffusely as the population is spread throughout the region. One-off locations do also lead to clusters but it is very clear that many of the clusters are heavily orientated to the size and scale of the agglomerations of population in the region. In terms of other significant but smaller groups, only about 5% of employment is in manufacturing which is about the same as in transport, with the transport activity concentrated around airports and central city hubs. Utilities, primary industry and construction are much more idiosyncratic in their locations, either spread fairly evenly in association with population or in very small one-off concentrations. In fact the scale of this activity is so small and the locational requirements so diffuse or special that we will treat these sectors rather differently as we illustrate below when we design the employment models.

In summary then, clustering is either caused by the agglomeration of activities in attractive business areas (overwhelmingly in central London), or the clustering of activities within and around large facilities such as airports and industrial parks. As we have noted, the most highly clustered activities are finance, with insurance closely behind, and these activities are dominant within Central London and thus good contenders to be aggregated together. Business services are significantly less spatially clustered and, while still centralised, are not concentrated to the same degree in central London, and thus should remain separate. In terms of clustering at large facilities, this can be seen in the air transport, which is highly clustered, overwhelmingly around Heathrow in outer London. Motor vehicle, electronic and textile manufacturing are moderately clustered, and are located in outer London and the wider region. The majority of manufacturing activities however do not show high degrees of clustering. The sub-regional geography also indicates some misclassifications in the proposed groups. Oil and gas is concentrated in central and inner London, and is surely service jobs rather than primary sector activities. Similarly the high degree of printing and publishing jobs concentrated in inner London shows that these are service jobs, rather than manufacturing. These groups need to be split into their component parts or simply classed in a service group, as the service proportion is likely to greatly exceed the manufacturing proportion. The utilities groups are highly clustered, and are located mainly outside Greater

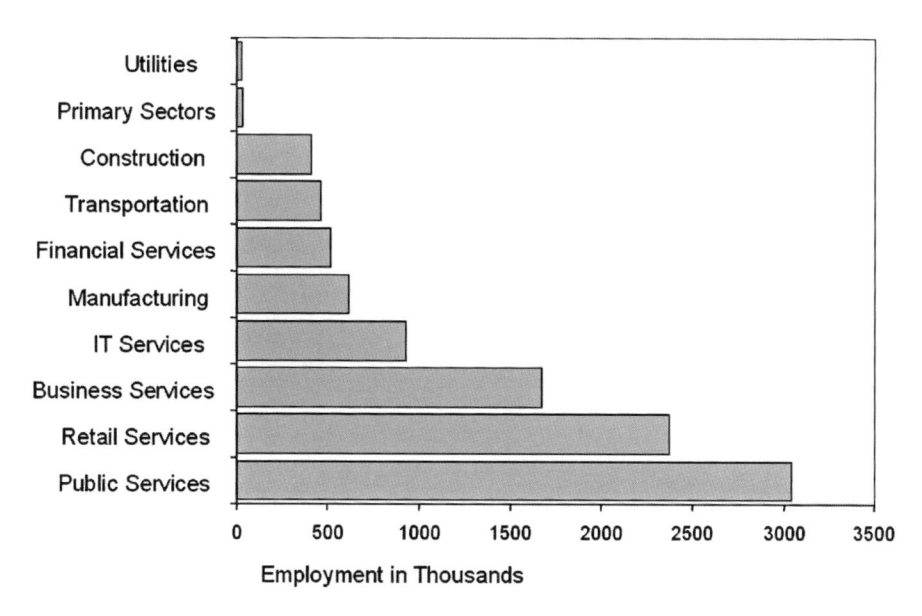

Fig. 5.4 The distribution of employment groups for the location model

London. Construction jobs are not clustered, and other than air and water transport, other transport jobs are also widely spread.

A process of to-ing and fro-ing in examining these patterns led us to aggregate the 42 sectors from the input–output model to 10 distinct sectors which we define in terms of their size as Public Services (30%), Retail Services (24%), Business Services (17%), IT Professional Services (9%), Manufacturing (6%), Financial Services (5%), Transportation (4%), Construction (4%), Primary Sectors (<0.5%), and Utilities (<0.5%). We show these in terms of employment totals in the bar graph in Fig. 5.4, from which it is quite clear that the 'services' categories completely dominate the region with 84% of the activity in these groups. Although we do not show this here for the region, this mirrors the polarisation of employment that we commented on above as implied by Fig. 5.3. As a structure for the ten sectors, it represents as good an aggregation reflecting all these diverse factors as we are likely to get. It is on this basis that we have developed the employment location model.

5.3.2 Specifying and Fitting the Floorspace Location Models

Of the ten sectors, four have very little spatial similarity to the other six which are the core sectors. In fact for each one of these four, they do not have any cross-correlations with any of the other nine sectors with values greater than 0.4 and most are below 0.1. Of the remaining six sectors, all the service sectors are highly

Table 5.1 Correlations (×100) between six key employment sectors and floorspace

	Manu-facturing	IT services	Financial services	Business services	Public services	Retail services
Manufacturing	100					
IT services	24	100				
Financial services	5	42	100			
Business services	18	*81*	*46*	100		
Public services	16	65	25	71	100	
Retail services	*33*	75	33	*77*	*67*	100
All-floorspace	47	75	49	78	65	81
Retail-floorspace	23	59	26	63	59	85
Office-floorspace	14	<u>79</u>	<u>65</u>	<u>89</u>	<u>71</u>	72
Industrial-floorspace	<u>66</u>	18	2	11	10	29

correlated with one another, in contrast to the manufacturing sector which is spatially quite distinct. We now define these sectors in terms of employment E_k where we use the index k to distinguish these new sectors from the original 42 input–output sectors m, n. In terms of possible explanatory variables, in particular floorspace, these six sectors have quite strong relationships with the three kinds of floorspace, each having a correlation of at least 0.66 with total, industrial, retail and office floorspace. Another interesting feature of this data, is that of the two accessibility variables measured using population potential from Hansen's (1959) measure for public and private transport networks, the correlations with these variables are all less than 0.2. This prompts us to compare these distributions with population itself and these correlations are also rather low, thus implying that the employment base is very different spatially from the distribution of population. The relevant comparisons between employment and floorspace categories shown as correlations are given in Table 5.1 where the key correlations are in bold italics (between the six sectors) and underscored for the floorspace comparisons.

As we noted in the previous sections, the employment model which is based on predicting the employment in each of these six sectors is divided into two stages. Clearly from Table 5.1, correlations between employment and floorspace are quite high and as floorspace is a critical supply side variable that can be manipulated to determine future scenarios that we might test, we decided to first build a predictive model of the three floorspace categories in Table 5.1, without the global category All-Floorspace which is in fact the sum of Retail-Floorspace, Office-Floorspace, and Industrial-Floorspace. This first model uses these various measures of floorspace as independent variables together with a series of other variables that reflect other key factors that appear important. In fact we used a stepwise procedure to generate the best set of variables for the three categories. The other variables are of two kinds; accessibility to existing office, retail and industrial floorspace (to generate clustering and agglomeration effects) where the attractor is the relevant floorspace variable, general accessibility to population, accessibility to particular infrastructure facilities such large airports based on network travel time to the facility; and then potential space for urban development, such as existing land for commercial uses which is linked to agglomeration and brownfields redevelopment potential, and open space with potential for development. This was calculated as total greenspace less protected greenspace crucial to existing and future planning policies.

Our initial models regressed the three categories of floorspace (which we now define as F_j^z, $z = 1$: retail floorspace, $z = 2$: office, $z = 3$ industrial) in different combinations of these variables and we immediately found that there were strongly contrasting patterns for urban and rural areas. These differences were difficult to capture in a single regression model with low values of the variance explained. This problem was tackled by introducing two sets of dummy variables in of each floorspace models. The objective of the first set is to differentiate wards with no commercial floorspace (isolated rural wards) from those with commercial floorspace (urban wards) and those with high commercial floorspace (town centre wards). The objective of the second set is to include the distinct conditions in Inner London and this is achieved by introducing three different subregional effects that partition the region into three distinct sets of areas based on nearness to the central core. The models which were ultimately selected incorporate these dummies and this does weaken the model's ability to generate completely new urban developments in rural areas. Arguably the majority of such development takes place at 'seed points' in small settlements that become towns, and on brownfield sites that are redeveloped and thus this should not be a major shortcoming.

The three models that have been fitted are variants of the generic structure that follows:

$$F_j^z = a_0 + \sum_{\substack{\ell=1,2,3 \\ \ell \neq z}} a_k F_j^\ell + a_4 B_j + a_5 D_j + a_6 A_j^z + a_7 T_j + a_8 G_j + \\ + a_9 H_j + a_{10} Q_j + a_{11} M_j + \sum_{\ell=1,2,3} s_\ell \delta_j^\ell + \sum_{\ell=1,2} f_\ell \delta_j^\ell \quad (5.20)$$

where the coefficients $a_o...a_{11}$ are the usual regression weights that imply a degree of significance for the variables to which they are ascribed, s_ℓ and f_ℓ are the coefficients of the dummy variables δ_j^ℓ which take on values of 0 or 1 for specific zones. The three dummies associated with s_ℓ, $\ell = 1, ..., 3$ are those for the three subregional effects noted above while the two dummies associated with f_ℓ, $\ell = 1, 2$ are those which determine thresholds on the relevant floorspace category. B_j, D_j, A_j^z, T_j, G_j, H_j, Q_j, M_j are respectively non-domestic buildings, domestic buildings, accessibility to floorspace of type z, accessibility by public transport to

Table 5.2 Estimates (and standard errors) for the three floorspace models

Independent variable	Retail floorspace $\{F_j^1\}$	Office floorspace $\{F_j^2\}$	Industrial floorspace $\{F_j^3\}$
Constant a_0	−3992.257	−13763.129	−21825.274
	(908.641)	(2517.158)	(2465.222)
Retail floorspace F_j^1	Na	0.771	−0.138
		(0.034)	(0.026)
Office floorspace F_j^2	0.125	Na	−0.162
	(0.0005)		(0.012)
Industrial floorspace F_j^3	−0.023	−0.278	Na
	(0.008)	(0.021)	
Nondomestic building B_j	56.984	226.331	450.912
	(5.640)	(15.232)	(8.434)
Domestic building D_j	25.062	Na	−23.834
	(4.198)		(8.604)
Access to retail floorspace A_j^1	0.070	Na	Na
	(0.004)		
Access to office floorspace A_j^2	Na	0.045	Na
		(0.003)	
Access to industrial floorspace A_j^3	Na	Na	0.125
			(0.0005)
Public transport accessibility T_j	Na	128.338	47.792
		(33.297)	(14.114)
Greenspace G_j	−0.149	−0.173	−0.364
	(0.018)	(0.050)	(0.037)
Time to heathrow H_j	Na	−121.947	Na
		(38.358)	
Time to all airports Q_j	27.001	−178.775	Na
	(7.939)	(58.570)	
Time to motorways M_j	Na	Na	−74.595
			(32.856)
Subregional effect core δ_j^1	−89153.674	94686.907	−62219.581
	(5057.273)	(15593.312)	(8014.538)
Subregional effect inner areas δ_j^2	−29533.271	−56162.792	−60027.959
	(2213.944)	(5268.918)	(3446.372)
Subregional effect outer areas δ_j^3	−6657.748	−12246.272	−27674.705
	(1065.701)	(2551.055)	(2273.093)
High impact floorspace for z	111907.227	69355.676	Na
	(2580.984)	(6664.709)	
No impact floorspace for z	−6583.713	Na	Na
	(958.500)		
Correlation squared	0.719	0.677	0.679
	(14712.458)	(41189.450)	(29902.481)

population, both based on Hansen's (1959) potential measure, greenspace, distance to Heathrow, distance to other airports, and distance to major motorway junctions. This model has been fitted to the three floorspace distributions and the results are presented in Table 5.2.

The performance of these models is quite good. Some 70% of the spatial variation over a very large region with many small zones is good enough for forecasting purposes where we are very largely concerned with specifying future changes in floor space in 'what if' contexts. Moreover we will use these models as inputs to the employment model where we will bring other variables to bear as well as these various constraints on ranges of values that we indicated in an earlier section. It is to these models that we now turn. Concern about error propagation is worth noting but the rather inelegant fitting and manipulation of these models is achieved in the context of a deeper learning about the region.

5.3.3 The Employment Location Models

As we argued above, we do not intend to develop separate models for all ten sectors of employment. Four of these sectors – the primary, utilities, construction and transport sectors are too specific in their locational requirements, too diffuse across the region, and/or simply too small to be significant for such forecasting. When we forecast the future locations of these sectors, we will simply scale them to their baseline values at 2005, the year at which the employment models have been calibrated and for which data exists, or we will pre-specify their values in terms of the scenarios we develop; or more likely a combination of these two strategies will be used.

The six sectors k that we formally model are all functions of their relevant floorspace which has been modelled in the previous section. We also add a number of key variables based on accessibility and travel time/distances to these models, developing a generic form that is applied differentially to each sector dependent on what we consider to be significant causal drivers of their respective locational patterns. The model is similar to that for floorspace and we state it as

$$E_{jk} = b_0 + \sum_{z=1,2,3} b_k F_j^z + b_4 A_j^{PT} + b_5 A_j^R + b_6 M_j + b_7 J_j + b_8 H_j + b_9 h_j + b_{10} Q_j$$
(5.21)

where $E_{jk}, k = 1, \ldots, 6$ or k = {financial services, manufacturing, IT services, business services, retail services, and public services}, $b_0 \ldots b_{10}$ are the respective regression coefficients, A_j^{PT} and A_j^R are the public transport and road accessibility potentials for the respective employment variables, J_j is distance/travel time to rail hubs, and h_j is the distance/travel time to A-Class roads. All other variables are as defined previously.

| Access by Public Transport | Accessibility to Airports | Accessibility to Motorways |

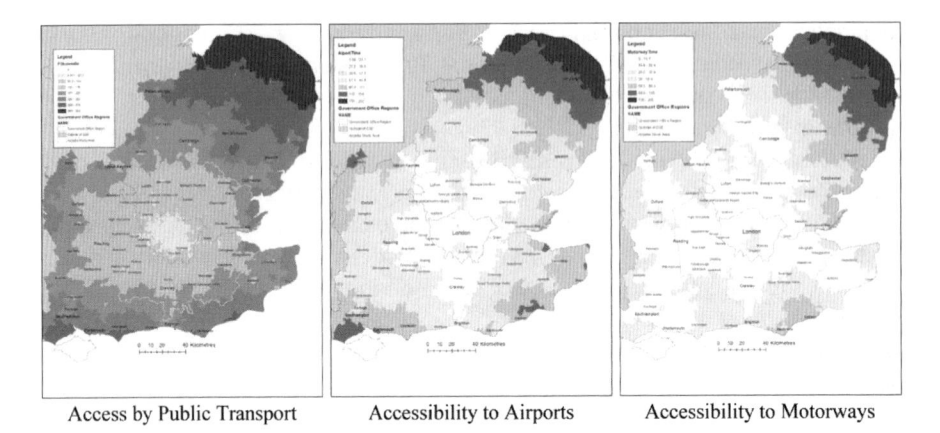

Fig. 5.5 Sample accessibility surfaces computed using various weighted indices

Even though the employment models are largely driven by floorspace from the previous model, and observed floorspace is used to fit the models in Eq. 5.21, these are also functions of the various accessibility potentials and direct distance measures to facilities. The generic Hansen (1959) measure A_j^k is defined for an activity k in zone j as

$$A_j^k = K \sum_i E_{ik} d_{ij}^{-2} \qquad (5.22)$$

where we show these as accessibilities to employment type E_{ik} and where the distance/travel time d_{ij} is that associated with the particular mode of travel, that is public transport PT or car/road transport R. To provide some sense of the distribution of these variables, we show three accessibility patterns for the Greater South East Region in Fig. 5.5. In Fig. 5.6, we also show the distribution of employment for the ten categories, six of which we are modelling using Eq. 5.21. It is clear that for the distributions we are not intending to model, the patterns are much more diffuse than those for the services and manufacturing that we are modelling. These latter distributions are very highly concentrated and the affect of the accessibility variables in Eq. 5.21 is to diffuse these concentrations a little when these variables are embedded into the employment model.

In Table 5.3, we present the results of fitting the six models. The performance of these models is good and we have included only those variables which are significantly different from their exclusion from the models at the 5% level. There has been considerable to-ing and fro-ing in developing these models, examining regional distributions such as those in Figs. 5.5 and 5.6, and then examining significance and the causal logic of activities and indices that might be related to one another. In terms of Table 5.3, each of the categories is associated with its most immediate floorspace – that is financial, business, IT and public services with office floorspace, retail with retail floorspace and manufacturing with industrial. Manufacturing is only associated with A-Class road and airport accessibility, while the other five services

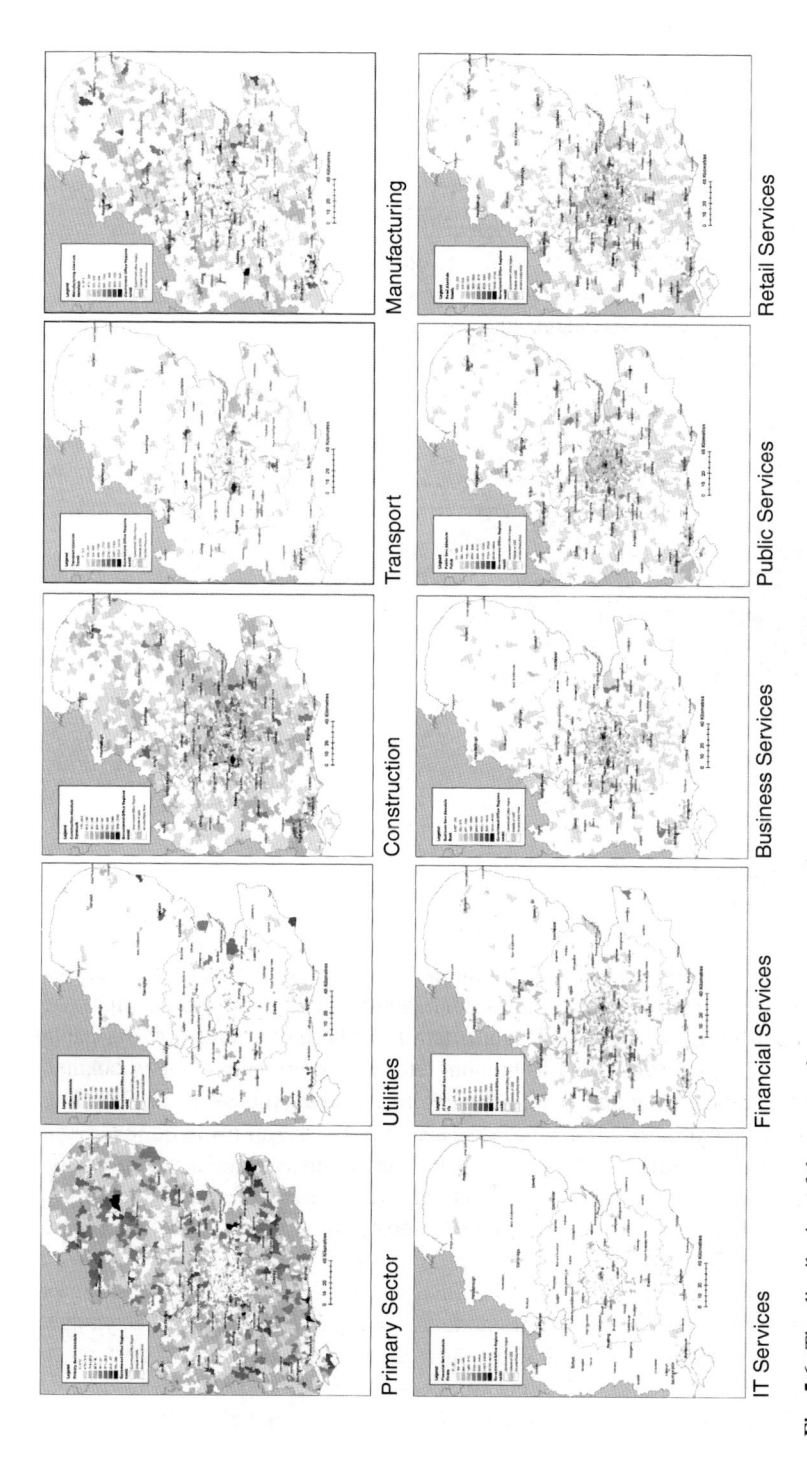

5 Simulating the Spatial Distribution of Employment in Large Cities: With... 99

Fig. 5.6 The distribution of the ten employment types across the greater south east region (readers should zoom in on these figures to extract the detail: the figures is located at http://www.complexcity.info/the-scale-project/)

Table 5.3 Estimates of the coefficients and performance of the employment location models (standard errors in (•))

	Financial services $\{E_{j1}\}$	Manufacturing $\{E_{j2}\}$	IT services $\{E_{j3}\}$	Business services $\{E_{j4}\}$	Retail services $\{E_{j5}\}$	Public services $\{E_{j6}\}$
Constant b_0	−405.715 (61.877)	44.992 (14.903)	333.043 (40.553)	268.647 (40.092)	113.798 (60.795)	1145.055 (70.777)
Retail floorspace F_j^1	Na	Na	Na	Na	0.054 (0.001)	Na
Office floorspace F_j^2	0.011 (0.000)		0.010 (0.000)	0.023 (0.000)	Na	0.019 (0.000)
Industrial floorspace F_j^3	Na	0.0005 (0.000)	Na	Na	Na	Na
Employment accessibility (within 25 min by public transport)	−4.408 (0.222)	Na	1.069 (0.117)	2.721 (0.185)	0.791 (0.095)	2.843 (0.251)
Employment accessibility (within 25 min by road)	0.461 (0.015)	Na	−0.086 (0.008)	−0.188 (0.013)	Na	−0.287 (0.017)
Time to motorway junctions M_j	23.519 (3.491)	Na	−4.617 (1.354)	−7.140 (1.990)	−13.383 (2.805)	−7.971 (2.732)
Time to rail hubs J_j	Na	Na	Na	Na	29.918 (6.472)	−34.275 (9.126)
Time to heathrow airport H_j	Na	Na	−1.656 (0.693)	Na	Na	Na
Time to nearest A-class roads h_j	Na	−4.709 (1.565)	Na	Na	Na	Na
Time to all airports Q_j	−5.593 (2.109)	0.870 (0.364)	Na	Na	−3.846 (1.705)	Na
Correlations squared	0.637	0.653	0.795	0.899	0.826	0.713

sectors are influenced by accessibility to motorways, and accessibility to public and car/road transport (with the exception of retail which is only public transport). Access to Heathrow airport appears to only influence IT services, while retail and public services are influenced by access to rail hubs. All this seems to make sense although there is substantial correlation between the various accessibility variables. In short, with the floorspace models, there is substantial implicit weighting through double counting. We make no apology for this, for we want to optimise the fit and as we will illustrate below, short of a perfect fit, we need to be very specific about how we use these models in prediction to reduce the error that is contained in the model estimates of the base year calibrations.

5.4 Using the Employment Models Within the Integrated Framework

5.4.1 Error Propagation in Comparative Static Models

Our model framework is in the comparative static tradition. The input–output, employment, and LUTI models are all estimated to a cross section of urban and spatial structure at the base year, in this case 2005. The various data that we use have culled from various series and censuses from 2001 to 2007 but we have normalised these to ground all the data in the common year. There is very little discussion of how to use cross-sectional forecasting models in predictive contexts. The notion of comparative statics which is an old concept in economics, is assumed to be one where a model of a system produces a static equilibrium and in forecasting, a new static equilibrium is assumed to occur at some point in the future where the actual change is the difference between the two equilibria. This, of course, assumes that whatever changes that take place in the future work themselves out to the new equilibrium by the end of the forecasting period. In short, it is assumed that any motions that take place between the equilibria are not required to be known for the prediction to be useful.

Lowry (1965) in his seminal article on model design says of comparative statics: "The process by which the system moves from its initial to its terminal state is unspecified. Alternatively, comparative statics may be used for impact analysis, where no target date is specified. Assuming only one or a few exogenous changes, the model is solved to indicate the characteristics of the equilibrium state toward which the system would tend in the absence of further exogenous impacts". In fact, Lowry (1964) in his **Model of Metropolis**, makes the further point that such comparative statics produces an instant metropolis, an emergent structure rather than the actual urban structure that we currently observe. It is on this premise that the baseline calibration of any comparative static model is 'what would happen' if the forces at work in the data describing the current urban structure were to work themselves out, rather than 'what has actually happened'. This has a profound impact on how such models should be used in prediction.

The problem with any model in terms of its calibration is that no calibration can give perfect estimates of the observed situation, whether it be a comparative static or a dynamic model. The problem is how the errors which are part of the calibration are to be treated in using the model in prediction. If one simply makes predictions with the same model that has been used in calibration, then these errors will be transmitted through to the future state. If the errors are greater than the differences that are associated with the future state, then what is error and what is new are completely confounded. Indeed, this is intrinsic to the entire prediction in that it will always be possible to find a forecasting period in which the overall change will be less than the overall error. It is somewhat remarkable that there has been so little discussion of this issue but in practice what is usually done is to generate differences rather than absolutes. In fact this is no easier with dynamic models that are fitted to differences than to static models that simulate the entire structure because both contain errors. It might even be argued that static models are preferable in that it is the difference between the two equilibria that is significant, and any prediction is then simply treated as a difference between the existing situation and the future state.

We can give this situation more clarity in terms of the input–output and employment models which are both comparative static. Adding time to the variables, we define the total population at the base year time t as $P(t)$ and total employment as $E(t)$. These quantities are external to the entire framework; in fact they can be predicted using aggregate demographic and employment models and one alternative we are considering is to predict population using the MoSeS model (Birkin, et al. 2009). However assuming we have predictions of these totals as $P(t+1)$ and $E(t+1)$ at time $t+1$, then we clearly know the increments

$$\Delta P(t) = P(t+1) - P(t) \\ \Delta E(t) = E(t+1) - E(t) \tag{5.23}$$

which may be positive of negative. These are net values, for population and employment may grow and decline in different areas and this represents an intrinsic problem to most modelling frameworks that accept exogenous totals of this kind. Nevertheless, the input–output model takes total employment at times t and $t+1$ and generates totals $E_k(t)$ and $E_k(t+1)$, the differences $\Delta E_k(t) = E_k(t+1) - E_k(t)$ also being either negative or positive.

The employment model distributes these totals to the zones of the system, but before we broach the problem of dealing with net totals, let us consider the predictions of the estimated model at the calibration year $E'_{ik}(t)$ and predictions at the same year but with input variables for the future year defined as $E'_{ik}(t:t+1)$. Now the sum of these employments across sectors and zones equals observed total employment at time t

$$\sum_i E'_{ik}(t) = \sum_i E'_{ik}(t:t+1) = E_z(t), \tag{5.24}$$

$$\sum_i \sum_k E'_{ik}(t) = \sum_i \sum_k E'_{ik}(t:t+1) = E(t), \tag{5.25}$$

and to model where the increment of employment is located, we simply take the ratios $E'_{ik}(t:t+1)/E'_{ik}(t)$ and apply these to the increment $\Delta E_k(t)$. Then the employment model for each sector k is

$$\Delta E_{ik}(t) = \Delta E_k(t) \frac{E'_{ik}(t:t+1)}{E'_{ik}(t)}, \tag{5.26}$$

and the new employment at time $t+1$ is

$$E_{ik}(t+1) = E_{ik}(t) + \Delta E_{ik}(t). \tag{5.27}$$

It is easy to show that these employments in Eq. 5.27 sum to the category totals produced by the input–output model. In short, the errors in the actual calibrated model are not transmitted to the new totals but only to the increment of employment due to the fact that we only allocated the increment using the model, not the original employment structure which is simply used as the baseline.

5.5 Problems with Predicting Total Aggregate Activity

The problems of specifying aggregate totals which begin the chain of prediction involve the fact that these totals reflect net rather than gross change. For example, a sector might be declining overall but growing in some places with decline elsewhere cancelling this out. This might be caused by simple differential growth in situ or in the case of population it can be a complex concatenation of changes in fertility, mortality and migration. However if we simply take the aggregate quantities and these are negative, then the largest negative values will be allocated to the areas which are most attractive to employment. In short, these kinds of models tend to work with positive quantities and when we feed them with negative, then the logic will simply be reversed. Strictly speaking, we need to separate out growth from decline – positive from negative. If we have some sense of this, then we probably need to input this distinction externally and for each quantity, $\Delta E_k(\tau)$, we need to break this into positive and negative components $\Delta E_k(\tau) = \Delta E_k^+(\tau) + \Delta E_k^-(\tau)$ and allocate these quantities directly using the logic which is embodied in Eqs. 5.26 and 5.27. In this case, we would then be somewhat more confident that the mechanisms of growth and decline are being dealt with appropriately in the model framework. But in the last analysis, the models need to be much more disaggregate spatial at all stages and the logic of the integrated input–output, employment and spatial interactions models developed by Echenique (2004) amongst others is unassailable. Models of the kind developed here tend to be based on the expediency of coupling models together which have been designed with different purposes in mind and in any further developments, the need to develop the integrated framework with strong coupling between the submodels would be critical.

5.5.1 Using the Model Framework for Scenario Generation

The integrated model is currently being used to explore and test the impact of three rather different scenarios for the future of the London region. We have a baseline scenario referred to as 'business as usual' which is essentially a trend projection based on current planning policies and infrastructure developments that are already being implemented. Current policies also imply a degree of compaction mirroring growth in the last two decades, and this is linked to the growth in services and the knowledge economy based in London and larger cities. The second scenario is one of 'decarbonisation' with low carbon cities being the dominant vision. This is implied in more concentrated, higher density forms and restricted car-based growth. In planning policy terms, it is an exaggerated version of 'business as usual, and it has implications for transport in terms of higher fuel prices/taxes promoting greener alternatives. The third scenario is one of 'deregulation' particularly with respect to compaction and the force of the greenbelt. It provides a direct contrast to the second.

All these scenarios imply different levels of employment. None of the ten categories of employment involve a decrease overall but this does not mean the problems with the net effects of change have been removed. In terms of the operation of the floorspace predictions, then starting with new estimates of floorspace which reflect these scenarios, we need to iterate Eq. 5.20 and also add new capacities, change accessibilities and add in anticipated changes in the building stock. In fact there are many locations that we need to specify in terms of these predictions and we will concentrate on making general changes across the board so that we might evaluate sufficiently different kinds of scenarios in terms of the anticipated physical changes. After we have made predictions of floorspace, we can make explicit changes in floorspace in the employment equations. In fact, we tend to reserve the employment model for handling one-off changes in physical and transport capacities while leaving the floorspace models to deal with more general changes which affect all zones within the wider region. In short, the floorspace and employment models have been designed so that we can intervene and formalise scenarios in this way through physical changes to the future region.

5.6 Conclusions and Next Steps

There are many problems in developing an integrated framework of models based on a loose-coupling of existing structures. In theory, an integrated strategy that weaves employment into population modelling with consistency at all scales is required but the problems of doing this at a disaggregate level and in a temporal, non-equilibrium context are formidable. The debate over cross-sectional or dynamic models or some fusion of both does not address the two issues that are relevant to using any model in prediction: the need to begin the process somewhere with assumptions about aggregate quantities and the need to deal with perpetuating errors in the calibrated outputs of models when they are used in prediction. Neither of these problems has been addressed much, somewhat remarkably given that there is nearly 50 years of sustained effort with these kinds of model. This must be due to the fact that the scale of these efforts has tended to end before their intensive use in prediction, often because of the sheer scale of the exercise in getting to the point where predictions are possible, and/or in terms of the organisations involved in building the model whose expertise and goals are often very different from those organisations involved in using it in prediction.

Computation has now reached the point where many of these problems can now be resolved. Models like this one can be quickly designed, changed, and implemented with stakeholders and other scientists in ways that enrich their use in prediction. We need a sustained attack on the problem of prediction in urban systems, given that during the time when these kinds of models have been developed, our view of the predictability of human systems and what predictive models actually mean has changed quite radically. The notion of providing predictions but for the shortest intervals of time, has changed quite dramatically. Models such as this and others in this book, now need to be used in conditional prediction of many kinds, to generate 'what if' scenarios and inform the debate about how these tools might be continually modified in response to what we learn about the problem and the future. In this context, there is a still a massive dilemma between ever more detailed models, and ever more scepticism about what we can predict and how we might use prediction in helping us inform these dialogues about the future. Integrated modelling is important in this but so also is the need to develop fast, visually accessible models that we can change quickly in response to our learning. This implies that the argument about simple, static, aggregate versus complicated, dynamic, disaggregate models structure is far from over: it is just beginning.

Acknowledgements This research was part funded by the ESRC Genesis Project (RES-149-25-1078), and the EPSRC Scale (EP/G057737/1) and Arcadia Projects (EP/G060983/1)

References

Batty M (1986) Technical issues in urban model development: a review of linear and non-linear model structures. In: Hutchinson BG, Batty M (eds) Advances in urban systems modelling. North Holland Publishing Company, Amsterdam, pp 133–162

Batty M, Vargas-Ruiz C, Smith D, Serras J, Reades J, Johansson A (2011) Visually-intelligible land use transportation models for the rapid assessment of urban futures, Working Paper 163, Centre for Advanced Spatial Analysis, University College London, http://www.bartlett.ucl.ac.uk/casa/pdf/paper163.pdf. Accessed 7 Sept 2012

Birkin M, Townend P, Turner A, Wu B, Xu J (2009) MoSeS: a grid-enabled spatial decision support system. Soc Sci Comput Rev 27(4):493–508

Echenique MH (2004) Econometric models of land use and transportation. In: Hensher DA, Button KJ, Haynes KE, Stopher PR (eds) Handbook of transport geography and spatial systems. Pergamon, Amsterdam, pp 185–202

Getis A, Ord JK (1992) The analysis of spatial association by distance statistics. Geogr Anal 24:189–206

Hansen WG (1959) How accessibility shapes land use. J Am Inst Plann 25:73–76

Junankar S, Lofsnaes O, Summerton P (2007) MDM-E3: a short technical description, Working Paper, Cambridge Econometrics, Cambridge, UK, available at http://www.camecon.com/. Mar 2007

Lowry IS (1964) Model of metropolis, memorandum RM-4035-RC. Rand Corporation, Santa Monica

Lowry IS (1965) A short course in model design. J Am Inst Plann 31:158–165

Macgill SM (1977) The Lowry model as an input–output model and its extension to incorporate full intersectoral relations. Reg Stud 11(5):337–354

Maoh H, Kanaroglou P (2013) A firm failure microsimulation model. In: Pagliara F, de Bok M, Simmonds D, Wilson A (eds) Employment location in cities and regions: models and applications. Springer, Berlin/New York, this volume

Moeckel R (2013) Firm location choice versus employment location choice in microscopic simulation models. In: Pagliara F, de Bok M, Simmonds D, Wilson A (eds) Employment location in cities and regions: models and applications. Springer, Berlin/New York, this volume

Putman SH (1983) Integrated urban models, policy analysis of transportation and land use. Pion Press, London

Putman SH (1991) Integrated urban models 2: new research and application of optimization and dynamics. Pion Press, London

Walsh CL, Dawson RJ, Hall JW, Barr SL, Batty M, Bristow AL, Carney S, Dagoumas AS, Ford AC, Harpham C, Tight MR, Watters H, Zanni AM (2011) Assessment of climate change mitigation and adaptation in cities. Urban Des Plann (Proc ICE) 164(2):75–84

Wilson AG (1970) Entropy in urban and regional modelling. Pion Press, London

Chapter 6
Complex Urban Systems Integration: The LEAM Experiences in Coupling Economic, Land Use, and Transportation Models in Chicago, IL

Brian Deal, Jae Hong Kim, Geoffrey J.D. Hewings, and Yong Wook Kim

Abstract Simulated alternative patterns of future land-use change at a fine cellular resolution can be a valuable tool in assisting urban decision making in a region. When coupled with top-down regional economic demand modeling techniques, these disaggregate, spatially-explicit models of urban transformation can help planners and policymakers understand how different public policy and investment choices will play out in the future, especially in conjunction with different economic and demographic trends. In this paper, we seek to advance inquiry into connecting economic, land-use and travel-demand simulation models through a qualitative assessment of the change in land-use and travel demand outcomes as a result of loosely coupling a macroeconomic model (CREIM), a fine scale regional dynamic spatial land use model (LEAM), and a zone based travel demand model (CMAP TDM) in the Chicago metro region. We then compare land use and travel demand simulations with and without coupling so as to isolate the effects of the coupling process. We conclude with discussions on the implications of these results on location variables for various land use types, and its potential implications for planning.

B. Deal (✉)
Department of Urban and Regional Planning, University of Illinois, 611 Taft Dr, 61820 Champaign, IL, USA
e-mail: deal@illinois.edu

J.H. Kim
Department of Planning, Policy, and Design, University of California, Irvine 206E Social Ecology I, Irvine, CA 92697, USA

G.J. Hewings
Regional Economics Applications Laboratory, University of Illinois, Champaign, IL, USA

Y.W. Kim
Department of Urban and Regional Planning, University of Illinois

6.1 Introduction

The importance of integrating economic, land use, and transportation systems models for forecasting and informing policy and investment choices in urban areas has been well established mainly in terms of the transportation/land use nexus (Hansen 1959; Lowry 1964; Kim 1983; Waddell 2002; Wegener 2004). Attempts to physically integrate transportation and land use models within a single analytic framework in an effective manner also have a long history (Lowry 1964; Crecine 1964; Garin 1966) although recent attempts have been less prolific (de la Barra 1989). What has been clear is that through a tighter coupling of modeling systems, planning academics and professionals hope to better simulate potential future changes in the urban fabric of our regions; produce more reliable and defensible policy options to cope with the changes; and to enhance the decision making abilities of our communities. Less intuitive however, is how these complex modeling approaches become integrated; both technologically and into a wide-ranging and public planning process.

Deal and Pallathucheril (2007) argue that simulations of future regional land-use change can be a valuable tool in helping planners and policymakers understand how different public policy and investment choices will play out in the future, especially in conjunction with different economic and demographic trends. These simulations provide valuable and reliable insights into the potential future implications of current decisions. Transportation engineers have long known the benefits of simulation modeling, developing the traditional four-step process to produce travel forecasts as early as the 1950s. They have also recognized that land use and transportation systems are not mutually exclusive. The four-step process for example, relies on future socio-economic and land use projections as the basis for determining potential 'zonal' demand. But until recently, this relationship between land use and transportation systems has been described in static terms – i.e., a static long term land use forecast is used to drive a dynamic transportation model, or a static transportation system is used to drive a dynamic land use model. The linking of these dynamic systems has been relatively cumbersome, costly and difficult to achieve – although we will argue, critical in understanding the dynamics of urban systems. When the mutual influence of land use and travel demand are effectively captured and conveyed, the resultant information is more reliable, believable, and defensible. These coupled simulations can better inform choices made with regard to both transportation and land use policies and investments.

The need for integration and feedback between dynamic land use and travel demand models has been recognized since at least the 1990s but efforts to jointly simulate future changes has only recently begun to produce meaningful results. Waddell et al. (2007) proffer three potential explanations: a preoccupation with expanding transportation infrastructure (and its accompanied funding) and a reluctance to confront possible negative consequences of this expansion; a lack of a solid theoretical basis for this integration; the complexity and massive data requirements of available tools. All three are plausible explanations, although perhaps quickly becoming obsolete.

Miller et al. (1998) describe a number of ways in which land-use change can drive travel demand models. Some use varying degrees of formal and quantifiable mechanisms for allocating activities to zones, including explicit representation of market systems. However, the approach that appears most prevalent involves transportation professionals exercising their judgment about potential activities resulting from future land-use changes within the spatial unit of analysis – traffic analysis zones.

Hunt and Abraham (2003) use spatial market representations as a way to describe the exchange of goods and services at different locations and between different actors located in different places. PECAS, an extension of the MEPLAN (Echenique et al. 1990; Hunt and Simmonds 1993) and TRANUS (de la Barra et al. 1984; de la Barra 1989) approach, uses an integrated approach to land use and transportation modeling. In PECAS, land and transportation markets are simultaneously modeled. Spatially disaggregated 'make' and 'use' input–output tables are used to distribute activity that is driven by region-wide economic conditions. Activities allocated to traffic analysis zones then drive a transport module. Temporal dynamics are simulated by interactions between successive time periods: activity allocations drive space development which in turn drives region-wide economic conditions and activity allocations in the next time period; utility (and negative utility) identified in the transport module drive activity allocations in the next time period. Hunt and Abraham identify two issues with PECAS implementations: lack of data at a fine enough resolution for calibration; and computational limitations on the number of zones that can be modeled. Both factors imply a potential coarseness of representation.

Waddell et al. (2007) describe how UrbanSim (Waddell 2002; Waddell et al. 2003), a land-use model, was integrated with a four-step travel demand model in the Greater Wasatch Front Region of north-central Utah. UrbanSim explicitly models individual agents in the land development process. It includes both the supply and demand side of the real estate market as well as the subsequent prices. Unlike a static equilibrium model, demand, supply, and prices are adjusted dynamically. UrbanSim is connected to the four-step model by projecting land use for 5 years; the 5 year projection spatially allocates new households and jobs: the new household and job information initiates the travel demand model; and the new state of the transportation network is now input back into UrbanSim for the next 5 year run.[1] This process produced significant changes in travel demand when compared with model runs that did not include dynamic land-use inputs: a 5 % increase in vehicle miles traveled (VMT) and vehicle hours traveled (VHT); a 16 % increase in total hours of congestion delay (TCD). However, the effects that this approach produced on land-use change are not examined in any great detail.

In this chapter, we seek to advance this area of inquiry by building on the experiences described above with a focus on assessing the land-use consequences of connecting economic, land-use and travel-demand simulations. The work

---

[1] Hunt and Abraham (2003) describe this kind of an approach as connection rather than integration of land use and travel demand.

presented here was done in collaboration with the Chicago Metropolitan Agency for Planning (CMAP), the Chicago region's designated Metropolitan Planning Organization. Prior to the completion of the work, CMAP used separate and static forecasting approaches. In this work, the regional economic structure and change is simulated using a regional macroeconomic model – the University of Illinois Regional Economics Applications Laboratory's Chicago Region Econometric Input–output Model (CREIM); future land-use change is simulated using an dynamic spatial land use model – University of Illinois, Department of Urban and Regional Planning, LEAM Laboratory's Landuse Evolution and impact Assessment Model (LEAM); while future travel demand is simulated using a typical four-step travel demand model – CMAP's Travel Demand Model (CMAP TDM). In this work, the two models were loosely coupled more or less in the manner described by Waddell et al. (2007): results from CMAP TDM are used as inputs into LEAM which is run for several annual time steps, the resulting land use output from LEAM is the basis for socio-economic inputs back into CMAP TDM, and the process iterates until the simulation is complete.

The following section provides a brief description of the 'loose coupling' of LEAM and CMAP TDM. We compare land use and travel demand simulations with and without coupling so as to isolate its effects. We conclude with discussion of the implication of these results on planning and planning support systems.

6.2 The Chicago Metropolitan Agency for Planning Travel Demand Model

The Chicago Metropolitan Agency for Planning is heir to a rich legacy of travel demand modeling procedures developed by its predecessor, the Chicago Area Transportation Study (CATS). CMAP now maintains a set of computational procedures that systematically estimates travel demand in the greater Metropolitan Chicago region. The CMAP procedures are typically referred to as the CMAP TDM Models.

The CMAP TDM use a four-step process (Fig. 6.1) to predict and evaluate future regional transportation demand under various system, policy and investment scenarios. CMAP uses TDM models to analyze and assist in the transportation planning process in the region. Its planning area comprises Cook, DuPage, Kane, Kendall, Lake, McHenry, and Will Counties and parts of Grundy County.

The CMAP TDM is implemented in EMME/2 (http://www.inro.ca), an interactive multi-modal transport planning modeling package. It offers a comprehensive set of tools for demand modeling and multi-modal network modeling. In the CMAP TDM, the nine-county Chicago region is represented as 15,147 trip generation zones (TGZs), each the size of a quarter-section (1/2 mile by 1/2 mile). Information about trends in households and employment in each TGZ is gathered from many sources and stored as an EMME/2 databank. Trip making behavior for the model is derived from this database of socio-economic and land use characteristics. Origin, destination and travel mode are based on trip making behavior and prevailing transportation conditions such as highway congestion and transit service.

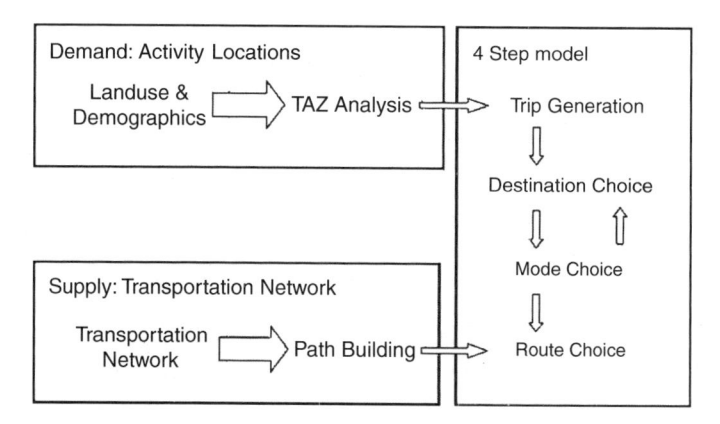

Fig. 6.1 The typical four-step model

Transportation system demand is used to determine which combination of highway and transit facilities are used in making a trip. All of these factors go into the EMME/2 model in order to produce vehicle equivalents on each link of the network.

The purpose of this chapter is not to describe the model in full. It is important to understand however, the traditional four-step travel demand models' approaches in order to discern data needs and potential integration with land use simulation models, such as LEAM.

6.3 The Landuse Evolution and Impact Assessment Model

Fundamentally, the LEAM model consists of two major organizational parts: (1) a land-use change model (LUC) – defined by a dynamic set of sub-models drivers that describe the local causality of change and enable easy addition and removal of variables and the ability to play out 'what-if' scenarios; and (2) impact assessment models that facilitate interpretation and analysis of land-use change depending on local interest and applicability – these help to assess 'so-what' questions and explicate the implications of a scenario.[2] The need in planning and policy-making to answer both 'what-if' and 'so-what' questions is a key basis for the LEAM framework.

LEAM uses a modified cellular automata approach where 30m × 30m cells evolve over a surface defined by biophysical factors such as hydrology, soil,

---

[2] More detailed descriptions of the LEAM framework is described elsewhere (Deal and Pallathucheril 2008) and as noted, it is not our intention to describe the model in full here. The intent here is to provide a very brief description of the basics of the model that will enable the reader to follow some of the reasoning behind its use-driven local applications.

geology and land forms; and socio-economic factors such as administrative boundaries, census districts, and planning areas. The probability of each cell change is decided not only by the local interactions of neighbor cells but also by global information available to the model. Therefore, cells within LEAM are intelligent agents that not only can capture local information, but also can sense and respond to regional or global information such as social and economic trends.

The probability for a certain type of development (e.g., for residential, commercial-industrial, and urban open space) can be expressed, as follows.

$$\text{IF } i \in \bigcup_{k=1}^{m} NG_{k,t}^{j} \text{ THEN } p_{i,t}^{j} = 0 \text{ ELSE } p_{i,t}^{j}$$
$$= \kappa \cdot f\left(CE_{i,t}, NE_{i,t}, ZO_{i,t}, AC_{i,t}\right) \quad (6.1)$$

where $p_{i,t}^{j}$ represents cell i's probability for j-th type of development at time t, and $NG_{k,t}^{j}$ indicates a set of land cells that cannot be developed for purpose j at t due to the k-th constraint. The constraints include both natural barriers to development (e.g., water, steep slopes, etc.) and regulatory actions for environmental or many other zoning purposes. Unless the development of a certain cell is prevented by any of NG, the probability is calculated based on a function $f(\cdot)$ that gives a probability value between 0 and 1 combined with a growth adjustment factor κ.[3] The probability calculation considers a broad range of factors: each cell's biophysical conditions ($CE_{i,t}$), neighboring cells' status, typically taken into account in cellular automata models ($NE_{i,t}$), socio-political and economic characteristics of the zone in which the cell is located ($ZO_{i,t}$), and its accessibility to various attractors, such as employment centers and public facilities ($AC_{i,t}$).

In this approach, the transportation network (and subsequent congestion coefficients) help to determine potential development probabilities by affecting the accessibility computation $AC_{i,t}$. As noted in the following sections, land use simulation outcomes can differ significantly depending on road capacities and congestion. This is consistent with regional observations of existing land use structures where the formation of land development patterns are significantly influenced by transport conditions.

The LEAM framework consists of a hierarchical structure with multiple scaled models incorporated. These models are loosely coupled in a modular framework where the information can be exchanged on-the-fly through aggregate or disaggregate approaches. This design strategy enables LEAM to integrate cellular micro models and regionalized macro socioeconomic models into a single model framework.

---

[3] κ is designed to ensure that the total amount of new land development follows the overall demand growth, forecasted by a regional economic model. See Deal et al. (2005) for detailed explanations of how this works.

Fig. 6.2 A LEAM residential probability map for the metropolitan Chicago region

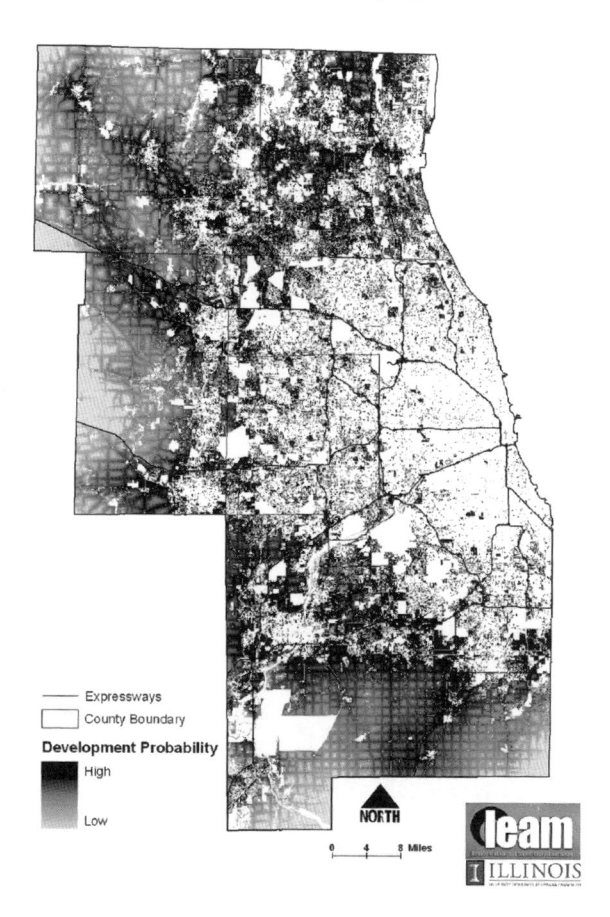

6.3.1 LEAM Simulations

Land use transformations are simulated using a Monte Carlo experiment in which potential land uses compete for space based on probability scores and a random number generator. In a given time step (for a specific type of land use), the regional demand for new development and the probability of land use change associated with each cell is determined regardless of whether the cell is transformed. Transformation occurs if the calculated probability is greater than a random number generated for each cell, so that even cells with low scores may be transformed (Fig. 6.2). For instance, an undeveloped cell will be transformed into commercial-industrial land at time t, if

$$p_{i,t}^C > Max(p_{i,t}^R, p_{i,t}^O, \rho)$$

where $p_{i,t}^R$, $p_{i,t}^C$, and $p_{i,t}^O$ indicate the cell's probabilities for residential, commercial-industrial, and open space purposes, respectively; and ρ is a randomly generated number for the cell. Then, the simulated commercial-industrial land development represents the spatial distribution of employment into the future.

The regional demand for land (residential, commercial, and open space) is a function of the generation of new households and jobs as projected over time using a regional econometric forecasting model (Israilevich et al. 1997). The conversion from regional economic forecasts to demands for new land development is accomplished using observed localized (county or census tract) averages for the ratio between developed land, household generation, and the types of jobs created. In lieu of sub-regional constraints on demand to determine spatial allocation (Wu and Martin 2002), the estimated demand serves only as a target for regional land allocation in each time period. Locations around the region compete to meet the targets, and market variables speed up or slow down development rates based on how well the demand was met, or was not met; this is captured by the growth adjustment factor κ mentioned above. A typical land use change map for the greater Chicago region is shown in Fig. 6.3.

Simulated outcomes are described in graphs, charts, text, and in map form and are used in engaging local dialogue and in analyzing the potential implications of the changes described. The environmental, economic and social system impacts of alternative scenarios can be modeled and tested (Deal and Schunk 2004). Scenario descriptions of alternative land-use policies, investments decisions, growth trends and unexpected events (among others) can be simulated, analyzed and compared for regional importance. LEAM's visual and quantitative representation of each scenario's outcome provides both an intuitive means of understanding and a basis for analyzing the implications of potential decisions. These representations act as a catalyst for discussion and communal decision-making.

A dialogue among stakeholders, planners, and policy makers completes the process by providing feedback on the value and relevance of the simulation outcomes. Dialogue and feedback are important parts of the LEAM framework, both mathematically and in local application. It has been the authors' experience that pure mathematics in complex urban dynamic models does not tell a complete story and sometimes not even a compelling one, about the local condition. Constant internal and external review and interaction are critical for producing salient land use change simulation outputs (Deal and Chakraborty 2010).

6.3.2 Economic Drivers

Regional economic change is an important driver of population and subsequent land use allocation in any region (See e.g. Briassoulis 2000; Veldkamp and Lambin 2001; Hewings et al. 2004). In other words, the number of people and number of jobs, representing the amount of economic activities in the region, are critical factors in determining how much land will be developed in the future. At the very basis of LEAM are population and employment forecasts. In basic simulations these are taken from known sources, such as US Census, and state or regional

Fig. 6.3 A typical land use change summary map for the metropolitan Chicago region. *Yellow* areas represent new residential developments in 2030. *Red* areas represent new commercial developments in 2030

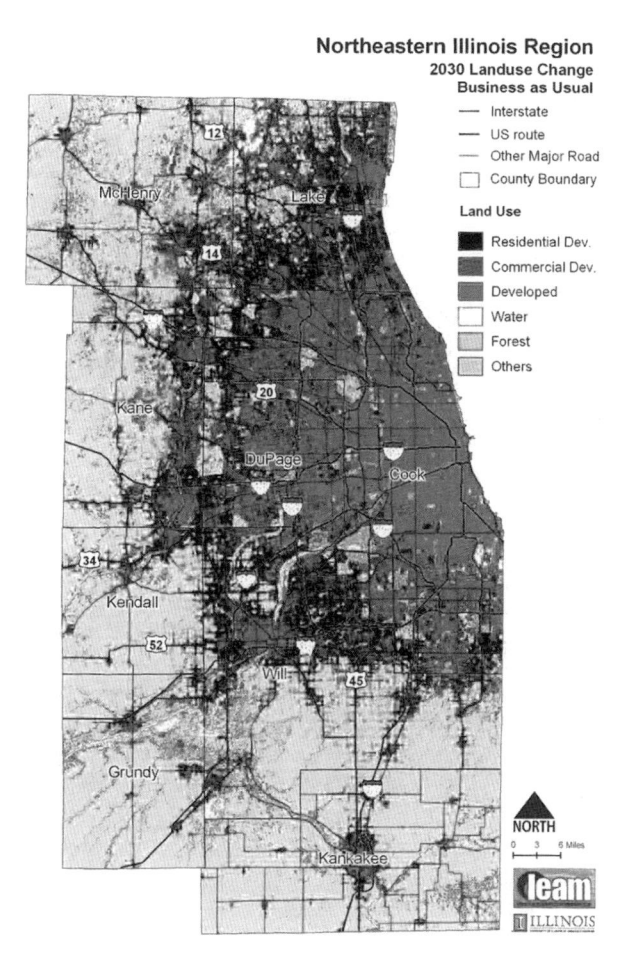

sources of information. In advanced simulations, these are derived from a regional economic model. Macroeconomic variables, (e.g., regional population and employment) are forecasted and used to estimate the residential/commercial-industrial land demand which partly control the total amount of new developments.

In this work we integrated an existing regional economic model and LEAM to provide a top down forecast of regional economic demand. Linking this to a spatial allocation model enables model users to understand how different public policy and investment choices will play out in the future, especially in conjunction with different economic and demographic trends or scenarios. The regional economic model used in this study is based on models developed at the University of Illinois Regional Economics Applications Laboratory (REAL). REAL, has developed a technique for building a Regional Econometric Input–output Model (REIM) for generating regional-level economic projections that consider not only the relationship between the regional economy and the national economy but also the region's own industrial structure. Since REIM reflects the dynamic changes in regional economic structure over time, the use of REIM enables us to get better forecasts of the regional economy, while maintaining its effectiveness in impact and policy analyses. Since 1989 REAL has developed (and continues to develop and apply) a REIM based model for projects in the Chicago region (CREIM – The Chicago Region Econometric Input–output Model).

The CREIM is a structural equation model that consists of five major blocs – output, employment, income, population, and final demands – tightly interconnected with each other. Among others, it projects regional employment by industry, used for LEAM simulations, by estimating the future trajectory of regional productivity in each industry and determining the industry's output (i.e., production) simultaneously, as below.

$$\ln\left(\frac{OUT_{s,t}}{EMP_{s,t}}\right) = g(\cdot) + \varphi_{s,t} \rightarrow EMP_{s,t} = \frac{OUT_{s,t}}{\exp\left[g(\cdot) + \varphi_{s,t}\right]} \quad (6.2)$$

where $OUT_{s,t}$ and $EMP_{s,t}$ represent the level of output and employment in industry s at time t, respectively. $\varphi_{s,t}$ is an error term. $g(\cdot)$ indicates industry-specific regional productivity function, that is typically characterized as a first or second-order autoregressive process influenced by some independent variables, mainly national economic indicators.

CREIM also generates regional population forecasts, another input for land use simulations. The model forecasts births and deaths and net migration in the Chicago metropolitan area based on historical trends to describe dynamic regional demographic changes. Furthermore, it sets working age population as a function of regional employment to reflect the economic-demographic interactions. In the CREIM framework, population also has an effect on regional employment indirectly through population – final demand variables (i.e., household consumption and government expenditures) – output – employment linkage. Israilevich et al. (1997) provide detailed explanations of the structure and formulation of CREIM.

The regional employment and population forecasts from CREIM are plugged into LEAM. The loosely coupled nature of the macroeconomic forecasting model (CREIM) and land use simulation model (LEAM) constitutes a one-way link – i.e., economic activity produces the push for new households and land area. In this framework land use patterns do not feedback to the macroeconomic variables. Reality however, might suggest that the performance of a regional economy is also affected by the availability of potential land in the region, government interventions in land use, the efficiency (or inefficiency) of the spatial pattern of land use in the area, and the amount of physical congestion that might cause differences in potential transaction outcomes (See e.g., Parr 1987; Cervero 2001; Kim 2011). This is a common limitation of most land use and regional economic models. New REAL/LEAM work tackles this issue to reflect the feedback effect from land use back to the regional economy (see Kim and Hewings 2009 and the related chapters in this book).

6.3.3 Transportation Networks and LEAM

As noted above, the structure and demands on the transportation systems within a study area is critical for both determining a specific cell's development probability and for simulating future land use patterns. More specifically, the development probability for each cell, at each time step, is based on three major attributes associated with transportation networks: road access, road carrying capacity, and road congestion. Road access represents cell proximity to roadway networks that influences travel times to important attractors. Since accessibility to employment, cultural, healthcare and other centers of interest are measured by each cell's travel time to the attractors, road access partly determines the cell development probabilities.

The probabilities also depend on road carrying capacity and road congestion that determine travel speeds on roadway networks, another critical component of travel times (accessibility) to attractors. Capacity is influenced by type, width, and speed limits. Wider roads, an increase in lanes, and faster speeds can sustain higher peak loadings than their narrower, slower counterparts. Congestion is based on road capacity and intensity of use at peak times for any direction; as congestion increases (decreasing speeds), accessibility measures decrease, causing potential development probabilities in that local area to decrease.

6.4 Coupling CMAP TDM and LEAM

In basic simulations (without coupling to a regional transportation model), LEAM uses posted speeds (or derived speeds at time 0) as a measure of the ease of access to important regional and local attractors. Furthermore, these measures do not (cannot) change over time. A loosely coupled land use TDM however, can provide important assessments of potential travel speeds that can be continually evaluated in response to changing land use conditions. In coupling the two models (Fig. 6.4), LEAM uses the speeds derived by the CMAP TDM simulations to calculate measures of accessibility.

The allocation of households derived in LEAM simulations provides important demographic and economic feedback for input into the CMAP TDM. The CMAP TDM uses demographic and economic variables calculated by TGZ to compute trip generation. Pre-coupling estimations may be less spatially and temporally explicit and therefore more difficult to estimate in terms of their responses to changes in the transportation network. Coupling with a spatio-temporal land use model like LEAM can inform the spatial demographic and economic inputs on an annual basis (if needed), at a fine resolution that can be aggregated to any spatial unit - including TGZs.

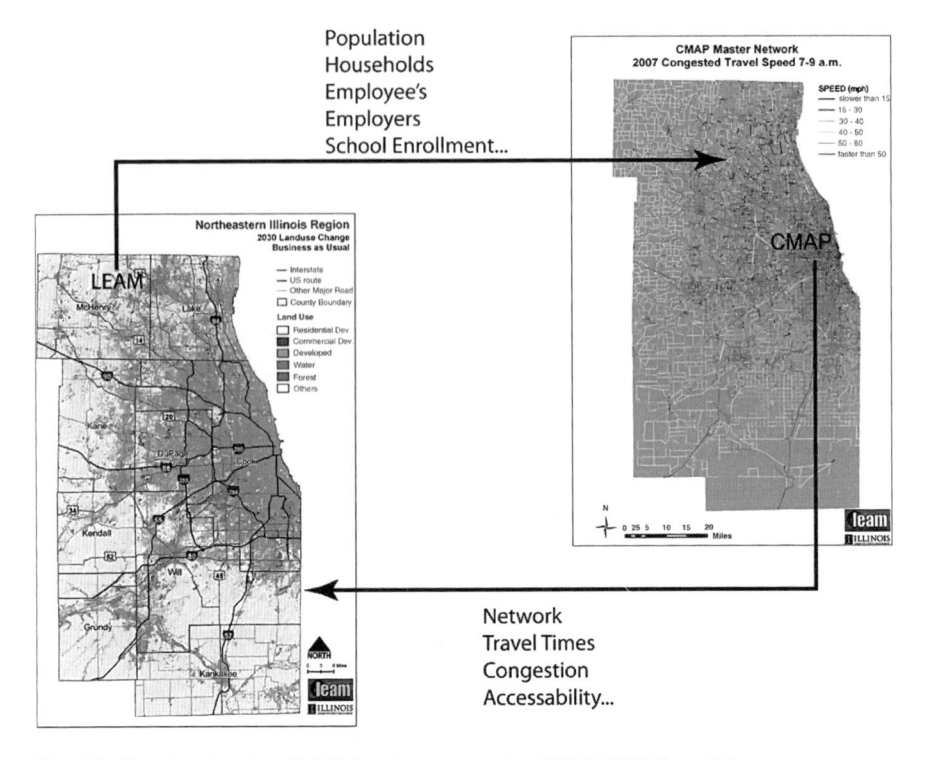

Fig. 6.4 Coupling Landuse (LEAM) and transportation (CMAP TDM) models

A fully integrated run follows the patterns described here (Fig. 6.5) at 5 year intervals. More frequent coupling may be needed or specific project years added (or subtracted), depending on the scenario modeled.

A more detailed description of the approaches described above follows.

6.4.1 From CMAP TDM to LEAM

Primary CMAP TDM model inputs include regional land use, economic (e.g., employers and employees), demographic (e.g., household size and income), and highway and transit network information (see Deal and Pallathucheril 2008 for more detailed inputs). Highway networks are directionally coded for divided highways and arterials and include any roadway functionally classified as a collector or higher. Transit networks include bus and rail systems. Each of the CMAP TDM models is calibrated (i.e., calculated coefficient values) and validated (i.e., base-year "forecasting") using household travel and on-board passenger data collected in 2002 and the region's concurrent socioeconomic data.

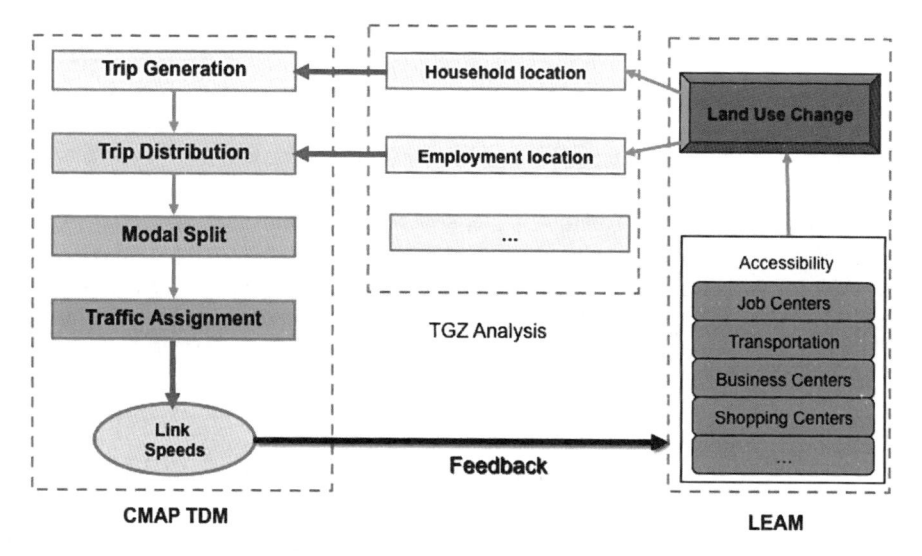

6 Complex Urban Systems Integration: The LEAM Experiences in Coupling... 119

Fig. 6.5 Feedback between coupled CMAP TDM and LEAM systems

In plugging the CMAP TDM into LEAM, some challenges are found. Among others, network representation was the most difficult and problematic variable to overcome, given the discrepancy between two different network files. Basic LEAM simulations use TIGER line data files produced by the US Census Bureau and its associated road classification system to represent the road network and assumed posted speeds. There are major differences in the way TIGER and CMAP networks are geospatially represented. TIGER data attempts to represent network locational information as direct geospatial coordinates – i.e., they try to place roads in space as accurately as possible. This enables geographic reference points and relationships to be accurately assessed, expressed and computed. In contrast, the CMAP network is a conceptual representation of the road network – it does not attempt to accurately reflect the geo spatial relationships between the road network and other geographies.

Since LEAM utilizes geo-referenced land cover data and accessibility measures in relational and geo specific ways, the actual road network geography is useful. It helps maintain consistency in both data input (geo-spatial relationships) and in graphic model output (that describe these relationships in map form). Also, because it utilizes raster based (grid cell structure) information, LEAM is not sensitive to direction of travel.

On the other hand, the CMAP TDM models produce rich data sets with numerous options, such as number of lanes, driving directions, time period restrictions, modal splits, and more – all useful information for formulating and expressing land use decisions and therefore useful for deriving more accurate LEAM simulations. Although unconstrained road conditions (posted speeds) provide valuable information, land use decisions may be biased toward constrained roadway conditions (i.e. realized speeds at peak times). It is this constrained network condition we sought in coupling with CMAP TDM models. The travel speeds of networks provided by CMAP TDM models, which must be a better assessment than the posted speeds, enable LEAM to estimate the accessibility (travel time) among locations considering the real constraints on networks (congestion), to generate a better probability surface, and to simulate future land use pattern more precisely.

How is it possible to reconcile the discrepancy between the two network files and transfer the CMAP TDM output describing congested network link speed to the LEAM road network that has different topology from CMAP master network? We resolved this issue using a travel speed zone to refer the CMAP TDM congested speeds to LEAM Tiger line network. In this process, the CMAP TDM output table data is joined to the CMAP master network using the common identifier of 'from node' and 'to node'. After joining the data, the congested speed for each node in the CMAP network is calculated using travel time and node length. A Travel Speed Zone (TSZ) (which can be conceptualized as a vehicleshed, much like a watershed) is created by developing a polygon that encapsulates each link from node to node. The TSZ represents the congestion data for that segment. The speeds in each zone are then projected to the LEAM Tiger line network by superpositioning the Tiger line network.

Then, LEAM probability surface can be constructed with the more realistic representation of congestions and spatially specific variables (e.g., employment centers, retail centers, etc.). Consideration is also given to the potential frequency of use of attractors (number of employees, size of the retail center, number of hospital beds, etc.). A cost-surface is derived from standard raster-based least-cost computations in which a cost surface is used to describe the time required (speed) to cross individual cells (based on land use, geography, road network, etc. – a river, for example would have an extremely large value). Gravity model computations are then applied to each attractor to compute the accessibility of any cell in the study area to the attractor in question (measured in terms of travel time in 15 s increments).

6.4.2 From LEAM to CMAP TDM

The LEAM land use change allocation approach locates likely new (or re-) development across a region. Aggregating this allocation by specific TGZ, in discrete time steps can provide the basis for estimating changes in future activity in the zones, addressing one of the current deficiencies in the typical four-step transportation model. This requires that information on each individual unit of land, in this case a 30 m × 30 m cell, is capable of characterizing different levels of socio-economic activity in different parts of the region. Hence, a residential cell that falls within a highly urbanized core might represent more potential households than another cell located in a sprawling exurban location. In order to inform sophisticated transportation models, each cell must be able to describe with some level of confidence, its demographic and economic potential.

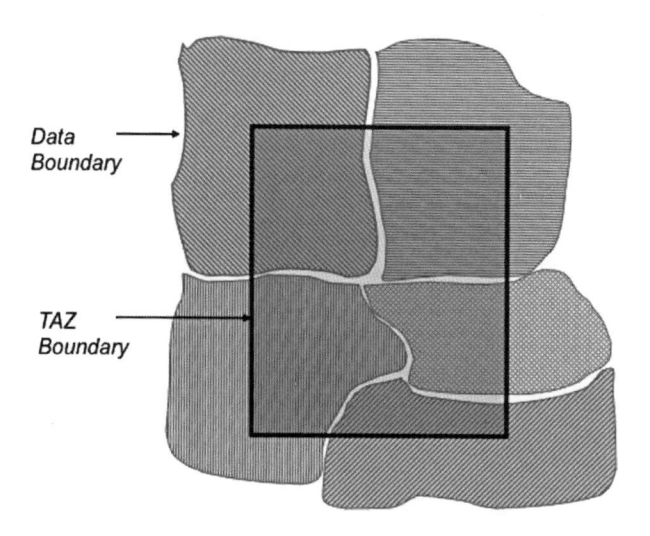

Fig. 6.6 The potential discrepancy between Census data and TGZ boundaries

The CMAP TDM, requires the value of various demographic and economic variables to be computed by TGZ, their spatial units of analysis (in this case quarter sections). These values are generally derived from census data. The shapes of Census geographic units: blocks, block groups, and tracts however, may not exactly match TGZ geography (Fig. 6.6) and need to be reallocated. Traditionally this is accomplished through an approximation based on area (e.g., if the TGZ covers 50 % of the census block area then 50 % of that block's population can be attributed to that TGZ). This assumes that population and jobs are uniformly distributed over the spatial geography of census blocks. We find this assumption is seldom accurate and can potentially generate large errors in assigning jobs and household information.

Because LEAM is able to generate geo-specific data at a fine resolution, we take an alternative approach to assigning TGZ information. Instead of assuming that population and jobs are uniformly distributed within a census unit, we allocate the demographic and economic information using a cell-based approach. Each cell spatially allocated in the LEAM approach is a carrier of a bevy of useful information: land use attributes (residential/commercial-industrial/undeveloped etc.), household size, income, employment characteristics, etc. Each individual cell is capable of being characterized with different levels of demographic and economic information. A residential cell that falls within a highly urbanized core represents more households, or a higher density population, than another cell located in a sprawling exurban location. Similarly, a commercial cell allocated to the urban core might represent a higher employee/sq ft ratio.

The translation of LEAM cellular output to CMAP TDM demographic and economic information starts with a base year estimation. A distribution of variable information (by household for example) is computed for each census block. This information is then randomly assigned to each corresponding cell in the block

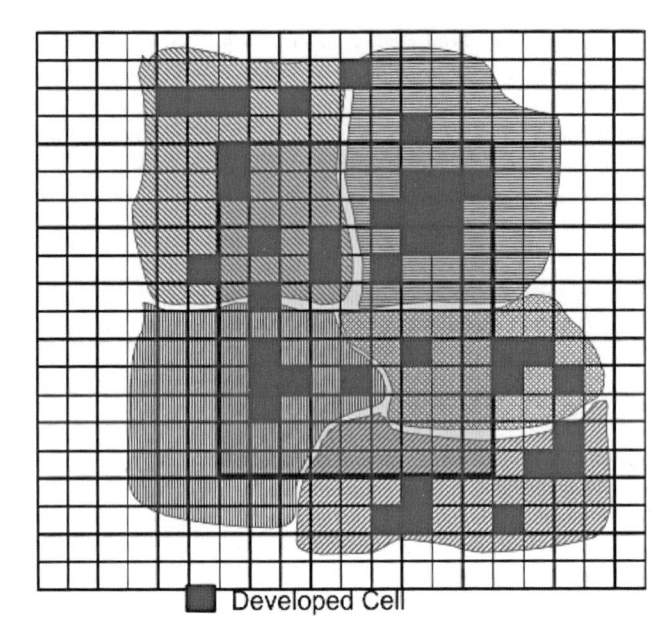

Fig. 6.7 The estimation of demographic and economic variables using a cell-based allocation technique

(i.e., residential or commercial).[4] As a result, each cell is assigned a particular value for each demographic variable (e.g., 0.5 households) desired (Fig. 6.7). Once each cell has been joined with the appropriate demographic values, we estimate the demographic variables for each TGZ by aggregating the values represented by residential cells within a specific TGZ. The estimation of economic variables is performed in a similar way, using commercial-industrial cells as the representation instead of residential cells.[5]

The logic for estimating future years is similar. The forecasted households are assigned to TGZs according to LEAM simulation allocations. In this process, we consider different households of different TGZs, measured in the base year estimation, with the consideration of overall density decrease over time in the region as a whole representing the combined effect of reduced family size and shifts in land consumption per household.[6]

---

[4] Though most zones have reasonable household densities, there are some zones having no valid household density in 2000, due to the discrepancy between Census and land cover data. For these zones, a constant density, which equalizes total number of households of the region to the forecasted regional total, is applied.

[5] For a more detailed description of this analysis see Sarraf et al. (2005)

[6] It should be noted that some variables, such as AH (Adults per household), CH (Children per household), and WH (Workers per household), are assumed to be constant between 2000 and 2035, since the changes in demographic structure are not analyzed.

Through the process of using the output of one model as the input for the other, LEAM and CMAP TDM are coupled into a more thorough and comprehensive economic/land use/transportation modeling environment.

6.5 Results and Analysis

Three different analyses were conducted for this work. A brief summary of these analyses follows. First, a LEAM land-use simulation was developed for a wider, nine county region to better match the CMAP study area geography (Fig. 6.8 left). This simulation used Tiger line network data with posted speeds. This regional model represents our most exhaustive LEAM effort to date with the important issue of redevelopment included in the model along and with some critical validation and calibration exercises completed. However, earlier versions of the model (e.g. LEAM) that were used to generate MY1 data for CMAP TDM activity generation, did not include redevelopment. Since the essence of this work was to present an approach to model coupling, we chose to rerun the LEAM model without redevelopment for use as our base condition in MY2035. We felt it important to use our best basis for comparing simulation outcomes and concluded that LEAM without redevelopment (Fig. 6.8 right) was the appropriate choice.

Next, the construction of LEAM simulations using the CMAP master network information as the basis for the road network interactions was tested. It was tested under both un-congested and congested speeds and compared to our established LEAM baseline (using Tiger based line data with posted speeds and no redevelopment). Network comparisons were also made with CMAP TDM results with and without the LEAM inputs. Figure 6.9 (left) shows the land-use change as simulated using proposed (uncongested) travel speeds from the CMAP TDM rather than the actual road network and posted speeds (which are higher than proposed simulated speeds). Figure 6.9 (right) compares this land-use pattern with patterns simulated using the actual road network and posted speeds (no redev). For each quarter section, the map indicates which simulation involved greater development. The differences between the two patterns appear quite marked. Using CMAP TDM travel speeds appears to move development between major expressways and arterials, while using posted travel speeds appears to move development further out along the expressways and arterials. The next set of maps explains why this may be the case.

Figure 6.10 (left) maps congested travel speeds as estimated using the CMAP TDM and LEAM inputs. Red segments in the network represent extremely slow travel speeds and green segments represent higher travel speeds. The adjacent image (Fig. 6.10 right) shows segments characterized by 'Posted' speeds. The map on the left shows that some segments, especially those located in the city of Chicago, have speeds less than 20mph. These differences in actual speeds on the regional network help explain the difference between the two land-use patterns shown above. The precipitous increase in friction on road travel significantly

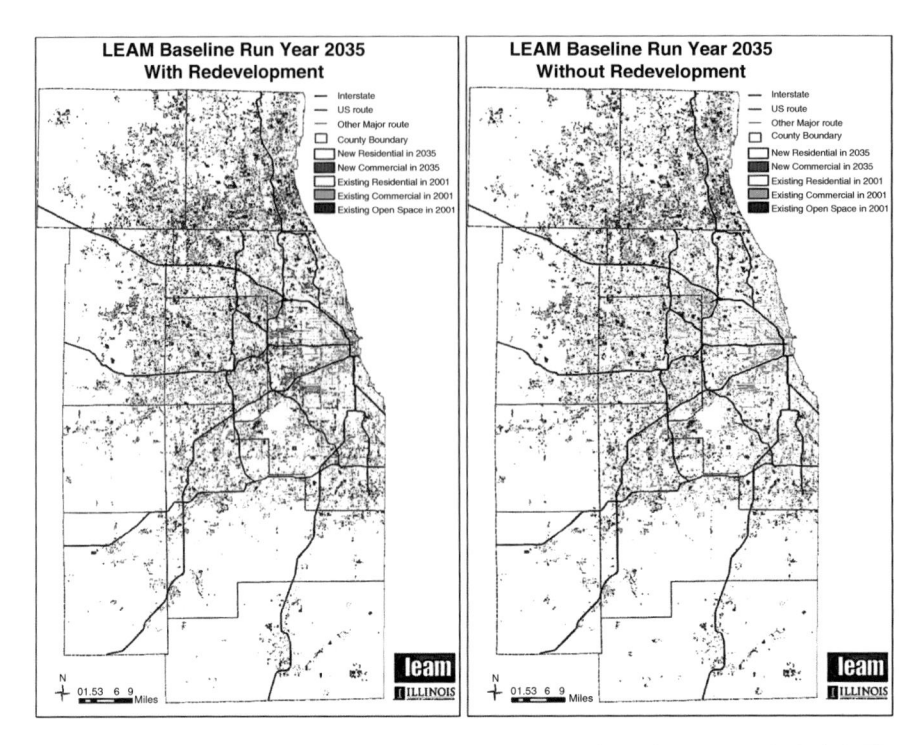

Fig. 6.8 The LEAM simulation with (*left*) and without (*right*) redevelopment included. The simulation without redevelopment is used for base lining and comparison purposes in this work

increases the attractiveness for development of cells further away from roads but involving less travel along roads and more travel across developed cells. In this situation, the least-cost computations that underlie many of the transportation-based LEAM probabilities tend to bypass roads and utilize other cells to get to attractors. This change in attractiveness tends to fill the spaces between expressways and arterials rather than move development further outwards as is the case when the cost of travel is represented by posted travel speeds on roads. The coupled LEAM simulation also shows much less development in Kankakee County and the south side of the region. Since the road network in this part of the region is very sparse, and when road segments are congested, this part of the region is rendered less competitive for new growth. If congested speeds are restricted to areas that are fully developed (such as the CBD), then this effect will likely be masked.

LEAM MY2035 simulation results using the CMAP TDM's congested speeds describe the effect of a highly congested roadway on LEAM simulations (Fig. 6.11 left). When compared to the base line condition (no redev) we find much of the new development has moved away from fringe urban centers (Kankakee) toward the metro Chicago center and also toward major roads (Fig. 6.11 right). This outcome is intuitive and probably more realistic than results using CMAP proposed speed or

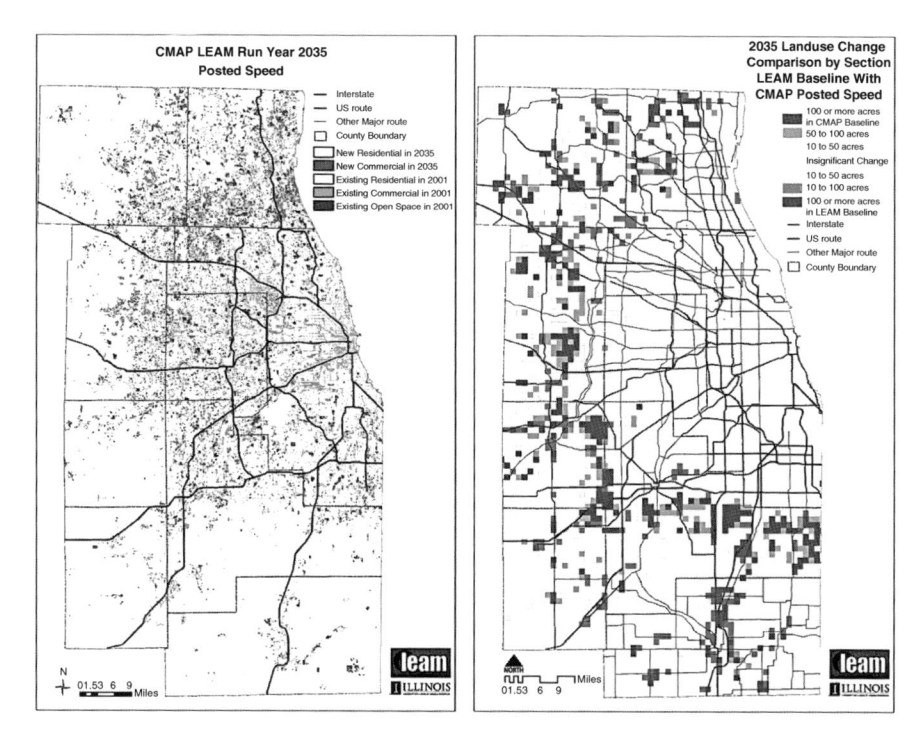

6 Complex Urban Systems Integration: The LEAM Experiences in Coupling... 125

Fig. 6.9 *Left*: LEAM 2030 projection using the CMAP master network and posted speeds. *Yellows* are new residential developments and reds new commercial areas. *Right*: Comparison map of LEAM 2030 projections using the CMAP master network with proposed speeds (*red*) and the LEAM baseline condition (*blue*)

LEAM travel speeds. It may also describe the advantages of the coupling of CMAP TDM congested speed to LEAM network.

As noted, the LEAM baseline includes posted speeds for its network condition. These speeds are an order of magnitude faster than the maximally congested speeds used from the CMAP TDM output. LEAM also uses a least-cost surface to analyze the potential travel times connected with various land cover forms (farmland is easy to cross – fast, water is hard to cross – slow, roads at posted speeds are very fast and easy to traverse). When maximally congested speeds are input into LEAM, the road network becomes more like a barrier to movement than a facilitator. This means that the travel time costs of migration between cells might be lower across cells that are not part of the road network. So development may not spread into the fringe as much as seek greener urban pastures. Figure 6.11 shows a significant decrease in the development to the south where the CMAP congested speeds may alter the attractiveness of exurban areas.

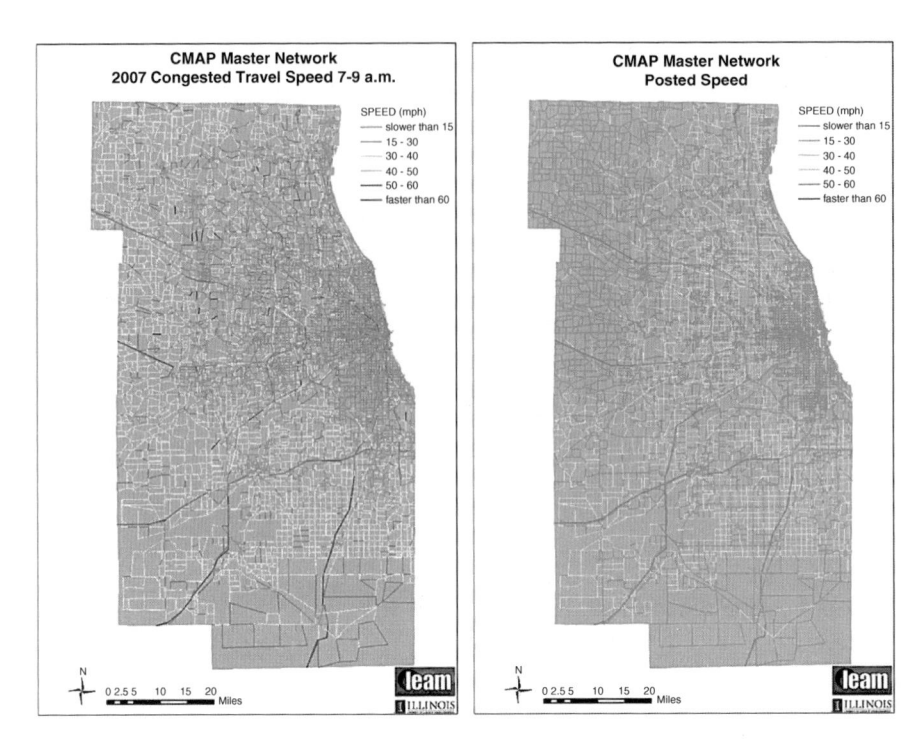

Fig. 6.10 *Left*: The CMAP master network with CMAP TDM derived most-congested speeds (*red*) and least congested speeds (*green*). *Right*: The CMAP master network with CMAP TDM derived congested speeds below 20mph (*red*) and above 20mph (*green*)

6.6 Discussion

As described earlier, differences between the two road networks is appreciable. As we see in Fig. 6.12 (left), the road network as derived from TIGER data is very dense. The CMAP Master Network is a reduced and abstract representation of this road network (Fig. 6.12 right). The differences between the two networks could have an effect on least-cost computations and thus alter the outcome at a finer grain of detail if not translated.

In this study, we also identified several problem areas in the coupling process. Comparing time 0 data developed using LEAM and actual TGZ counts, we discovered that input datasets for the two models are not perfectly consistent. In addition, we found the computational time required to complete a four-step model run of the magnitude of CMAP TDM to be overly cumbersome (20–30 h runtimes). LEAM utilizes a parallel processing approach to distribute the required computations and improve run-time efficiency. Although the Chicago region contains over 25 million cells with each performing hundreds of calculations at each time step, we have been able to reduce our computational run-time from days, to hours and now minutes using several strategies including parallel processing of

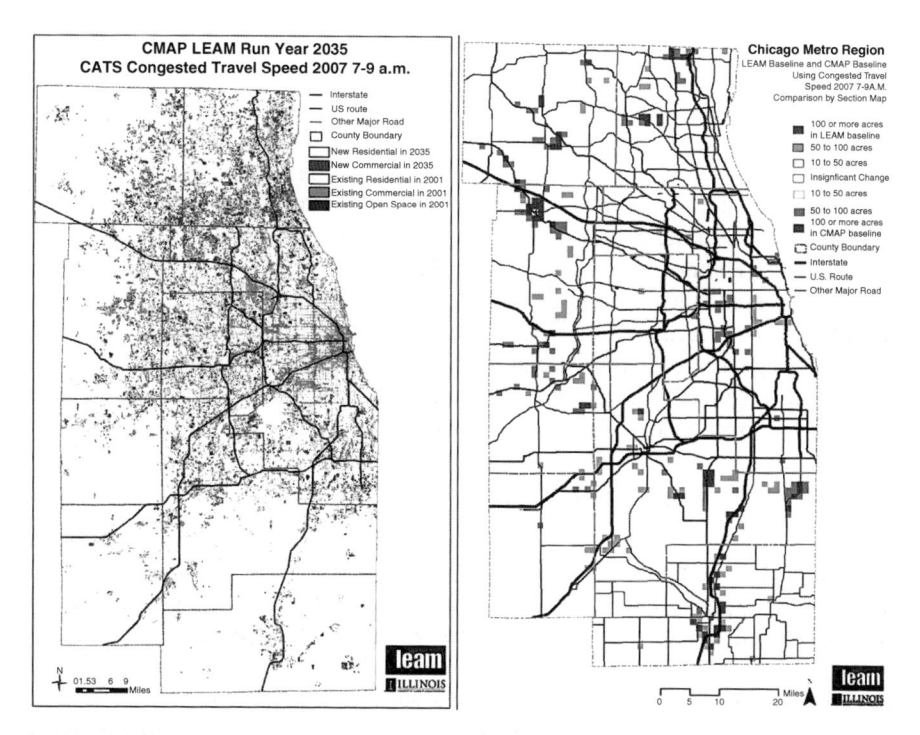

Fig. 6.11 *Left*: LEAM 2035 projection using the LEAM network and CMAP TDM congested speeds. *Yellow* areas are new residential developments and reds are new commercial developments. *Right*: Comparison map of LEAM 2035 projections using the CMAP TDM congested speeds (*red*) and LEAM baseline (*blue*). Note the decrease in development activity to the south of the study area

the computations to up to 128 processors. The distribution of the typical four-step model is more complex however, due to the use of tables and best-fit algorithms not easily translated onto separate processors. Citilabs has begun to address this problem with a quasi-distributed computing solution in which the steps are separated onto four processors increasing run-time efficiencies.

The project also revealed some more direct challenges. Though the CMAP TDM estimates the network with various options, such as number of lanes, driving directions, time period restrictions, and so on, the most difficult and problematic part is the representation of its network geography into the real-life road network geography. LEAM uses actual network geography in order to represent realistic outcomes. Given the discrepancy between the two network files, it is challenging to transfer the CMAP TDM output describing network conditions to the LEAM road network. Furthermore, to better translate LEAM output to CMAP TDM input, we need to know several things including: (1) how to improve initial land cover data for more precise initial estimation of demographic and economic variables, (2) how to properly deal with the zones having no valid household or employment densities at present according to the given input datasets, (3) how to illustrate the overall

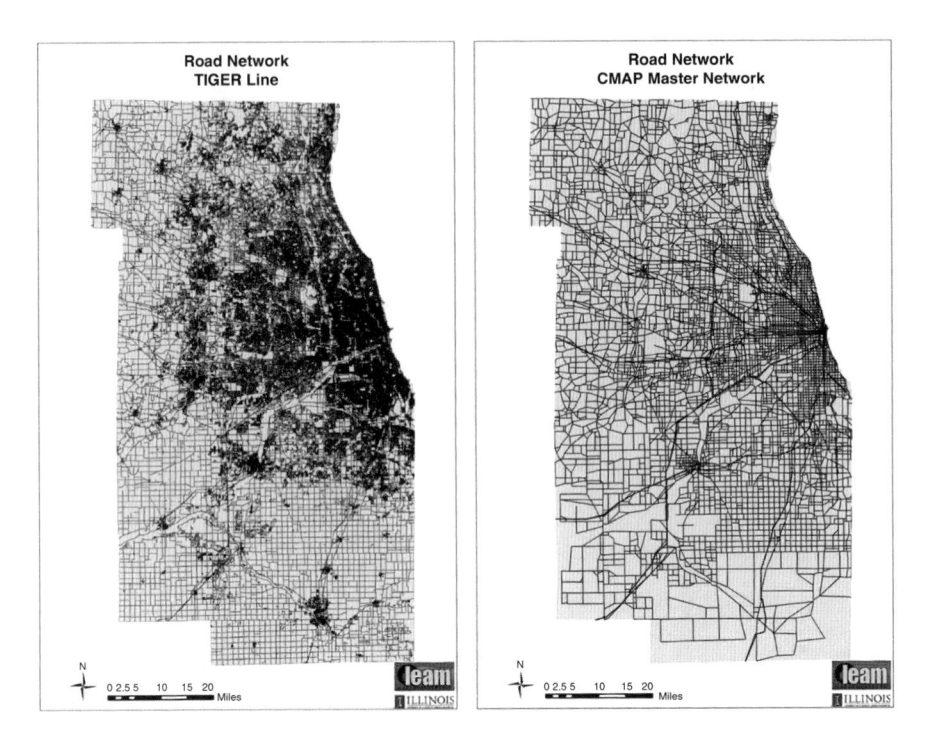

Fig. 6.12 *Left*: Tiger line data. *Right*: The CMAP master network

household and employment density changes more effectively, and (4) how to forecast and consider the changes in demographic structural variables (AH, CH, WH) over time.

6.7 Conclusions

The need for integration and feedback between dynamic land use and travel demand models has long been recognized. When the mutual influence of economic, land use and travel demand models are effectively captured and conveyed, the resultant information is richer, and may in fact be more reliable, believable, and defensible.

This work attempts to advance the art and the science of integrating land use and transportation models with a focus on the potential land-use consequences of a loosely coupled system. We describe a process of coupling a regional macroeconomic model – CREIM, a dynamic spatial land use model – LEAM, and a traditional four-step transportation model – CMAP TDM, and discuss the implications on land use in the region. The results of this coupling indicate that notable differences occur in land use patterns when average, posted, and minimum peak speeds are used as inputs. Average and peak speeds appear to be more effective in representing resultant behaviors, although the results also suggest that people's perceptions of congestion might be driven less by the worst conditions on the road and more by an overall sense of traffic conditions.

This project illustrates how competitive travel demand and land use simulation models can be linked without sacrificing their advantages and can actually produce more illustrative results. Though CREIM, LEAM and the CMAP TDM are loosely coupled in this example, the process has helped us identify the potential for connections between approaches, established the bridge, and accomplished a cycle of actual data transfers and coupled simulations. By iterating this cycle, we will be able to achieve a dynamic and systematic integration.

The policy implications of this work suggest several things. Our outcomes suggest that people do in-fact make land use decisions based on, among other things, travel times. This suggests that computing congestion-based coefficients are critical for analyzing the land use implications of transportation decisions. Unfortunately, most transportation models use static depictions of land use to determine demand and geographic locational variables – which we show are linked to the way the transportation system works and is therefore a moving target. Linking dynamic models of land use change with transportation models helps to relieve this problem.

Our work also delineates the critical differences between roads dominated development patterns and infill development. Other land use models might rely too heavily on transportation structure for assessing probable future land use outcomes. These network based without including congestion will fail to adequately project areas that are attractive for infill development. These differences in locational outcomes have important implications for public transit, school locations, and other density driven land use decisions. Our approach is able to capture and assess these critical dynamics.

Future work includes: improving and codifying the coupling technique described here for a more seamless and consistent model integration; the development of dynamic frameworks for assigning evolving demographic and economic characteristics to households and businesses in cells; the use of more efficient techniques to better generate transportation-based probability surfaces. Additional work is also needed in the areas of model efficiency and version control systems. Using three complex modeling environments is difficult to manage and accounting for modifications and their implications in scenario planning exercises requires additional control tools. Additional and new approaches calibrating and testing these complex models is also needed. We also need to include critical systems feedbacks, especially in terms of the affects that land use patterns have on the structure and efficiency of a regional economy.

In conclusion, we argue that the coupling of economic, land use, and travel demand models is critical for developing a better and more reliable understanding of the dynamics of urban systems. We feel that the work described here, while adding to the literature, demonstrates a very rough and elementary cut at the complexity of the problem and the potential of these coupled tools for improving planning and decision making in a region.

References

Briassoulis H (2000) Analysis of land use change: theoretical and modeling approaches, The Web Book of Regional Science, Regional Research Institute, West Virginia University http://www.rri.wvu.edu/WebBook/Briassoulis/contents.htm

Crecine JP (1964) TOMM (time oriented metropolitan model), Pittsburgh, PA: CRP Technical Bulletin, No.6, Dept. of City Planning

Cervero R (2001) Efficient urbanization: economic performance and the shape of the metropolis. Urban Stud 38:1651–1671

de la Barra T (1989) Integrated land use and transport modelling. Cambridge University Press, Cambridge

de la Barra T, Perez B, Vera N (1984) TRANUS-J: putting large models into small computers. Environ Plann B: Plann Des 11:87–101

Deal B, Chakraborty A (2010) Cyber-physical planning support systems: advancing participatory decision making in complex urban environments. Int J Oper Quant Manag 16(4):353–367

Deal B, Pallathucheril V (2007) Developing and using scenarios. In: Hopkins LD, Zapata MA (eds) Engaging the future: forecasts, scenarios, plans, and projects. Lincoln Institute for Land Policy, Cambridge, pp 221–242

Deal B, Pallathucheril V (2008) Simulating regional futures: the land-use evolution and impact assessment model (LEAM). In: Brail RK (ed) Planning support systems for cities and regions. Lincoln Institute for Land Policy, Cambridge, pp 61–84

Deal B, Schunk D (2004) Spatial dynamic modeling and urban land use transformation: a simulation approach to assessing the costs of urban sprawl. Ecol Econ 51:79–95

Deal B et al. (2005) LEAM technical document: overview of the LEAM approach. Landuse Evolution and Impact Assessment Modeling Laboratory, University of Illinois at Urbana-Champaign

Echenique MH, Flowerdew AD, Hunt JD, Mayo TR, Skidmore IJ, Simmonds DC (1990) The MEPLAN models of Bilbao, Leeds and Dortmund. Transp Rev 10:309–322

Garin RA (1966) A matrix formulation of the Lowry model for intra-metropolitan activity location. J Am Inst Plann 32:361–364

Hansen WG (1959) How accessibility shapes land use. J Am Inst Plann 25(2):73–76

Hewings GJD, Nazara S, Dridi C (2004) Channels of synthesis forty years on: integrated analysis of spatial economic systems. J Geogr Syst 6:7–25

Hunt JD, Abraham JE (2003). Design and application of the PECAS land use modeling system. In: Proceedings of the 8th international conference on computers in urban planning and urban management, Sendai, Japan, May 2003

Hunt JD, Simmonds DC (1993) Theory and applications of an integrated land use and transport modelling framework. Environ Plann B: Plann Des 20:221–244

Israilevich PR, Hewings GJD, Sonis M, Schindler GR (1997) Forecasting structural change with a regional econometric input–output model. J Reg Sci 37:565–590

Kim TJ (1983) A combined land use-transportation model when zonal travel demand is endogenously determined. Transp Res Part B 17(6):449–462

Kim JH (2011) Linking land use planning and regulation to economic development: a literature review. J Plann Lit 26(1):35–47

Kim JH, Hewings GJD (2009) Integrating the fragmented regional and subregional socio-economic forecasting: a spatial regional econometric input–output framework. Paper presented at the 56th Annual North American meetings of the Regional Science Association International. San Francisco, Nov 2009

Lowry IS (1964) A model of metropolis. Technical Report, RM-4035-RC, Rand Corporation, Santa Monica

Miller EJ, Kriger DS, Hunt JD (1998) Integrated urban models for simulation of transit and land-use policies (TCRP Web Document 9). Transit Cooperative Research Program, National Academy of Sciences, Washington, DC

Parr JB (1987) Development of spatial structure and regional economic growth. Land Econ 63:113–27
Sarraf S, Pallathucheril VG, Donaghy K, Deal B (2005) Modeling the regional economy to drive land-use change models. Paper presented at the 46th annual conference of the Association of Collegiate Schools of Planning, Kansas City, Nov 2005
Veldkamp A, Lambin EF (2001) Predicting land-use change. Agric Ecosyst Environ 85(1–3):1–6
Waddell P (2002) Urbansim: modeling urban development for land use, transportation and environmental planning. J Am Plann Assoc 68(3):297–314
Waddell P et al (2003) Microsimulation of urban development and location choices: design and implementation of UrbanSim. Netw Spat Econ 3:43–67
Waddell P et al (2007) Incorporating land use in metropolitan transportation planning. Transp Res Part A 41:382–410
Wegener M (2004) Overview of land-use transport models. In: Hensher DA et al (eds) Handbook of transport geography and spatial systems. Pergamon/Elsevier Science, Kidlington, pp 127–146
Wu F, Martin D (2002) Urban expansion simulation of Southeast England using population surface modeling and cellular automata. Environ Plann A 34(10):1855–1876

Chapter 7
Employment Location Modelling Within an Integrated Land Use and Transport Framework: Taking Cue from Policy Perspectives

Ying Jin and Marcial Echenique

Abstract For over four decades the integration of land use and transport models has contributed to systematic studies of economic change and employment location in cities and city regions around the world. This paper discusses one continuously running work stream that originated as a spatial model of urban stock and activity (Echenique et al. 1969) and later became encapsulated in the MEPLAN land use and transport modelling package (Echenique, 2004). One central feature of this approach is its emphasis upon simultaneously solving the employment location model with the production, trade, residential location and transport demand models for any specific year. Fast improving data availability and computer power have enabled an increasingly more complex series of models under this framework to find extensive use in supporting real policy decisions of major infrastructure investment, urban expansions and regeneration. In turn, practical policy needs have guided the priorities of model development.

The paper first reviews the key ideas leading up to the formulation of the MEPLAN package. It then outlines the main applications and the experience gained from them. Current developments of this approach, particularly to incorporate advanced social accounting matrices and recursive spatial equilibrium modelling are reported next. It concludes with an assessment of the strengths and weaknesses of the approach and its future potential from a policy perspective.

7.1 Introduction

Being able to predict the consequences of employment location, and ultimately to predict employment location itself as an integral part of the urban system, has been a major preoccupation ever since the start of applied urban modelling. The Lowry

Y. Jin (✉) • M. Echenique
The Martin Centre for Architectural and Urban Studies, University of Cambridge, Cambridge, UK
e-mail: ying.jin@aha.cam.ac.uk

model (1964) forecasts the location of working residents based on that of employment required by basic industries, and then predicts the location of the service industry employment based on the consumer demand of the residents. Echenique et al. (1969) extends this model to cover physical constraints upon location arising from buildings, town planning and the transport network, and to predict the distribution of employment and population within a city according to the functional relationships among urban activities subject to those constraints.

Echenique et al. (1969) was the herald of a continuous stream of operational models that were developed at Cambridge University Centre for Land Use and Built Form Studies (later the Martin Centre for Architectural and Urban Studies), the consultancy and software firm Applied Research Cambridge and the planning and transport consultancy firm Marcial Echenique and Partners (ME&P). The approach has further contributed, directly or indirectly, to software development and model applications at a number of other consultancies and academic institutions and continues to do so today.

The early models used analogies to Newton's gravity law to describe the location behaviour of businesses as well as households. The modelling of location choice then adopted spatial interaction models based on entropy maximisation (Wilson 1970; Batty 1976). Over the years the modelling approach became progressively grounded in micro-economic theory in their representation of production, consumption, trade, location choice and supply of land, buildings and transport services, whilst retaining its strength in incorporating the planning constraints. By 1975 an integrated land use and transport modelling framework has taken shape; The MEPLAN software package was developed in the early 1980s at ME&P which encapsulated the suite of theoretical models in a context-free manner, which in turn greatly facilitated the subsequent development of large scale, integrated land use and transport models (Echenique 2004; ME&P 1992a).

A close integration into the MEPLAN framework of a number of key economic and engineering models such as the Input–output Tables (Leontief 1951, 1986), random utility theory for location choice (McFadden 1974) and transport demand modelling (Domencich and McFadden 1975; Ben-Akiva and Lerman 1985; Daly and Zachary 1978), and traffic flow simulation (Sheffi 1985) have made the applied models established within the MEPLAN framework effective and practical tools for exploring and assessing urban land use and transport development options at the geographic scales suitable for cost-benefit and wider impact analyses of policy initiatives.

In particular, the incorporation of the Social Accounting Matrix (SAM), which is an extension of the Leontief Input–output Tables, at the national or regional level in the MEPLAN models have made it possible to forecast the location of all employment including Lowry's basic industries (Echenique et al. 2011). The incorporation of hybrid Cobb Douglas-constant elasticity of substitution (CES) production functions for industries have facilitated the representation of producer behaviour under monopolistic competition and of urban agglomeration effects; furthermore, through recursive spatial equilibrium modelling the non-linearity of employment growth effects and hysteresis of the development process are represented in the current research models (Jin et al. 2011). The recursive spatial equilibrium framework provides opportunities to link general equilibrium modelling with the emerging dynamic urban models for policy appraisal (for an in-depth review of the connections among the models and their roles in policy analysis, see Batty (2005; 2009)).

This paper first reviews the key ideas leading to the formulation of the MEPLAN package in Sect. 7.2. It then outlines in Sect. 7.3 the main applications and the experience gained in the process. Current developments of this approach, particularly in incorporating advanced social accounting matrices, new production functions and recursive spatial equilibrium modelling are reported in Sect. 7.4. It concludes with an assessment of the strengths and weaknesses of the approach and speculates on its future potential from a policy perspective.

7.2 Key Ideas Encapsulated in MEPLAN

Employment location is represented in the integrated land use and transport models as an outcome of complex interactions and feedbacks arising from individual business and personal decisions. For all private and public sector producers that employ labour, the decisions regarding where to locate are subject to (1) what inputs they use for production, (2) from where they choose to obtain the inputs and (3) where the demand is for their products and services – whilst (1) is modelled through production functions, (2) and (3) are based on advanced spatial interaction modelling that takes account of all direct and indirect costs of production, sales market penetration and transport. These are key to employment location modelling in the research and policy models in the lead up to the formulation of MEPLAN. Because of the limited space here we focus on presenting the key ideas as encapsulated in MEPLAN, rather than tracing back the historic development of the ideas.

Following Echenique (2004), the key ideas in MEPLAN that are relevant to employment location modelling may be presented in terms of the industries' production functions and their links to the wider economy, producers' locational choice, impacts of transport and other spatial costs, constraints and prices, and feedback loops towards spatial equilibrium. This section discusses these components in turn.

7.2.1 Production Functions and Links to the Wider Spatial Economy

MEPLAN employment location modelling starts from the premise that in a modern economy the private and public producers (who ultimately determine employment location) are closely connected – not only to the population who are both suppliers of labour and consumers, but also to one another, to government expenditure, investment and foreign trade. Representing these functional relationships is therefore considered the very first step towards employment location modelling in MEPLAN.

It is thus unsurprising that the MEPLAN models have made extensive use of the Leontief Input–output Table (Leontief 1951, 1986), which has been and still is the most commonly available operational model of the economy and which can meet, in its specific way, the need of MEPLAN models for production functions and wider economic linkages. For a given country or region and year, the Input–output Table provides a linear fixed production function for each type of industry, which accounts for each unit of output (typically in monetary value), the amount of intermediate inputs such as raw materials and business services from domestic producers, imports from foreign producers, labour, capital, government subsidies and taxes, and other assorted expenditure. For a given level of household consumption, government expenditure, public and private investments and demand for exports, the Input–output Table can predict for the country/region in question the total demand for goods and services from each industry, and at the same time the total demand from that industry for labour inputs.

Since the Input–output Tables are created primarily for aspatial, macro-economic analysis, the application of Input–output Tables in MEPLAN models implies a number of necessary adaptations in order to provide appropriate simulation results for land use and transport policy assessment. One common adaptation is to attribute the production and consumption of all goods, services, expenditure, investment and trade to specific model zones, typically administrative divisions of the study area. The adaptations of particular variables that are directly relevant to employment location modelling are summarised in Table 7.1, along with the other adaptations which may have an indirect effect on employment modelling. It is clear that the adaptations transform the Input–output Table into a spatial social accounting matrix with activities being measured in both physical and monetary units.

Typically a MEPLAN model makes use of an adapted Leontief type production function, where the demand for q types of capital, labour, buildings and raw material inputs are worked out through predetermined coefficients:

$$K_j^n = a_j^{qn} X_j^n \qquad (7.1)$$

$$L_j^{wn} = a_j^{wn} X_j^n \qquad (7.2)$$

$$B_j^{kn} = a_j^{kn} X_j^n \qquad (7.3)$$

$$Y_j^{mn} = a_j^{mn} X_j^n \qquad (7.4)$$

where X_j^n is the output of industry n in zone j, and the production technology is represented by the use of input shares $a_j^{qn}, a_j^{wn}, a_j^{kn}$ and a_j^{mn} respectively for q types of capital, w types of labour L, k types of buildings B, and m types of intermediate inputs Y.

The demand for each type of intermediate inputs and labour that arises from all industries n at a production zone j are summed respectively, e.g. $L_j^w = \sum_n L_j^{wn}$.

Table 7.1 Typical adaptations of Input–output models in MEPLAN models

	Variable	Standard input–output table	Typical MEPLAN applications
Directly relevant to employment location modelling	Labour inputs	Represented by lump-sum monetary values of labour inputs with no differentiation of the types of labour inputs	Labour inputs are converted into average number of employed or self-employed persons per unit of output, often by skill/occupation levels. The inputs are often disaggregated by model zone to reflect differences in productivity, industry composition, etc
	Business premises and building stock	Does not identify business premises and building stock as a distinct input	Added as a new category, with corresponding reduction in average input amounts from industries such as estate property services. Such inputs are also variable by zone owing to the property mix, physical inertia and prices
	Consumer goods	Producers deliver goods and services directly to consumers	Can use retail as an intermediary, i.e. retail industry buys retail goods from the other industries and households buy from retail industry
	Delivery of freight	Goods are delivered directly to customers	Can represent intermediate logistics via distribution centres and depots away from the production and sales sites
Other adaptations	Household consumption	As lump-sum monetary values of household consumption	Are converted into two components: households by socio-economic category and average expenditure per type of household for each type of goods and services. This links consumption to production and labour demand
	Transport inputs	Transport inputs are distributed among transport service industries, and among the other products and services (where transport costs cannot be separated from the recorded costs of goods and services)	For each industry, a calibrated amount of average transport inputs is subtracted and the transport and other spatial costs from the transport model (see Sect. 7.2.3) which are specific to each model zone are added. By definition, the modelled transport inputs are variable by production zone and the size of trade catchment

Since the 1950s specific Input–output Tables have been regularly produced by countries and administrative regions for the purposes of economic accounting and forecasting. Their use in the MEPLAN applications has greatly enhanced the empirical underpinnings in representing the functional relationships and made the models consistent with macro-economic analysis of national, regional and urban economies.

MEPLAN models typically follow the definitions of existing Input–output Tables which operate at the level of industry rather than any individual firm. This on the one hand makes it practical in producing empirical tables given existing data, and often overcomes the idiosyncrasies that exist in the firm-level observations. On the other hand, however, the use of industry level data does not readily allow modelling at the level of individual firms. In cases where e.g. the location of one very large firm is the object of study, bespoke Input–output model parameters will have to be developed outside a MEPLAN model.

Technically the most complex challenge in applying Input–output Tables in MEPLAN models has been the updating the fixed and non-substitutable input shares e.g. a_j^{qn}, a_j^{wn}, a_j^{kn} and a_j^{mn} through time. Recent developments of operational computable general equilibrium (CGE) models which allow input shares to vary subject to prices and physical constraints have opened a new pathway to modelling the functional relationships (See Sect. 7.4).

7.2.2 Producer's Locational Choice

By definition producer's locational choices are spatial, and the MEPLAN models represent the spatial interaction through sspatialising the average functional relationships from an underlying Input–Output Table. The resulting choices among the locations of production, work and consumption are modelled through discrete choice models that are based on random utility theory (McFadden 1974; Domencich and McFadden 1975).

While it has been demonstrated that MEPLAN can handle complex movements of trade through contemporary supply chains from the original supplier to the ultimate user via alternative logistical stages using a series of discrete choice models (WSP Policy & Research, 2005b), it is beyond the word limit of this paper to incorporate that mechanism here. A simplification that has often been made and does not affect the generality of the model is to represent the sourcing of intermediate goods as one step sourcing (i.e. directly from the supplier to the user) using a multinomial logit model that is commonly used in spatial interaction models:

$$Y_{ij}^m = Y_j^m \left\{ S_i^m e^{-\lambda^m (c_i^m + d_{ij}^m + \omega_{ij}^m + \Omega_i^m)} \Big/ \sum_i S_i^m e^{-\lambda^m (c_i^m + d_{ij}^m + \omega_{ij}^m + \Omega_i^m)} \right\} \quad (7.5)$$

where Y_{ij}^m is a trade flow of goods type m sourced from i to j, S_i^m is a term that represents production capacity for goods type m in zone i, C_i^m is the mill price of goods type m in zone i, d_{ij}^m is the door-to-door cost (i.e. transport plus logistics) of moving a unit of m from i to j, ω_{ij}^m and Ω_i^m are constants for residual attractiveness respectively of the trade link for m from i to j, and of production of m in i, λ^m is a scale parameter that reflects the observation error term that underlies the logit formula. All parameters and constants are to be determined by empirical calibration. Goods m may be a product or service.

For commuting, a similar logit formulation is used for the flow of each type of labour w as commuters:

$$L_{ij}^w = L_j^w \left\{ S_i^w e^{-\lambda^w (c_i^w + d_{ij}^w + \omega_{ij}^w + \Omega_i^w)} \bigg/ \sum_i S_i^w e^{-\lambda^w (c_i^w + d_{ij}^w + \omega_{ij}^w + \Omega_i^w)} \right\} \quad (7.6)$$

The demand for production of goods in the supplier zone is $Y_i^m \equiv \sum_j Y_{ij}^m$ and for labour in the zone of residence $L_i^w = \sum_j L_{ij}^w$. Note production zone i can be either domestic or foreign – for foreign zones border costs from/to i must be included in the door-to-door cost. The delivered price of goods and services in zone j in monetary terms is therefore a trade volume weighted average of production and spatial costs $p_j^m = \sum_i \{Y_{ij}^m (c_i^m + d_{ij}^m)\} / \sum_i Y_{ij}^m$ where p_j^m reflects the spread of the market catchment. The utility of the goods given the spatial choice is $\hat{p}_j^m = \frac{1}{\lambda^m} \ln \sum_i S_i^m e^{-\lambda^m (c_i^m + d_{ij}^m + \omega_{ij}^m + \Omega_i^m)}$. Similarly the average wages in the workplace zone is $p_j^w = \sum_i \{L_{ij}^w (c_i^w + d_{ij}^w)\} / \sum_i L_{ij}^w$ and the utility of labour is $\hat{p}_j^w = \frac{1}{\lambda^w} \ln \sum_i S_i^w e^{-\lambda^w (c_i^w + d_{ij}^w + \omega_{ij}^w + \Omega_i^w)}$.

7.2.3 Transport and Other Spatial Costs

The flows of goods in value X_{ij}^m generate freight haulage, while the flows in business services generate traffic of distribution vans and business travel. The freight haulage, vans and firm-to-firm business traffic can be generally expressed as

$$F_{ij}^{mT} = \Phi^{mT} X_{ij}^m \quad \text{and} \quad F_{ji}^{mT'} = \varphi^{mT'} X_{ij}^m \quad (7.7)$$

where F_{ij}^{mT} is the outward traffic type T, and $F_{ji}^{mT'}$ is the return traffic generated by value flow m within a typical working day from i to j, Φ^m and φ^m are the average trip rates respectively for the outward and return traffic. Note $F_{ji}^{mT'}$ only applies to those movements of goods and service delivery that generate return traffic (such as empty returns of trucks and vans).

The spatial cost of moving goods and services from i to j for trade m is therefore:

$$d_{ij}^m = \Phi^{mT} c_{ij}^T + \varphi^{mT'} c_{ji}^{T'} + \tau_{ij}^m + \Lambda_{ij}^m \quad (7.8)$$

where $c_{ij}^T, c_{ji}^{T'}, \tau_{ij}^m$ and Λ_{ij}^m are respectively the generalised transport cost per unit of T from i to j, the returning generalised transport cost per unit of T arising from F_{ij}^{mT}, non-transport spatial costs (such as permits for movements, lodging expenses, etc. associated with the movement) per unit of m from i to j, and the generalised border costs per unit of m which are only > 0 for i's that are in foreign territory.

Analogously, the trade of labour is converted to the number of commuting trips per typical working day:

$$F_{ij}^{wT} = \Phi^{wT} X_{ij}^w \quad \text{and} \quad F_{ji}^{wT'} = \Phi^{wT'} X_{ij}^w \tag{7.9}$$

The spatial costs for the trade of labour are the total commuting costs per unit of labour, i.e. $d_{ij}^m = \Phi^{mT} c_{ij}^T + \varphi^{mT'} c_{ji}^{T'}$.

In all cases, generalised costs of transport contain the monetary costs such as road vehicle operating costs and the tariffs of other transport services plus a monetary valuation of travel time and the user perception of non-monetary aspects of the transport services such as reliability and quality of service.

7.2.4 Constraints and Prices

The MEPLAN models make a clear distinction between costs and prices: while the costs are accounted for from the producer's perspective as an accumulation of the expenditure on the inputs used, prices are estimated through the supply–demand mechanisms and market clearing in the model, where the prices may rise if demand is higher than supply constraints, and vice versa.

The constraints that are used in the model are typically land, buildings and infrastructure supply (particularly transport) which are defined together with the pricing and regulatory regimes. The incorporation of supply constraints leads to the generation of an economic rent in Marshallian terms, which can be either positive or negative depending on the level of demand relative to supply. For land and buildings, the prices are modified by

$$p_j^n = p_j^{n*} + \frac{1}{\beta^n} \ln(X_j^n/K_j^n) \tag{7.10}$$

where p_j^n is the new price of supply type n in zone j taking account of the original price of n, p_j^{n*}, the demand for n, X_j^n, and capacity restraint of n, K_j^n. β^n is a step size parameter to be set iteration stability of the model.

The new price p_j^n is then fed into the model equations that determine the demand for n, X_j^n in the next iteration of the model, such that the demand that is generated within the model adjusts toward meeting the constraint K_j^n. The adjustments usually take place through modifying the input share coefficients, i.e.

$$a_j^{mn} = b^{mn} + \overline{a}_j^{mn} f(p_j^n) \tag{7.11}$$

where b^{mn} is the minimum amount to be consumed per unit of production, $\overline{a}_j^{mn}$ is the additional amount to be consumed when $p_j^n=0$, and $f(p_j^n)$ is a function of the new input price of n at the current iteration. In a vast majority of cases, the price and demand adjustment mechanisms will ensure that total demand meets the constraint. However, either the existence of the minimum consumption term b^{mn} or an exogenous demand input into the model may imply that in some cases total demand will never meet the constraint.

For transport network capacity constraints, the economic rent usually takes the form of time penalties, generated using traffic engineering principles. For example, congestion penalties on the highway links are generated using speed-flow curves for vehicular traffic. In contrast to conventional transport models, MEPLAN road congestion penalty equations include terms for both travel times and the valuation of travel times, both of which may be raised as a function of the traffic volume to capacity ration to represent respectively delays in travel time and reduced reliability. Similarly crowding on public transport services are modelled through delays on boarding and alighting, and raised valuation of on-board travel time in the case of overcrowding.

7.2.5 An Integrated Model

It is clear from the above that employment location choice is represented in the MEPLAN models as a part of the complex interactions in a spatial market, and the model solutions for employment location are worked out simultaneously with residential location and transport demand at a spatial equilibrium subject to land use and transport supply constraints. In particular, the focus of the employment location model is to represent how industries exploit input cost advantages including the variation of spatial costs for production in a series of their own trade-offs, while responding to choices made by other producers and the population in the study area, and by trade partners outside the study area. This is probably best seen in a typical simulation sequence of the integrated land use and transport model which we summarise below following the general format set out by Echenique (2004).

The spatial equilibrium solutions are worked out for a specific year. All quantities are defined as total flows for the year and all prices are average prices. The model is initiated with exogenous inputs which are (1) final demand which is determined outside the model, typically exports to outside the study area, private and public investment in the study area, and government expenditure on goods and services,[1] (2) land, buildings and infrastructure supply (particularly transport),

---

[1] Note that, in contrast to Input–output Tables, consumer final demand from the household sector is usually generated endogenously within the model.

together with the pricing and regulatory regimes, (3) expenditure propensities for all types of households, and (4) in cases where the location of non-working households are not modelled (which is often the case) the assumed zonal distribution of such households. At the initiation of the model, intermediate production is zero; all prices of goods and services are initiated as 1.0, in a way that is consistent with Input–output Tables. After the initiation, the model is solved by iterating through the following steps with a number of feedback loops.

Step 1 is to estimate the total demand for each industry at the zone of consumption – this includes exogenous demand and any intermediate demand that is generated after Iteration 1.

Step 2 is to start with consumption of each industry at each zone, and estimate the pattern of trade flows from all relevant production zones that satisfy this consumption through a spatial interaction model. The trade patterns depend on the cost of production and non-monetary attractiveness at each production zone and transport cost, all of which may be updated after Iteration 1.

Step 3 tots up the total trade demand at each production zone for each industry

Step 4 uses the input shares to estimate the demand for intermediate inputs, capital, labour and buildings. Typically the input share for labour is allowed to vary, therefore generating employment demand at production zones that is consistent with the prices, industry composition, and productivity. The input shares of intermediate inputs, capital and taxes/subsidies can be either kept constant or allowed to vary. The adjustment of input shares is carried out at Step 8 in line with price variations.

Step 5 estimates the commuting pattern of each type of labour, and the formation of working households at the residential end of the commuting journeys.

Step 6 estimates the demand for goods, services and non-commuting personal travel from both working and non-working households at each zone of residence, and distribute them to zones where such demands are met. The household demand for goods and services are to be added to total demand for each industry after Iteration 1 (see Step 1).

Step 7 compares the demand for constrained land use supply such as land and buildings against the supply available, which generates the variations in prices in order to clear the land and building markets. This leads to adjustments of all production and consumption prices.

Step 8 adjusts all the variable input shares and household consumption propensities in line with prices.

Step 9 adjusts the patterns of trade among all zones of production and consumption in line with the new prices, e.g. all being equal, production in zones of lowered prices will increase.

Step 10 translates all matrices of trade, commuting and other personal travel into passenger and freight movements between trip origins and trip destinations for a day (usually a typical working day outside the holiday periods, in order to reflect the needs of traffic planning). The origin-destination movements include both the out and return trips necessary for completing the movements.

Step 11 distributes the origin-destination movements among the available door-to-door modes of transport for each time periods of a day (typically morning peak, inter-peak, afternoon peak, off peak, etc.). The trips on each modal network (typically road and rail alternatives, and for large regional and national models waterway and air may also be included) are then assigned onto available routes for each time period of the day. The choice among door-to-door modes and routeing alternatives is modelled through appropriate discrete choice models.

Step 12 compares the passenger and vehicular traffic assigned on each road, rail and intermodal link against the service capacity constraint on these links. This comparison generates time delays and increases in the valuation of travel time which represent e.g. overcrowding or reduced reliability, using speed-flow curves on road and crowding functions on train services. The time delays and rises in travel time valuation are then fed into the updating of generalised transport costs (which includes monetary costs, a monetary valuation of the travel times, reliability and service quality), and ultimately into the updating of spatial costs at the next iteration.

It is apparent that all steps above except Steps 5 and 6 and part of 10 are directly connected to employment location modelling. Steps 5 and 6 and part of 10 are related to the modelling of households, which also exert an indirect influence on the location of production, jobs, and the premises of businesses and institutions.

There are many feedback loops present among the steps above which are solved iteratively in model runs. They may be summarised as:

Feedback A: it updates intermediate and factor inputs – the process is analogous to solving the multiplier effect in a standard Input–output Table; the key difference is that the multiplier effects are solved here spatially, i.e. together with trade flows between model zones.

Feedback B: it updates the variable input shares subject to changes in input prices, which are derived from cost and generalised cost changes in the transport and trade markets, and price changes arising from land and building constraints.

Feedback C: it updates the spatial patterns of trade, commuting and other personal travel subject to changes in prices of production, consumption and transport.

Feedback D: it updates the door-to-door mode, time period of day and routeing choices of passenger and freight movements, and through the iterations achieve user equilibrium between transport demand and supply on the multimodal network for each time period of the day.

Feedback E: it further updates the monetary and generalised costs of trade, commuting and other personal travel, which then influences feedback loops C, B and A above.

Taken together, these feedback loops solve the integrated land use and transport model at the point of spatial equilibrium, subject to model inputs such as exogenous final demand, supply of land, buildings and infrastructure (particularly transport), pricing and regulatory regimes, household consumption preferences and exogenous location choice of non-working households. These model inputs may be used to define alternative scenarios that represent initiatives of land use policy, infrastructure investment, pricing and regulation. In addition, the model inputs can also define specific combinations of demographic and globalisation trends, technological development, and macro-economic forecast at the national and regional levels, which are generally outside the control of the policy makers.

The model outputs under different model scenarios may be compared through comparative statics for each modelled year. This framework of scenario comparison is often extended to modelling a number of points in time, typically in 5–10 year increments to coincide with intermediate policy horizons. When the model is further extended in its representation of producer and consumer behaviour, the inter-temporal framework may be considered as that of comparative recursive dynamics (see Sect. 7.4).

7.3 Policy Applications

The overarching aim of the integrated land use and transport models reviewed above has been to produce operational tools for policy analysis and decision support. Employment location modelling is placed among the simultaneous interactions between diverse land use and transport markets. It can be seen as having been directly influenced by the practical and policy oriented modelling ethos of Lowry (1964). In the early stages of model applications, bespoke pieces of software were written for specific urban and regional studies. By the time the MEPLAN software was designed and built in the early 1980s, the aim was to create a general purpose modelling software suite that could fully address the needs of a diverse portfolio of urban, regional and national scale land use and transport studies that have been carried out worldwide. As a result, MEPLAN was written in a context free manner which in theory can cope with a wide range of application styles. However, the MEPLAN model applications continue to be focused closely on concurrent policy needs and empirical models that are supported by available data sources.

This practical and policy focus would in part explain why the integrated land use and transport models have found extensive use in supporting policy decisions of major infrastructure investment decisions, new urban expansions and city-wide regeneration. Table 7.2 summarises the main model applications that we are aware of in the lead up to and including the MEPLAN models.

It is clear from Table 7.2 that the MEPLAN models have found extensive applications in testing combined land use and transport planning policy packages in both developing and developed countries – the strengths of the models lay in the system-level assessment of both the impacts of employment location upon the urban system, and the impacts of major policy initiatives (particularly in land use and transport) upon employment location.

As a rule the very early models followed the Lowry tradition by treating basic industry employment as an exogenous input. This changed as the Input–output Tables were incorporated in the model (e.g. in the model of Sao Paulo Metropolitan

Table 7.2 Main pre-MEPLAN and MEPLAN models and their applications

Year	Methodological development	Application	Main reference
1967	Simple static spatial model of urban stock and activity	Reading, UK	Echenique et al. (1969)
1968–1970	Indicators for comparing planning and design of British New Towns	Stevenage & Milton Keynes, UK	Echenique et al. (1972)
1972	Land use and transport model	Caracas, Venezuela; Santiago, Chile	Feo et al. (1975), de la Barra et al. (1975)
1974	Integration with cost-benefit analysis	Sao Paulo, Brazil	Flowerdew (1977), Echenique (1983b)
1975	Land markets and prices	Tehran, Iran	Hirten and Echenique (1979)
1976	Input–output model for economic and commodity flows	Sao Paulo, Brazil	Williams and Echenique (1978), Echenique (1986)
1977	Floorspace module and logit mode choice module	Bilbao, Spain (version 1)	Geraldes et al. (1978), Echenique (1983b)
1978	Modelling floorspace market segments	Sao Paulo, Brazil	Echenique (1983a), Echenique (1983b)
1981	Detailed representation of public transport & urban freight traffic	Bilbao, Spain (version 2)	Devereux et al. (1982)
1981	Financial model for new town development	Guasara, Venesuela	Echenique (1983b)
1982	Detailed traffic model	Bilbao, Spain (version 3)	Marsden (1984)
1980–1985	Development of a context-free software package MEPLAN	Cambridge, UK	ME&P (1992a), Echenique (2004)
1986	Impact of land prices on labour market	Cambridgeshire, UK	Echenique et al. (1987)
1987	Housing by tenure	Leeds, UK	Echenique et al. (1990)
1989	Public transport congestion, region-wide planning strategies	London and surrounding regions, UK (Version 1)	ME&P (1989)
1990	Transport and urban development	Naples, Italy	Hunt (1994)
1990	Detailed retail sector model comparing planned and market provisions	Beijing, China	Jin (1990)
1992	Transport and urban development strategies	London and surrounding regions, UK (Version 2)	ME&P (1992b), Williams (1994)

(continued)

Table 7.2 (continued)

Year	Methodological development	Application	Main reference
1993	Regional economic impacts of the Channel Tunnel Rail Link	UK & Western Europe	Rohr and Williams (1994)
1993	Integrated land use and transport strategies at the macro-region level	Santiago, Chile (Macro Zona 1)	ME&P (1993) Echenique et al. (1994)
1995	Integrated land use and transport strategies at the metropolitan level	Santiago, Chile (Macro Zona 2)	ME&P (1995)
1996	Integrated land use and transport strategies at the national level	Santiago, Chile (Macro Zona 3)	ME&P (1997)
1999	Integrated spatial economic and transport model (EUNET1.0)	Manchester-Leeds corridor, UK; Helsinki region, Finland	ME&P (2000), LT Consult (1999)
2000	Land use and transportation infrastructure	Sacramento, California, US	Abraham and Hunt (2000)
2000	Alternative sub-regional strategies for long term growth ('Cambridge Futures')	Cambridge sub-region, UK	Echenique et al. (1999) Echenique and Hargreaves (2001)
2002–2005	Alternative city region land use and transport strategies for medium to long term growth	London and surrounding regions, UK (Version 3)	Jin et al. (2002); ME&P (2002), WSP (2005a)
2005	Industry location and freight traffic (EUNET2.0)	Great Britain, UK	WSP (2005b)
2010	Alternatives of suburban land use and transport development (EPSRC SOLUTIONS project)	London and surrounding regions, Cambridge sub-region & Tyneside, UK	Echenique et al. (2009), Echenique et al. (2010)

Note: This table builds on and extends Table 1 of Hunt and Simmonds (1993)

Region, see Williams and Echenique 1978), where the location of all types of employment started to become endogenous to the model. It became the mainstay with subsequent models since 1976, although occasionally the Lowry approach was allowed to return, notably in Version 3 of the London and South East Regional Model (ME&P 2002; Jin et al. 2002) where an existing official employment forecast for all industries except local retail and education services was adopted as an input for testing the coherence of the UK government's housing and infrastructure investment policies.

The land use and transport modelling research at the Martin Centre in Cambridge under the leadership of Marcial Echenique has led to the formation of a number of other modelling packages in addition to MEPLAN. For example, the main developers of TRANUS (de la Barra, 1989), DELTA (Simmonds, 2004) and PECAS (Hunt and Abraham, 2005) have all carried out research at the Martin Centre. Of these the broad modelling framework of TRANUS remains to be closest to MEPLAN.

7.4 Current Model Developments

A new development phase to extend the modelling methodology is underway in the on-going UK Research Council funded projects. The focus of the new developments is again guided by what is considered to be the basic needs of a policy analyst in dealing with urban development and restructuring in the foreseeable future.

Although the fundamental strength of integrated land use and transport models in understanding the system-wide effects remain, the policy context in the next decade is likely to be increasingly focused on attracting business and foreign direct investment, productivity growth, and sustainable urban development. While the backdrop in the developed economies is largely retrofitting existing urban areas with limited urban extensions, in the emerging economies the main challenge is often managing dramatic urban expansion and the formation of mega-city regions at an unprecedented scale.

7.4.1 New Trends in Policy Analysis

Large scale developments tend to have wide impacts upon all aspects of the city region, the ramifications of which persist through time. Models that represent system-wide effects are obviously desirable. But system-wide, integrated land use and transport models are resource-intensive to develop, use and maintain. New trends such as decentralisation in decision making will have major impacts on how urban and infrastructure development proposals are assessed. Sectoral allocation of funding from central government may well diminish, e.g. for housing or transport. Urban development policy initiatives will have to compete head-to-head with spending in other areas of a local authority's remit. Although this may reinforce the needs for a broader understanding and rationalisation of urban development initiatives by local authorities, there may be fewer opportunities for economies of scale in funding major model development projects upon which large comprehensive models depended in the past.

Models that are based on micro-economic theories and behavioural sciences are better equipped to deal with emergent patterns of economic activity resulting from decentralised decision-making than the outcomes of centralised planning. Nevertheless, the models will need to be made more relevant to 'business plan' style decision-making. Also, since any model-based assessment of a large development proposal will be contended by different communities and specialists, the models must be transparent and empirically robust.

Furthermore, there are changes afoot in the funding and management of major infrastructure projects. Traditionally, asset utilisation assessment for a particular piece of infrastructure, forecasting of its future use and assessment of its strategic impacts had been kept fairly separate in many countries. The UK is a good example of this, where asset utilisation was assessed by the owners and operators; the future demand for a particular new infrastructure project was estimated as part of the business plan of the scheme promoter; proposed infrastructure projects were appraised by central and metropolitan governments in terms of their national and local impacts.

Significant changes have been made to the infrastructure planning process e.g. since the UK Labour government's proposal of a new Infrastructure Planning Commission (IPC) of 2008, and the recent modification of that proposal in the implementation by the current coalition government. First, the new infrastructure planning system which has recently come into force requires the promoters of major infrastructure projects to demonstrate, in their planning applications, the strategic fit of the projects and project impacts through detailed analytical evidence. Secondly, the government has promised to complete scheme review and approval process within a relatively short period (up to 1 year at the IPC/Planning Inspectorate and 3 months with the Government Minister) once a planning application is submitted. However, these changes imply that the evidence gathering and impact analysis (which includes demand forecasting) will have to be front loaded – in other words it would be up to the scheme promoters to demonstrate, inter alia, future demand for the project and the strategic and local impacts including those on employment location.

The new infrastructure planning system in the UK is fairly unique so far. Given that the aim of the new system is to speed up the planning and consultation process for major infrastructure projects, the new planning process has been keenly observed by governments and scheme promoters in many other countries, particularly in the emerging economies such as China and India, where the interest is focused on how a streamlined system can be equipped with evidence based, robust analysis of the business case and wider impacts.

It is early days yet to judge whether and to what extent this new infrastructure planning system will be successful. It can be argued that its success would in part depend on whether a robust new methodology could be developed to support the operators and scheme promoters in carrying out a robust assessment of the wider impacts. This is clearly a new challenge, since the vast majority of operators and scheme promoters have at best used narrowly defined sectoral forecasting models, which are often not capable of forecasting impacts such as employment location.

All the changes in the policy context point towards increased technical sophistication on the one hand, and reduced technical complexity in model use on the other. First, the technical sophistication allows more realistic modelling of the behaviour of firms and institutions, such as monopolistic competition and urban agglomeration effects, and the dynamic feedbacks through time. Secondly, given the increased decentralisation in decision making and project planning, there is more intense pressure to provide robust forecasts for option development, proposal screening, and outline business case studies for projects that aim to promote employment growth and productivity. Thirdly, there is also pressure to do the above with very limited budget and short turnaround time.

7.4.2 New Modules for the Model

We are developing new modules which extend the capability to represent employment location as well as the wider land use and transport system.

Our on-going work in the UK EPSRC ReVISION Project has developed a parsimonious integrated land use and transport model for the UK that can be used for strategic assessment purposes at the regional and sub-regional level (Echenique et al. 2011). This model is based on an advanced Social Accounting Matrix for the UK that adopts the features that are reported in Table 7.1. Furthermore, the model also extends its modelling of infrastructure supply and demand to energy, water, sewerage, solid waste as well as transport, thus covering probably the fullest range of infrastructure within any integrated land use and transport model.

On modelling producer behaviour, our EPSRC Energy Efficient Cities Project has implemented a new set of production functions that represent monopolistic competition and urban agglomeration effects (Jin, Echenique and Hargreaves 2011). These production functions are inspired by new models of the spatial economy that range from Krugman (1991), Bröcker (1998) to Anas and Liu (2007), where a hybrid functional form is adopted with an overarching Cobb-Douglas production function for groups of inputs coupled with CES functions for substitution among varieties within an input group e.g. where there is significant choice among input varieties. Its general functional form is

$$X_j^n = A_j^n (K_j)^{v^n} \left(\sum_w V_j^{wn}(L_j^w)^{\theta^n} \right)^{\frac{\delta^n}{\theta^n}} \left(\sum_k V_j^{kn}(B_j^k)^{\zeta^n} \right)^{\frac{\mu^n}{\zeta^n}}$$

$$\times \prod_m \left(\sum_{u \in m} V_j^{un}(Y_j^u)^{\varepsilon^{mn}} \right)^{\frac{\gamma^{mn}}{\varepsilon^{mn}}} \quad (7.12)$$

Where X_j^n is the output of industry n in zone j. The use of a Cobb-Douglas production function for inputs groups of capital K, labour L, buildings B and intermediate goods and services Y implies constant internal returns to scale of production through defining the sum of cost share parameters for the respective input groups, $v^n + \delta^n + \mu^n + \sum_m \gamma^{mn} = 1$. For w varieties of labour, k of buildings, and m of intermediate inputs respectively, a CES function is used to represent the substitution effects, the elasticities of substitution among the inputs to one output n being governed by positive parameters θ^n, ζ^n and m number of ε^{mn}'s. Parameters V_j^{wn}, V_j^{kn} and V_j^{un} specify input biases that are to be empirically determined. A_j^n is a function that represent zone and product specific Hicksian-neutral total factor productivity (TFP) effects.

Under constant return to scale, the unit price of production for industry n in zone j, p_j^n, will be equal to average and marginal cost (Anas and Liu 2007):

$$p_j^n = \frac{\rho^{v^n}}{A_j^n (v^n)^{v^n} (\delta^n)^{\delta^n} (\mu^n)^{\mu^n} \left(\prod_m (\gamma^{mn})^{\gamma^{mn}} \right)} \left(\sum_w (V_j^{wn})^{\frac{1}{1-\theta^n}} (W_{*j}^w)^{\frac{\theta^n}{\theta^n-1}} \right)^{\frac{\delta^n(\theta^n-1)}{\theta^n}} \times$$

$$\times \left(\sum_k (V_j^{kn})^{\frac{1}{1-\zeta^n}} (R_j^k)^{\frac{\zeta^n}{\zeta^n-1}} \right)^{\frac{\mu^n(\zeta^n-1)}{\zeta^n}} \prod_m \left(\sum_{u \in m} (V_j^{un})^{\frac{1}{1-\varepsilon^{mn}}} (p_{*j}^u)^{\frac{\varepsilon^{mn}}{\varepsilon^{mn}-1}} \right)^{\frac{\gamma^{mn}(\varepsilon^{mn}-1)}{\varepsilon^{mn}}} \quad (7.13)$$

where ρ^{v^n} is the real interest rate for the use of capital in industry n, and W_{*j}^w, R_j^k and p_{*j}^u are respectively the average workplace wage (including commuting costs) of labour type w, average rent of buildings of type k and the average delivered price of intermediate input of type u in zone j; furthermore, the unit demand for intermediate input type $u \in m$ in zone j by industry n can be expressed as value-based variable demand coefficients:

$$a_j^{un} = \frac{V_j^{un} p_{*j}^u}{\sum V_j^{un} p_{*j}^u} \gamma^{un} \quad \text{for} \quad u \in m \quad (7.14)$$

where p_{*j}^u is the average price of the delivered goods type u in zone j, and the intermediate inputs of sub-type $u \in m$ for producing n in zone j are therefore

$$Y_j^{un} = a_j^{un} X_j^n \quad (7.15)$$

The demand for capital, labour and buildings that is consistent with the variant Cobb-Douglas-CES production function that we adopt is, for industry n at the production zone j (cf. Eqs. 7.1, 7.2, 7.3):

$$K_j^n = \left(\frac{1}{\rho} \right) v^n p_j^n X_j^n \quad (7.16)$$

$$L_j^{wn} = \frac{(V_j^{wn})^{\frac{1}{1-\theta^n}}(W_j^w)^{\frac{1}{\theta^n-1}}}{\sum_w (V_j^{wn})^{\frac{1}{1-\theta^n}}(W_j^w)^{\frac{\theta^n}{\theta^n-1}}} \delta^n p_j^n X_j^n \tag{7.17}$$

$$B_j^{kn} = \frac{(V_j^{kn})^{\frac{1}{1-\zeta^n}}(R_j^k)^{\frac{1}{\zeta^n-1}}}{\sum_{k'} (V_j^{kn})^{\frac{1}{1-\zeta^n}}(R_j^k)^{\frac{\zeta^n}{\zeta^n-1}}} \mu^n p_j^n X_j^n \tag{7.18}$$

Urban agglomeration effects are usually estimated using a production function methodology (see reviews in Rosenthal and Strange 2004 and Melo et al. 2009). This is also the focus of recent empirical literature. For example, Graham and Kim (2008) use an empirical analytical framework to investigate effects of agglomeration economies based on a translog production-inverse input demand system and a UK firm-level company financial information dataset. Because the translog function represents non-homothetic production technology, they are able to examine the effects of urban agglomeration on total factor productivity (TFP), partial factor productivity, factor prices and factor demands. The results highlight the importance of the TFP effects, i.e. the total shift in levels of output that arises from agglomeration. More recently, Graham et al. (2009) further estimate the effects of urban agglomeration when distance decay effects are incorporated. They suggest that urban agglomeration for firms can be measured in the general form $M_j^n = \sum_i L_i^{w''}(l_{ij})^{-\beta}$ where M_j^n is a measure of the accumulated economic mass for industry n in location j, $L_i^{w''}$ is the total size of employment of type w'' that is relevant to industry n and within distance band i, l_{ij} is the average distance from location j, and β is a distance decay parameter. They include a variable representing the economic mass in an empirical production function for the industries. A change in TFP from scenario A to scenario B can then be calculated as $A_j^{nB}/A_j^{nA} = (M_j^{nB}/M_j^{nA})^\pi$ or

$$A_j^{nB} = A_j^{nA}(M_j^{nB}/M_j^{nA})^\pi \tag{7.19}$$

where A_j^{nA} and A_j^{nB} are the TFP term (see Eq. 7.12) for scenario A and B respectively, and π is a parameter governing the elasticity with respect to economic mass changes.

In addition, the modelling of households and property development markets has also been improved as part of the new model development in terms of household location modelling (for further details, see Jin et al. 2011).

7.4.3 Recursive Equilibrium Structure for Longer Term Forecasts

A particular challenge for modelling large scale land use and infrastructure development is the requirement for explorative forecasting for several decades into the future. This is both a result of the long working lives of major infrastructure, and the inter-generational obligations on matters such as mitigation of long term environmental impacts (Wegener 2011).

For a given point of time, the CGE approach coupled with spatial interaction modelling (particularly the city level applications such as Anas and Liu 2007) has already demonstrated that industrial production, household consumption, and the interactions within the transport, property and land markets can be successfully brought together under a general equilibrium framework, which provides a unique point of reference among the alternative urban development predictions.

However, the durability of physical building stock, the development process, the planning system and the wider decision-making all imply that many of the variables that have a major influence upon the results of a general equilibrium model, such as the supply of floorspace and transport networks, do not respond instantaneously to price signals. Furthermore, many of the industries and households do not react instantaneously to price signals. Long term general equilibrium models can forecast an end state where certain supply constraints and inertia are removed, but for policy makers in the city region, it is of great relevance to know e.g. whether and how the pace of change in building and transport supply affect the economic, social and environmental outcomes. Policy assessments therefore require an explicit reference to how city regions evolve through time.

The discussions above suggest that there is scope to combine the features of (1) spatial equilibrium of location and travel behaviour which tend to respond fairly quickly to prices and constraints, and (2) dynamic modelling of building and transport supply, demographics and life cycle events, technical change and foreign trade conditions which are inert and only partially responsive to policy actions within any given time period. The combination of spatial equilibrium and dynamic modelling can provide unique predictions for a city region at distinct points in time with clearly defined and stated boundary conditions, so long as the boundary conditions are recursively linked between those points in time.

We should note that such practice is not new in essence: for example the embryonic forms can be observed ever since the early applications of the MEPLAN models. However, enhancing the capabilities of general equilibrium modelling and dynamic modelling has and will continue to make a difference to the theoretical and empirical robustness of the predictions and policy assessment.

Figure 7.1 below outlines the main information flows within and between the recursive spatial equilibria sequence that start from a Base Year t to a forecast year t + 10 years and then t + 20 years. The Base Year is a calibration year where the exogenously input assumptions such as macro-economic and demographic variables, transport supply, floorspace supply and land supply are input as known variables. The city region is represented in five categories of markets: the product markets where industries interact, the labour and consumer markets where households interact with industries, the transport markets, the property markets and the land markets. The known geographic distribution of land use, urban activities and travel are input into the model in order to calibrate the model

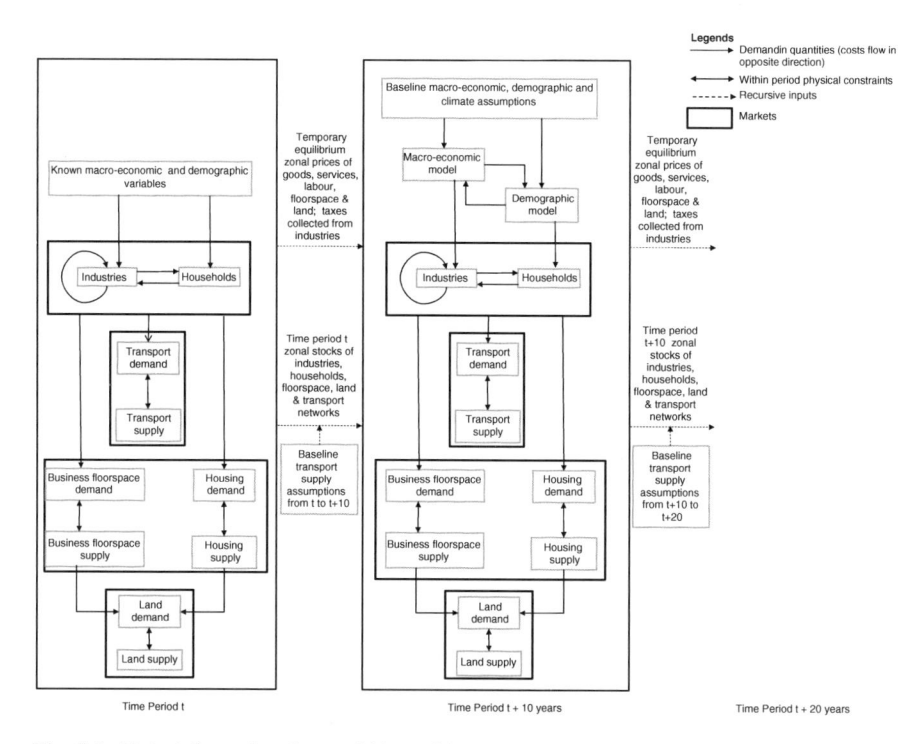

Fig. 7.1 Main information flows within and between recursive spatial equilibria. Note: In this diagram where solid single headed arrows indicate a 'demand', double headed arrows represent constraints, and dashed line arrows are recursive inputs from the previous year. (Source: Jin, Echenique and Hargreaves, 2011)

parameters uniquely by assuming system equilibrium regarding where industries and households locate and how goods and people move between locations.[2]

Moving to the first forecasting period t + 10 years, the zonal stocks of industries, households, floorspace, land and transport networks and services of the Base Year become the initiation values of the model. Baseline economic and demographic assumptions that are made outside the model are modified through comparing the modelled prices of production, cost of living, labour supply, productivity, and tax revenue with the same variables used by the Baseline scenario. Similarly, the baseline transport supply scenario is modified based on a comparison of the modelled tax revenue against the baseline transport investment assumptions. The five main markets in the model are then run in two broad steps. At Step 1, land supply, floorspace supply for business floorspace and housing, and transport supply respond to price signals from the previous time period subject to stock

---

[2] Note the choices may be subject to stock inertia, e.g. only a proportion of non-working households move within each year and the housing occupied by those who do not move are not available for relocating households.

inertia. Then at Step 2, the locations of industries, households and the derived transport demand arising from the locational process are run to market equilibrium subject to the supply constraints established at Step 1. The stocks, prices and tax revenues are output and to be used recursively by t + 20 years, and so on.

7.5 Conclusions

The integrated land use and transport models presented in this Chapter are in the main developed to provide practical policy analysis tools for planning infrastructure and land use development at urban, regional and national scales. The models have been used for over 40 years to forecast employment location, almost always as an integral part of the forecast of wider urban change. The wider impacts of employment location include, on the one hand, its consequences on the cost of production, location of residents, demand for land and infrastructure services, etc., and on the other, the feedback over time from the urban system on employment location.

This integration links employment location modelling directly to practical policy questions. As a result, these models have found extensive applications in testing policies that aim to influence employment location through transport and land investment, regulation and pricing.

The models have become progressively grounded in micro-economic theory in the explanation of the behaviour of firms and individual workers (whether employed or self-employed), in a way that is consistent to the representation of the behaviour of the wider urban population. In the model, the employment location equations are solved simultaneously with equations of production, trade, residential location and transport demand. The model forecasts are subject to an overall spatial equilibrium within an appropriate study area, which typically consists of one or more self-contained regional labour market catchments plus main locations of external trading partners.

The model predictions are therefore outcomes of a spatial equilibrium forecast of the location of employment as well as of production, trade and the population. The models also predict the prices of production, labour and consumer goods at the spatial equilibrium. The quantities and prices forecast at an explicitly defined spatial equilibrium provide a transparent reference point for assessing the costs and benefits of policy interventions, and explore scenarios that involve spatial disequilibrium.

A key contribution of the models discussed in this Chapter relate to their incorporation of land, buildings and infrastructure supply constraints in spatial equilibrium modelling in a way that is directly connected to the policy instruments available to governments at different geographic levels. These constraints are modelled together with investment, regulatory and pricing options that can influence employment location either directly or indirectly. It has been shown time and again in the model applications that indirect interventions such as supply of residential land or major transport infrastructure can often have as much influence upon employment location across a city region, particularly with a medium to long term perspective. This highlights the importance of the integrated modelling approach.

A key challenge to the application of an integrated land use and transport model is naturally the much increased burden for data collection, empirical calibration and the related complexity of setting up spatial equilibrium modelling. This requires higher modelling costs and specialist skills which can be a barrier to potential users who lack resources, particularly at the first sight.

However, as the modellers gain experience in this type of modelling, it has become clear that integrated modelling can require fewer observations than first appears to produce consistent land use and transport predictions (Hunt and Simmonds 1993). This is because an integrated model can often use well established economic theories regarding producer, labour and consumer behaviours to infill the gaps in the data. Modelling an integrated system can also help identify the inconsistencies in empirical observations of different urban sectors. Learning from the past experience has also helped to develop more succinct models that focus on the key relationships (Echenique et al. 2011). Fast development in IT and software has also helped to lower the costs of model development whilst enabling the users to access enhanced simulation models.

Acknowledgements Both authors wish to acknowledge the funding support of the UK Engineering and Physical Sciences Research Council (EPSRC) through the Energy Efficient Cities (EECi) Project (www.eeci.cam.ac.uk) and Marcial Echenique wishes also to acknowledge the support of the EPSRC ReVISIONS Project (www.regionalvisions.ac.uk). They would like to thank the reviewers for constructive comments on the paper. The usual disclaimers apply and the authors alone are responsible for the views expressed and any remaining errors.

References

Abraham J, Hunt JD (2000) Policy analogies using the Sacramento MEPLAN land use transportation infrastructure model. In 80th annual meeting of the Transportation Research Board, Washington, DC
Anas A, Liu Y (2007) A regional economy, land use, and transportation model (RELU-TRAN©): Formulation, algorithm design, and testing. J Reg Sci 47:415–455
Batty M (1976) Urban modelling. Cambridge University Press, Cambridge
Batty M (2005) Cities and complexity: understanding cities with cellular automata, agent-based models, and fractals. MIT Press, Cambridge, MA
Batty M (2009) Urban modeling. In: Kitchin R, Thrift N (eds) International encyclopaedia of human geography, vol 12. Elsevier, Oxford, pp 51–58
Ben-Akiva M, Lerman S (1985) Discrete choice analysis. The MIT Press, Cambridge
Bröcker J (1998) Operational spatial computable general equilibrium modeling. The Annals Reg Sci 32:367–387
LT Consult (1999) EUNET case study: a land-use and transport model for the Helsinki Region. LT Consult, Helsinki
Daly A, Zachary S (1978) Improved multiple choice models. In: Hensher D, Dalvi M (eds) Determinants of travel choice. Saxon House, Sussex

de la Barra T (1989) Integrated land use and transport modelling. Cambridge University Press, Cambridge

de la Barra T, Echenique MH, Quintana M, Guendelman J (1975) An urban regional model for the central region of Chile. In: Baxter RS, Echenique MH, Owers J (eds) Urban development models. Construction Press, Lancaster, pp 137–174

Devereux LS, Echenique MH, Flowerdew ADJ (1982) Bilbao land-use and transport model. Presented at the meeting of the International Study Group of Land-Use/Transport Interaction, IIASA, Schloss Laxenburg, Vienna

Domencich T, McFadden D (1975) Urban travel demand: a behavioural analysis. North Holland, Amsterdam

Echenique MH (1983a) The Sao Paulo metropolitan study: a case study of the effectiveness of urban system analysis. In: Batty M, Hutchinson B (eds) System analysis in urban policy-making and planning. Plenum, New York, pp 243–270

Echenique MH (1983b) Urban and regional policy analysis in developing countries. In: Chatterjee L, Nijkamp P (eds) Gower, Aldershot, pp 115–158

Echenique MH (1986) The practice of modelling in developing countries. In: Batty M, Hutchinson B (eds) Advances in urban system modelling. Elsevier, Amsterdam, pp 275–297

Echenique MH (2004) Econometric models of land use and transportation. In: Hensher DA, Button KJ (eds) Transport geography and spatial systems, vol 5, Handbooks in Transport. Pergamon/Elsevier Science, Kidlington, pp 185–202

Echenique MH, Hargreaves AJ (2001) Cambridge futures 2: what transport for Cambridge. University of Cambridge, Cambridge, The Martin Centre

Echenique MH, Crowther D, Lindsay W (1969) A spatial model for urban stock and activity. Reg Stud 3:281–312

Echenique MH, Crowther D, Lindsay W (1972) A structural comparison of three generations of new towns. In: March L, Martin L (eds) Urban space and structures. Cambridge University Press, Cambridge, pp 219–259

Echenique MH, Simmonds DC, Starr CM (1987) A MEPLAN model of Cambridgeshire. In: Proceedings of the PTRC Summer annual meeting, PTRC, London

Echenique MH, Flowerdew ADJ, Hunt JD, Mayo TR, Skidmore IJ, Simmonds DC (1990) The MEPLAN model of Bilbao, Leeds and Dortmund. Transp Rev 10:309–322

Echenique MH, Jin Y, Burgos J, Gil A (1994) An integrated land-use/transport strategy for the development of the central regional of Chile. Traffic Engineering and Control, September, pp 491–497

Echenique MH et al (1999) Cambridge futures. Cambridge University Press, Cambridge

Echenique MH, Hargreaves AJ, Jin Y, Mitchell G, Namdeo A (2009) Solutions project strategic London case study. University of Cambridge, Cambridge. See www.suburbansolutions.ac.uk. Accessed 1 September 2012

Echenique MH, Barton H, Hargreaves AJ, Mitchell G (2010) SOLUTIONS final report: sustainability of land use and transport in outer neighbourhoods, See www.suburban-solutions.ac.uk. Accessed 1 September 2012

Echenique MH et al (2011) A land use spatial interaction model based on random utility theory and social accounting matrices. Symposium for applied modelling 2011, Cambridge

Feo A, Herrera R, Riquezes J, Echenique MH (1975) A disaggregated model for Caracas. In: Baxter RS, Echenique MH, Owers J (eds) Urban development models. Construction Press, Lancaster, pp 175–202

Flowerdew ADJ (1977) An evaluation package for a strategic land-use transportation plan. In: Bonsall P, Dalvi Q, Hills P (eds) Urban transportation planning. Abacus, Tunbridge Wells, pp 241–258

Geraldes P, Echenique MH, Williams IN (1978) A spatial economic model for Bilbao. In: Proceedings of the PTRC Summer annual meeting, London: PTRC

Graham DJ, Kim (2008) An empirical analytical framework for agglomeration economies. Annal Reg Sci 42:267–289

Graham DJ, Gibbons S, Martin R (2009) Transport investment and the distance decay of agglomeration benefits. Centre for Transport Studies, Imperial College, Mimeo

Hirten JE, Echenique MH (1979) An operational land-use transport model for the Tehran region, Iran. Transport Research Circular No. 199, pp 6–7

Hunt JD (1994) Calibrating the Naples land use and transport model. Environ Plann B 21:569–590

Hunt JD, Abraham JE (2005) Design and implementation of PECAS: a generalised system for the allocation of economic production, exchange and consumption quantities. In: Lee-Gosselin MEH, Doherty ST (eds) Integrated land-use and transportation models: behavioural foundations. Elsevier, St. Louis, pp 253–274

Hunt JD, Simmonds DC (1993) Theory and application of an integrated land-use and transport modelling framework. Environ Plann B: Plann Des 20:221–244

Jin Y (1990) Locational propensities under state provision and market conditions: retailing in Beijing 1978–1988. Ph.D. dissertation, Department of Architecture, University of Cambridge, Cambridge

Jin Y, Williams IN, Shahkarami M (2002) A model for London and its surrounding regions, European transport forum, Cambridge. Available at http://www.etcproceedings.org/paper/a-new-land-use-and-transport-interaction-model-for-london-and-its-surrounding. Accessed 1 September 2012

Jin Y, Echenique MH, Hargreaves AJ (2011) A spatial recursive equilibrium model that incorporates the dynamics of large scale land use development and restructuring. Symposium for applied modelling 2011, Cambridge

Krugman P (1991) Geography and trade. MIT Press, Cambridge

Leontief W (1951) The structure of the American economy 1919–1939, 2nd edn. Oxford University Press, Oxford

Leontief W (1986) Input–output economics, 2nd edn. Oxford University Press, New York

Lowry I (1964) Model of metropolis, Memorandum RM-4035-RC. Rand Corporation, Santa Monica

Marsden A (1984) A model solution to Bilbao's traffic congestion. The Survey 163(4799):4–6

McFadden D (1974) Conditional logit analysis of qualitative choice behavior. In: Zarembka P (ed) Frontiers in econometrics. Academic, New York, pp 105–142

ME&P (1989) A land-use/transport model for London. Report to the Department of Transport, London

ME&P (1992a) Technical introduction to MEPLAN. Marcial Echenique and Partners (ME&P), Cambridge

ME&P (1992b) MEPLAN for London and the South East. TRRL Project No. N04290. Marcial Echenique and Partners (ME&P), Cambridge

ME&P (1993) A land use and transport model for Macro Zona Central of Chile (in Spanish). Marcial Echenique and Partners (ME&P), Cambridge

ME&P (1995) A land use and transport model for Santiago Metropolitan Area (in Spanish). Marcial Echenique and Partners (ME&P), Cambridge

ME&P (1997) A national model of land use and transport for Chile (in Spanish). Marcial Echenique and Partners (ME&P), Cambridge

ME&P (2000) EUNET case study – the trans-pennine model: final report for HETA, DETR. Department of Environment, Transport and the Regions, London

ME&P (2002) LASER enhancement project: final report. Marcial Echenique and Partners (ME&P), Cambridge

Melo P, Graham DJ, Noland RB (2009) A meta-analysis of estimates of urban agglomeration economies. Reg Sci Urban Econ 39:332–342

WSP Policy & Research (2005a) Wider South East Regional Research Study: compendium of model tests using the London and South East Land Use and transport model (LASER3.0). London: Department for Transport. (Lead author with ID Elston and others). See http://webarchive.nationalarchives.gov.uk/20070129210246/http://www.dft.gov.uk/pgr/economics/rdg/laserenhancementproject. Accessed 1 September 2012

WSP Policy & Research (2005b) The EUNET2.0 freight and logistics model - Final Report. Regional Pilot for Economic/Logistic Methods. DfT Contract PPAD 9/134/24. London: Department for Transport. See http://webarchive.nationalarchives.gov.uk/20070129210246/ http://www.dft.gov.uk/pgr/economics/rdg/theeunet20freightandlogistic3126. Accessed 1 September 2012

Rohr C, Williams IN (1994) Modelling the regional economic impacts of the channel tunnel. Environ Plann B 21:555–568

Rosenthal S, Strange WC (2004) Evidence on the nature and sources of agglomeration. Rev Econ Stat 85:377–393

Sheffi Y (1985) Urban transportation networks: equilibrium analysis with mathematical programming methods. Prentice-Hall, Englewood

Simmonds DC (2004) Introduction to the DELTA Package. Cambridge: David Simmonds Consultancy. See http://www.cix.co.uk/~davidsimmonds/main/pdfs/DELTAintro.pdf. Accessed 1 September 2012

Wegener M (2011) Transport in spatial models of economic development. In: de Palma A, Lindsey R, Quinet É, Vickerman R (eds) Handbook in transport economics. Edward Elgar, Cheltenham

Williams IN (1994) A model of London and the South East. Environ Plann B: Plann Des 21(5):535–553

Williams IN, Echenique MH (1978) A regional model for commodity and passenger flows. In: Proceedings of the PTRC Summer annual meeting, PTRC, London

Wilson AG (1970) Entropy in urban and regional modelling. Pion, London

Chapter 8
Integrating SCGE and I-O in Multiregional Modelling

Model Development for Swedish Planning

Christer Anderstig and Marcus Sundberg

Abstract Economic activities can be modeled at different levels of aggregation. Different levels of detail regarding spatial or temporal resolution, or levels of sectoral aggregation are appropriate depending on the question at hand. In cases where changes on the micro scale affects what happens at the macro level, and vice versa, an integrated approach is required. In this paper a modeling framework is presented, where focus is placed on the interactions between production and employment. The aggregate spatial computable general equilibrium model STRAGO is interacted with the highly detailed input-output model system rAps. Interregional and intersectoral relations of production, including agglomeration, are represented in the aggregate model, providing a coarse description of production by which the rAps model system is constrained. Such constraints will affect where individuals may find a suitable job. At the same time the aggregate model is dependent on the labour supply, provided by rAps, in determining production. An application of the proposed modeling framework is presented where future projections are compared to historical data.

8.1 Introduction

At the regional level, the assessment of future developments of population, employment and the labour market and the economy are of critical importance for planning in several fields. One obvious example is national transport planning. Decisions

---

C. Anderstig (✉)
WSP Analysis & Strategy, 12188 Stockholm-Globen, Sweden
e-mail: christer.anderstig@wspgroup.se

M. Sundberg
Division of Transport and Location Analysis (TLA), KTH Royal Institute of Technology, Teknikringen 10, 100 44 Stocholm, Sweden
e-mail: marcus.sundberg@abe.kth.se

about infrastructure investment should be based on principles of economic efficiency and cost benefit analysis. Such calculations need projections (forecasts) of travel and freight volumes for detailed geographic zones in the whole country. And, since transport is derived demand, projecting the regional distribution of population, employment, production and income is a prerequisite for estimation of transport demand, and the subsequent cost benefit analysis.

Similar conditions apply to other sectors. It is therefore a matter of importance that planning authorities and other actors at the regional level have a common view as for future regional development. As far as possible this view should be consistent with present official long term economic projections, as presented by the Ministry of Finance or the like. From a Swedish viewpoint this approach has prevailed in the past. From the 1970s and for about two decades regional economic forecasting was performed in conjunction with national economic long term projections (i.e., the Long-Term Surveys, prepared by officials at the Ministry of Finance). This reflects the common view at that time, namely that it is feasible to make descriptions of future economic development by means of econometric modelling. This view no longer prevails. "Gradually the feasibility of detailed forecasts for the development of economic sectors has been called in question, at the same time as the need of planning has been judged to be of less interest.[1]"

Today, social planning at the regional level is different and with different aims compared to the situation 20 or 30 years ago. But, the need for scenarios, projections and forecasts has increased. Here is a dilemma. The conditions for forecasting, in the sense of detailed predictions of future economic development, have become even more uncertain (if they ever existed). Nonetheless, projections of economic development are required in long term planning.

For transport planning in particular, a final requirement is the location of residential population and employment defined at a detailed zone level. But, in such modelling efforts it is reasonable to presuppose that the location of households is influenced by accessibility to jobs and places to buy goods and services, and that the location of firms is influenced by accessibility to workers and customers. Further, decreasing travel costs imply that the interaction costs within the spatial economy will be reduced. With lower interaction costs agglomeration of economic activity increases, creating benefits through economies of scale.

Thus, modelling and projecting location for transport planning purposes should take transport-induced location effects and agglomeration effects into account. These notions are now pivotal in numerous approaches within land-use transport interaction modelling.

Applied land-use transport interaction modelling approaches are normally defined at the spatial scale of an urban region, subdivided in a fine-scale zone system. The constraining total volumes of jobs (employment) and residences (population) at the regional level are normally determined exogenously, and it is

---

[1] The Long-Term Survey 2003/04; http://www.regeringen.se/sb/d/2379/a/13207

not always clear that these constraints at the regional level are consistent with constraints at an aggregate (multiregional) level.

In this paper we present an integrated model system by which transport-induced location effects and agglomeration effects are described at the regional level, in a multiregional context. The integrated model framework comprises a recently developed spatial computable general equilibrium (SCGE) model, STRAGO,[2] in combination with a multiregional I-O model, rAps. The idea of integrating models, working at different levels of spatial and sector detail, is in line with the thoughts of Wilson (2010) in terms of "Layers of disaggregation". While Wilson discusses the role of different models at different levels of disaggregation primarily in an urban context, our application is related to the (dis)aggregation from the national to different regional levels.

In the SCGE model both prices and quantities are described, whereas the I-O model is working in terms of values (fixed prices). By including prices in the model framework it is possible to describe how producers and consumers substitute between goods due to changing prices. Further, the importance of transport costs and agglomeration effects has recently been emphasized with respect to regional development, and these factors can be handled in SCGE. One of the strengths of SCGE models is the use of underpinning microeconomic theory for modelling the economy. The SCGE framework represents a theoretically and mathematically consistent approach to describing the economy, and it allows us to perform policy assessments and investigate scenarios within this consistent framework, where, e.g., the interactions between the economy and transports can be studied.

However, for applied projection and planning purposes SCGE models cannot provide results in enough spatial and sectoral detail. The advantage of combining the two model types is that the I-O model permits a high degree of disaggregation, whereas the strength of SCGE models is the potential to describe the behaviour of different actors, as pointed out above. Thus, analyzing a policy implying e.g. reduced transport costs, will shift the spatial distribution of demands for goods and services in the SCGE framework, yielding new spatial production patterns. On a finer regional scale, as represented in the rAps model system, these changes in spatial production patterns will affect regional labour supply and employment. By feeding these changes in labour supply back to the SCGE model, we may capture the interactions between production and employment.

As commissioned work for national authorities this integrated model framework has been used in two full-scale applications. It has been used to give projections of regional development to year 2030, consistent with national scenarios presented in the Long-Term Survey 20008.[3] A later development was to provide the transportation authorities with scenario data to year 2030, to be used as input to their national transportation models.

---

[2] STRAGO, Swedish Trade of Goods. The model has been developed by Marcus Sundberg, see Sundberg (2009).
[3] See ITPS (2009).

8.2 The Model System: An Outline

An overview of the model system is given in Fig. 8.1. The regional models STRAGO and rAps are working with assumptions (exogenous factors) which include constraints generated from national models. Exogenous factors are applied to all three model steps. Results generated from the national CGE model are treated as assumptions in the modelling at regional level. Annual change in production, productivity and exports by industry exemplify factors which are taken into account in the regional models, aiming at results which are consistent with assumptions at the national level.

STRAGO is used in a middle step to disaggregate results at the national level to Sweden subdivided into nine larger regions. In the multiregional input–output model, rAps, production by industry and region are further disaggregated under the regional constraints provided by STRAGO. The geographic distribution of production generates a regional labour supply and employment in rAps, which is fed back to STRAGO.

The idea of using a SCGE model for regional disaggregation of results by industry at the national level has been discussed by Lundqvist et al. (2004). Another example of regional disaggregation, with consistency between regional and national level, is found in Dixon and Rimmer (2004), describing a top-down approach.

8.2.1 STRAGO

STRAGO represents a SCGE model which describes Sweden, divided into nine domestic regions and the rest of the world. At present 14 different sectors/industries are represented, sectors which mainly correspond to producers of goods classified into STAN-groups.[4] In addition there is a sector 'Other[5]' and a transportation sector. The nine regions are mainly defined by the classification of NUTS 2 regions for Sweden[6], see Fig. 8.2. A brief technical description of the model is provided in the Appendix A, for a more detailed description, see Sundberg (2009).

---

[4] STAN is the grouping of goods which is used in the Swedish freight model Samgods. This grouping has been chosen in STRAGO to get information about transport costs for various goods. The STAN sectors are: Agricultural products, processed lumber, processed wood products, foodstuffs, crude petroleum, petroleum products, iron ore and metal waste, metal products, paper and pulp, earth/stone and building material, chemicals, and finally manufactured industrial products.

[5] The sector "Other" comprises the main part of all service production (i.e. a major part of the economy).

[6] The eight NUTS 2 regions in Sweden are: Stockholm, East-middle, Småland & islands, South, West, North-middle, Mid-north, Upper-north. The ninth region defined in STRAGO is Örebro & Västmanland, originally a part of East-middle.

Fig. 8.1 Outline of the model system STRAGO-rAps

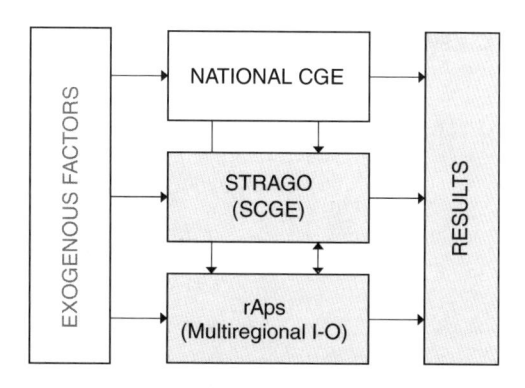

The theoretical underpinning of the model is mainly derived from the description of monopolistic competition by Dixit and and Stiglitz (1977), a framework widely used in trade modelling, e.g. Ethier (1982), and in economic geography, see Krugman (1991). For a pedagogical technical description of a SCGE model under perfect competition, see Bröcker (1998).

Each region includes a number of economic actors, creating the economic activity of the region: household, firms and transport agents, see Fig. 8.3. Both households and firms buy goods, either to be used for consumption or as intermediates to production. Moving goods from the location of production to where it is consumed is handled by an explicit transport sector.

In this application of the SCGE model the households in a region are represented in an aggregate way, through one representative household per region, acting under static but adaptive expectations.[7] At each point in time the household decides how much of the budget to be spent on consumption and investment respectively, where investment results in different future production and consumption potentials. The household earns its income by supplying labour and capital to firms within the region of residence. The main purpose of the STRAGO model within our combined modelling framework is to describe regional production patterns. Labour supply consequences are not explicitly modelled at this instance, but rather in the rAps model. Therefore the labour supply in STRAGO is set to equal that generated by rAps.

Production of goods is performed by a number of monopolistically competitive firms represented in each region. Each firm is producing a single variety, or type of an output. The production is carried out using a technology where both primary and intermediate inputs are used. The primary inputs represent production factors provided by households, such as labour and capital, while intermediate inputs are services and goods purchased from the different sectors and regions.

---

[7] The model can also be formulated in a way where the households are acting under perfect foresight.

Fig. 8.2 NUTS 2 regions in Sweden

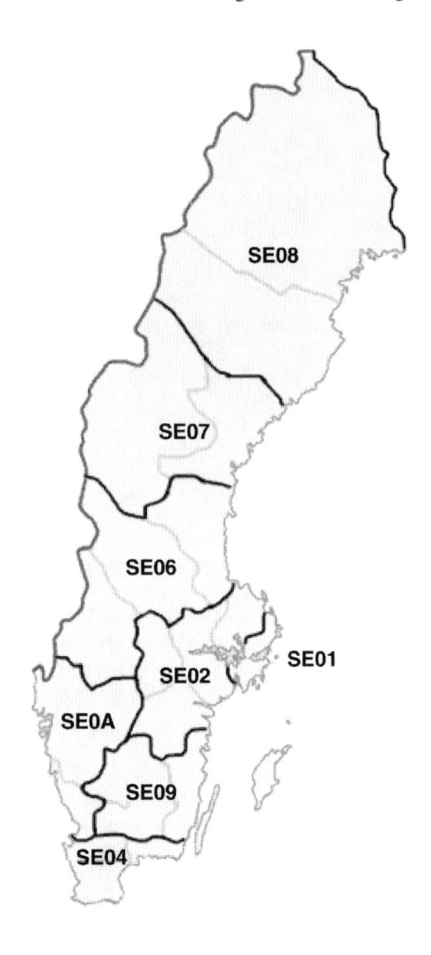

Inputs used for production, consumption or investment are provided from different regions. Hence, the demand will generate flows of goods between regions, i.e. interregional trade and a corresponding demand for transport services.

As indicated in the figure above, for the various industries there is import to, and export from, each region. In the model a complete trade structure is represented, where each region may use imported goods from every other region and industry, or use production from the own region to satisfy the demand. Trade with other regions implies transportation, and the transportation cost is rising with the distance between regions. This gives an incentive for the economic actors in a given region to substitute goods in a way that results in trade between proximate regions, if possible.

It is assumed that one of the driving forces behind interregional trade is the so called "love of variety", Dixit and and Stiglitz (1977). This property implies that it is attractive to trade with regions with a large supply of goods of a given type. Together with the existence of transportation costs this provides a basis for agglomeration of different activities, i.e. activities tend to be located in geographic proximity and different regions may specialize in different activities, see

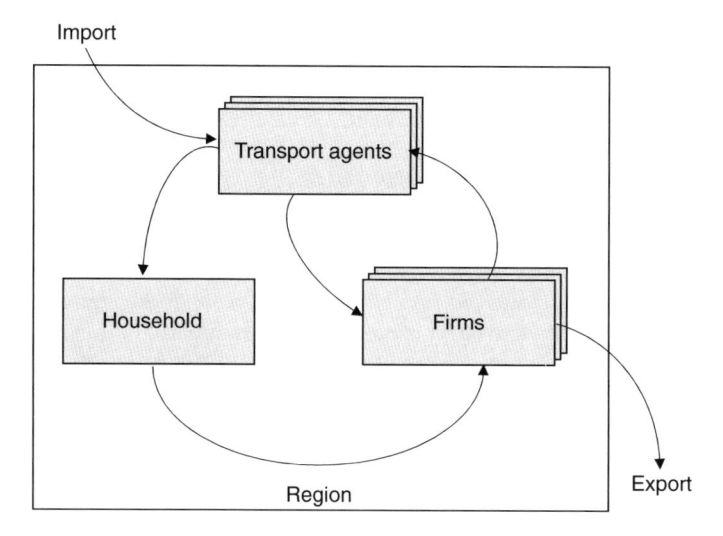

Fig. 8.3 Schematic illustration of a region in STRAGO

Krugman(1991). This can be grasped by looking at a given firm: the firm will benefit from being located close to its suppliers, and close to many suppliers, and at the same time close to the consumers, basically avoiding transport costs both on the input side of production and the output side. By using a mixture of standard monopolistic competition and the Armington (1969) approach to trade, where goods are demanded with respect to origin of production, the model provides a parameterization between perfect and monopolistic competition. The degree of agglomerative forces represented in the model may thus be controlled.

The analysis of long term scenarios, required for planning purposes or cost-benefit analyses, is typically dependent on one's view of the economic progress. Working under the constraining national scenarios, stating which sectors will grow and which will not, has region specific consequences, since different regions are specialized in different economic activities. Adding agglomerative forces to this development, implies that regions already relatively specialized in an industry will attract a disproportionally large part of growth in that particular sector. Thus agglomeration will affect the spatial production patterns, and thus the possibilities of employment in the different regions.

In STRAGO it is assumed that the supply of production factors is fully utilized. Thus, unemployment is not represented in the current version of the model. This is the main reason for the link going from rAps to STRAGO, as the labour market is being modelled explicitly in rAps. In this respect the consistency between the two models means that employment in rAps, aggregated to NUTS 2 regions, should be fully replicated in STRAGO. Thus, our approach is not strictly top-down since we exploit the disaggregate representation of labour markets provided by rAps, and use the labour market implications as restrictions to the workings of STRAGO.

8.2.2 rAps

rAps is basically a comprehensive regional model made up of various modules, and its theoretical basis could be described as eclectic. The multiregional input–output model is working by linking regional models for each of 72 labour market regions in Sweden, called FA-regions ("Functional Analysis regions"). In the linking procedure, two kinds of interregional flows are dealt with: interregional export and import of goods and services, and interregional in- and out-migration. In the multiregional model these flows are balanced, implying that the sum of all regional exports of each product (defined by industry) is balancing the sum of all regional imports, and that the sum of all in-migrants (defined by categories) is balancing the sum of all out-migrants. The balancing of these flows is handled through an extra region, a "pool", which acts as an intermediary with respect to all interregional trade and migration.

The regional model works with detailed data: population is grouped by age (100), sex (2), native country (3) and education (12). Labour supply is grouped similarly, but by age classes (8). Production is classified by 49 industries, and the corresponding labour demand is specified by education.[8] Labour demand is represented by input coefficients which change according to exogenous assumptions concerning productivity growth rates by industry.

The model is driven by exogenous demand (export, investment and parts of public consumption) directed towards production in various industries in the specific region. These demand components are changing in time according to assumptions about industry specific growth rates, based on assumptions at the national level, e.g. according to assumptions in scenarios made in the national Long-Term survey.

The construction of the regional model includes linkages both at the regional and sub-regional (local/municipal) levels, connections which can be grouped into five distinct modules, see Fig. 8.4.

The model calculations start in module (1) Population, by estimating the number born, who have died and progressed in education from one year to next year. This projection of population change is completed by estimated net migration, which is affected by the development at the regional labour and housing market in the previous year.[9]

Next, output from the Population module is input to module (2), Labour market, where preliminary estimates of the labour supply, L^S, and commuting are made.

---

[8] For a definition of industries in rAps, see Appendix B.

[9] Population is calculated at the level of municipalities (290) and the result is summed up to the level of regions (72). At the regional level in- and outmigration is based on estimated equations of migration propensities, determined by demographic attributes (age group, education, sex and native country), national migration propensity (change of national employment and change of foreign migration), regional labour market (change of regional versus national employment in previous year) and change of regional house price index previous year. Thus the influences from labour and housing markets are made exogenous by the lag structure. This makes computations tractable and also seems a reasonable way to model migration behaviour.

8 Integrating SCGE and I-O in Multiregional Modelling

```
                    Prod  X = AX+C(X) + I + Ex-Im(X)
                    Emp   L^D = L^D (X)
```

                              ┌──────────────┐
                              │ (3) Regional │
                              │   economy   │
                              └──────────────┘

L^S + Commuters =   Households
L^D + Unemployed   New dwellings
                                                           House price

--> Migration(*)   ┌──────────┐   ┌──────────┐   --> Migration(*)
                   │(2) Labour│   │(4) Housing│
                   │ market │   │ market   │
Supply L^S = L^S (Pop)└──────────┘   └──────────┘
In- & outcommuters
preliminary calculation

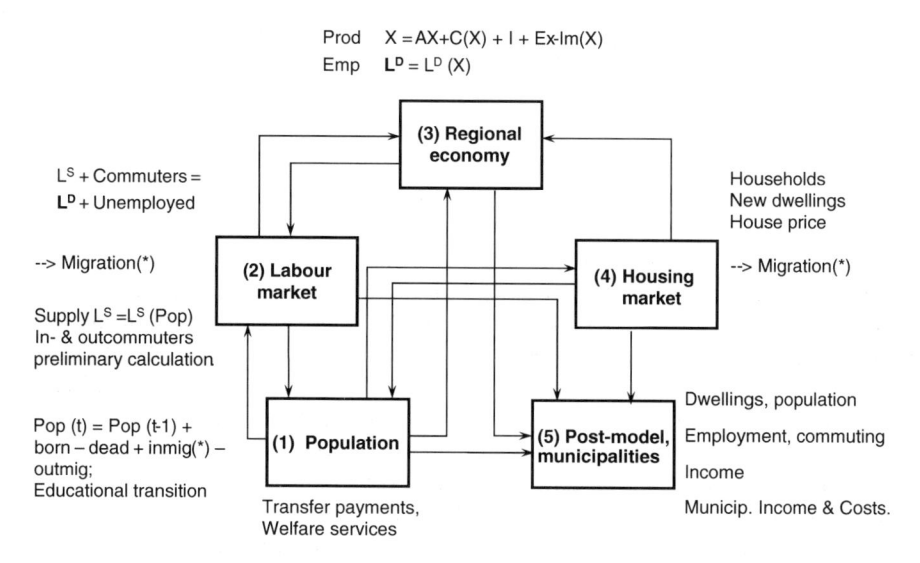

                                                          Dwellings, population
Pop (t) = Pop (t-1) +   ┌──────────┐   ┌──────────────┐ Employment, commuting
born – dead + inmig(*) –   │(1) Populat│──▶│(5) Post-model│
outmig;   │   ion   │   │ municipalities│ Income
Educational transition   └──────────┘   └──────────────┘

                       Transfer payments,   Municip. Income & Costs.
                       Welfare services

Fig. 8.4 rAps, five modules in the regional model

These estimates are partly affected by the development of the regional labour market in previous year.[10]

The third module (3), Regional Economy (Input–output) gives estimates of production, X, value added, income and labour demand, L^D. Intermediate demand, import, the major part of income, and household consumption are endogenously determined, whereas export, investment and part of public consumption are exogenous demand, as mentioned above. However, one part of investment demand, new dwellings, is estimated in the fourth module (4), Housing market. A preliminary estimate is based on data from the previous year.

Estimates of local public consumption are based on population figures. The connection from (1), Population to (3), Regional economy, also refers to estimates of various transfer payments which are treated as an exogenous income, in addition to the income generated by regional production. Another exogenous income is generated from net commuting, estimated in module (2). When labour demand has been estimated in (3), the preliminary estimates in the Labour market module (2) are supplemented by estimated unemployment, which is partly determined by the change in labour demand from previous year.[11]

The next step is to balance the results from the Labour market (2) with the results from Regional Economy (3). For each category (defined by education) regional

---

[10] The preliminary estimate of labour supply is based on estimated equations of labour force participation rates, determined by demographic attributes and the specific unemployment rate.

[11] The unemployment rate is determined by use of estimated equations including demographic attributes, the change in regional employment in previous year, and the change in national unemployment in previous year.

labour supply plus net commuters must equalize the sum of employed and unemployed. To achieve this balance the labour force participation rate and/or the commuting propensity is adjusted. Adjusting net commuting means, however, that also a part of exogenous income is adjusted, with repercussion on household consumption, production and labour demand in Regional economy (3). A stable "equilibrium" is achieved by a number of iterations between the two modules.

Module (4) Housing market gives estimates of house prices, new households and new dwellings. The regional house price index is determined by an estimated equation where the regional dwelling stock per inhabitant and change in regional income are the main determining factors. From this module there is a feedback to module (1), Population, with respect to how the housing stock is assumed to affect the distribution of regional population to municipalities at the local level. The last module (5), Post-model municipalities, provides, inter alia, estimates of employment and commuting at the municipality level. This module is termed 'post-model' since it is breaking down the results from the regional level, while these results do not serve as input to any other module.

What has been described above is the operation of rAps in a "stand alone" version. In the combined model framework the calculations start by running this multiregional model, subject to the constraints with respect to total national employment, growth of exogenous demand and labour productivity by industry. The resulting total employment by region (72) is aggregated to total employment by STRAGO-region (9), to provide STRAGO with a functional estimate of labour supply, given that unemployment is not represented in the current version of STRAGO. The communication between STRAGO and rAps is described below.

8.2.3 Calibrating the Models

Calibration of the model involves to adjusting the parameters of the model in such a way that both exogenous data constraints as well as consistency requirements from our modelling framework are achieved.

8.2.3.1 STRAGO

Exogenous conditions are imposed in terms of I-O data, which describes the intersectoral trade, consumption, exports, and factor use in the form of labour and capital for the whole of Sweden. A calibration requirement for the model is that all these data are replicated. In addition to the use of I-O data, production/consumption-data which describes the regional distribution of production for the various sectors is used. Furthermore, data on the transport intensities for different types of goods is used to calibrate the transport demand for the various sectors, which stems from the trading activities where goods are involved. This basic calibration of the model provides us with a relevant starting point for the model, where the national and regional structure of production, contained in the data, is replicated.

Also, we perform a consistency calibration with the aim that the model should provide results which are consistent to those of the national model. That is, we take into account the results of the national model, generated in a base scenario and ensure that key variables from the national model are replicated. For example, for the base scenario we require that the sectoral trends in exports, gross output and productivity, provided from the national model, also are found in STRAGO.

The calibration we have discussed so far implies that, as far as possible, we use either existing data or the requirements of consistency between models to determine the parameters of the model. In the case of the model's elasticity parameters, we set those parameters as guided by existing literature. The elasticity parameters determine the different actors' ability to substitute between different goods.

The values used for various elasticities in the model have been chosen, partly in view of retaining as large degree of consistency as possible to the national model, partly based on the estimates available in the literature, see, for example Donnelly et al. (2004), Beaudry and Wincoop (1996) and Ardelean (2006).

8.2.3.2 rAps

Calibration of rAps is mainly about adjusting the parameters of the model in such a way that exogenous conditions in terms of data are satisfied. This means, for example, that the population is changing according to the rate of change for fertility and mortality, and foreign in- and outmigration, used in the national scenario. Another example is the calibration to reproduce the official long term macroeconomic scenario with respect to growth of productivity by industry, and also with respect to the rate of change in exogenous demand by industry (exports, gross investment and government spending). Another kind of calibration is performed with respect to what can be judged as reasonable limits for labour participation rates and employment rates. Since the model is operating without explicitly considering regional variation in labour costs this calibration must be performed outside the model, by adjusting parameters affecting labour demand.

8.2.4 Communication Between Models

Model results from the different model steps – national model, STRAGO and rAps – must be translated into constraints to the other models to achieve consistency between models. Since the different model steps are operating with different classifications of goods and industries there is also a need of keys to transform results from one classification to another. An overview of the communication between model steps is presented in Fig. 8.5. The national model provides constraints to both STRAGO and rAps, whereas the regional models are providing constraints to each other.

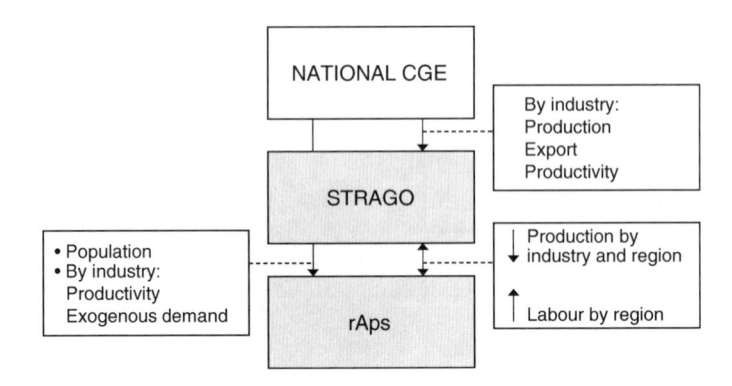

```
                    ┌─────────────────────┐
                    │   NATIONAL CGE      │         ┌──────────────────┐
                    └─────────┬───────────┘         │ By industry:     │
                              │ ─────────────────── │ Production       │
                              ▼                     │ Export           │
                    ┌─────────────────────┐         │ Productivity     │
                    │      STRAGO         │         └──────────────────┘
┌──────────────────┐│                     │         ┌──────────────────┐
│ • Population     ││                     │ ──────▶ │ Production by    │
│ • By industry:   │└─────────┬───────────┘         │ industry and region
│   Productivity   │─ ─ ─ ─ ─ ┼ ─ ─ ─ ─ ─           │                  │
│   Exogenous demand         │                     │ Labour by region │
└──────────────────┘┌─────────▼───────────┐◀─────── └──────────────────┘
                    │       rAps          │
                    └─────────────────────┘
```

Fig. 8.5 Communication of model results for consistency between model steps

The national model is providing the regional models with results referring to population, productivity, exogenous demand and production. The obvious reason is the ambition to get consistency between assumptions at national and regional level. This aim means, e.g., that growth rates by industry in the national model should be replicated in the regional models, when summing over all domestic regions.

Regional aggregation between the models is quite straightforward, simply aggregating by summing results from subregional levels. Regional disaggregation is provided by the STRAGO model and rAps respectively. A slightly more complex issue is sector aggregation and disaggregation. It turns out in our application that the national model has a rather fine representation of sectors, such that national restrictions can be provided for both STRAGO and rAps through aggregation. For the purpose of aggregation a transformation key is used.

For a base year we have a given transformation key matrix A ($m*n$, $m < n$), Results on a disaggregated sector level, given by a vector x, are transferred into aggregate sector values y through

$$y = Ax.$$

Corresponding to the aggregation key, we also have a disaggregation key $\tilde{A}$ ($n*m$, $m < n$) such that

$$x = \tilde{A}y$$

Provided that we have national restrictions on the disaggregated sector levels at different points in time x_t, we may generate the corresponding aggregated y_t through

$$y_t = Ax_t.$$

Since rAps is working with a finer sector representation than STRAGO we need to be able to disaggregate the production results from STRAGO. Yet, simply applying the disaggregation scheme $\tilde{A}$ directly will in general not produce results

# 8 Integrating SCGE and I-O in Multiregional Modelling

which are in concordance with the nationally provided restrictions. Rather, given x_t and y_t we now want to find time-specific transformations $\tilde{A}_t$ such that

$$x_t = \tilde{A}_t y_t.$$

For this purpose, a RAS procedure incorporating row and column adjustment of the original disaggregation key has been used. That is $\tilde{A}_t = R_t \tilde{A} S_t$, where R_t and S_t are diagonal matrices. We find these matrices through the following restrictions

$$x_t = R_t \tilde{A} S_t y_t$$

$$1 = 1 R_t \tilde{A} S_t.$$

That is, we find disaggregation keys which recreated the national restrictions, and the column sums of the key have to sum to one. This means that we are only allowed to distribute proportions of what is contained in x_t into y_t, and these proportions sum to one. Through this procedure we find disaggregation keys for transforming results provided by STRAGO into restrictions for rAps, which are consistent with the restrictions provided from the national level.

The communication between rAps and STRAGO includes the following steps. First, rAps is used in a preliminary run, mainly to generate employment, which aggregated from FA-regions to NUTS 2 regions is used as labour supply input to STRAGO. Second, STRAGO is applied to get a regional picture of the results at the national level. Third, rAps is used in a second run, where output from STRAGO – production by industry and NUTS 2 region – is used as constraints. A scaling procedure is applied where region- and industry specific scale factors are defined, telling how much production from the first step should be adjusted to satisfy the constraints from STRAGO in the final year. These scale factors are implemented by adding a yearly exogenous demand factor to the model.

The resulting regional distribution of employment may diverge from the initial input to STRAGO. If this is the case an iterative procedure can be applied until the regional distribution of employment in the two models harmonize.

8.3 Results from the Combined Framework

The regional allocation of employment resulting from rAps in the first run is mainly driven by the regional industry structure in the base year, and the growth in exogenous demand and labour productivity by industry, according to assumptions in the national model. As for the agglomeration advantages in metropolitan regions and regions with large labour markets, these advantages also/only reflect effects of the industry structure. For instance, fast growing industries according to the national model (e.g. knowledge intensive business services) are concentrated to

Table 8.1 Average annual employment growth as difference to national average. Actual growth 1985–2009 and projection 2005–2030 with rAps unconstrained and rAps constrained by results from STRAGO

NUTS 2	Annual employment growth		
	1985–2009 (%)	2005–2030 Preliminary (%)	Constrained (%)
Stockholm	0.54	0.38	0.54
East middle	0.08	0.00	0.20
Småland	−0.22	−0.49	−0.43
South	0.07	−0.01	−0.03
West	0.13	−0.10	0.10
North middle	−0.61	−0.29	−0.58
Middle north	−0.67	−0.11	−0.43
Upper north	−0.43	0.11	−0.52
Örebro + Västm	−0.49	0.02	−0.49

Stockholm and other metropolitan regions. There is also an effect of market size on the size of the multiplier, which tends to be in favour to larger regions. But the results from rAps in the first step are not derived from a model which explicitly is taking transportation costs (or other costs) into account, nor are agglomerative forces being modeled. The regional allocation of production resulting from STRAGO in the second step can be grasped as a corrective in these respects.

Given the altered regional allocation of production in STRAGO, whereas regional labour supply is constrained according to data from rAps, the resulting growth of labour productivity by industry differ from the assumptions in the national model. There may be reasons to expect some regional variation in labour productivity growth by industry. However, it seems more reasonable to see this variation mainly as an indication for migration incentives. While migration is not being modeled in STRAGO, migration is included in rAps. Thus, in the second run with rAps, employment is constrained by the allocation of production from STRAGO, and constrained by labour productivity growth from assumptions in the national model in the same way as in the first run. As a result there will be a relocation of population and labour force, driven by the altered relocation of employment.

From a projection of regional development 2005–2030, the average annual growth in regional employment is presented in Table 8.1. Both the results from the first (preliminary) run and the second run (constrained by STRAGO) are presented. As a comparison the actual annual growth for 1985–2009 is also presented. Since the national employment growth during this period was almost zero, while the national growth 2005–2030 is expected to be 0.36 % per year, regional growth is presented as differences to the national average.

It is interesting to note that rAps constrained by STRAGO predicts a relocation of employment much closer to the historical pattern than in the preliminary, unconstrained case. Table 8.2 reports how production (aggregated into three aggregate sectors, mainly agriculture, manufacturing and services) is redistributed

Table 8.2 Redistribution of production between NUTS 2 regions when comparing preliminary distribution in rAps with distribution constrained by results from STRAGO

NUTS 2	Economic sector		
	Primary (%)	Secondary (%)	Tertiary (%)
Stockholm	0.3	−0.2	1.9
East middle	1.0	0.8	0.5
Småland	−0.3	0.8	0.2
South	0.3	0.3	−0.5
West	−0.4	5.3	0.0
North middle	0.9	1.5	−0.9
Middle north	−0.5	0.2	0.1
Upper north	−0.5	−0.1	−0.7
Örebro + Västm	−0.7	−8.5	−0.5
Sum	**0.0**	**0.0**	**0.0**

between NUTS 2 regions, when comparing the preliminary results from rAps with the results where rAps is constrained by the results from STRAGO.

For example, the redistribution is expected to mirror the inclusion of agglomeration factors in STRAGO. Accordingly, it can be noted that production in the tertiary sector (services) is redistributed to the largest labour market region, Stockholm.

8.4 Concluding Remarks

The outcome and experiences from working with this integrated model framework is promising. The strength of SCGE models is the description of the behaviour of different actors. However, this description makes use of non-linear models which turn out to be intractable when disaggregated. Hence most operational SCGE models are highly aggregated in terms of number of industries and/or number of regions. The advantage of I-O models is that they permit disaggregation in both these respects.

Analysis at the regional level, dealing with regional consequences of national scenarios, effects of policy changes, effects of shocks in sectors of the industry etc., needs modelling tools which have a good theoretical underpinning. But the modelling tools must also be able to work at a fairly detailed level to serve its purpose from the policy and planning point of view. In this respect it thus seems productive to further develop the model framework presented above.

Appendix A STRAGO, Model Specification

In this application we run STRAGO in a recursively dynamic version, in which decisions are made under static but adaptive expectations. We provide a short description of the model. For a more extensive description, see Sundberg (2009).

All variables in the description below contain a time index, but this has been suppressed to simplify the presentation; time indices are expressed explicitly in the decisions that are dynamic. Unless otherwise stated, Greek letters denote exogenous parameters of the model, which are typically calibrated to data. Furthermore, let superscripts i and j refer to sectors, while the subscripts r and s reflects regions.

Production under monopolistic competition and demand for inputs. We state the monopolist's profit maximization problem, where p_r^i reflects the price of inputs, ϕ represent fixed input requirements, and χ is the marginal input requirement per unit of produced output, we have

$$\max_{x_r^{Mi}} \pi_r^{Mi}(x_r^{Mi}) = p_r^{Mi}(x_r^{Mi})x_r^{Mi} - p_r^i(\phi - \chi x_r^{Mi})$$

p_r^{Mi} is the price charged by the monopolist, and x_r^{Mi} the quantity produced. Profit maximization together with a zero profit condition, implies the following price and quantity for each monopolist in a sector i and region r

$$p_r^{Mi} = \frac{\varepsilon_r^i}{\varepsilon_r^i - 1} \chi p_r^i,$$

$$x_r^{Mi} = (\varepsilon_r^i - 1)\frac{\phi}{\chi},$$

where the price is given by the marginal cost of production times a markup, which depends on the perceived elasticity of demand[12] in a region. The marginal cost is given in turn by a Constant Elasticity of Substitution (CES) unit cost function, using different types of goods as inputs, as well as labour and capital

$$p_r^i = \left\{ \sum_j \alpha^{ji}(q_r^j)^{1-\rho_i} + \sum_k \alpha^{ki}(w_r^k)^{1-\rho_i} \right\}^{\frac{1}{1-\rho_i}}.$$

The prices payed for different inputs reflects both production costs as well as transport costs. Given the unit cost function, the corresponding demands for intermediates and primary inputs (labour and capital) are given by Sheppard's lemma

$$a_r^{ji} = \alpha^{ji}\left(\frac{p_r^i}{q_r^j}\right)^{\rho_i} \quad \text{and} \quad b_r^{ki} = \alpha^{ki}\left(\frac{p_r^i}{w_r^k}\right)^{\rho_i},$$

describing the input demands per unit of the firm's aggregate use of inputs. In each region and sector there are a number, or mass, of active monopolists, this number is denoted by n_r^i.

---

[12] The elasticity of demand is defined through $\frac{1}{\varepsilon_r^i} = -\frac{dp_r^{Mi}(x_r^{Mi})}{dx_r^{Mi}}\frac{x_r^{Mi}}{p_r^{Mi}}$.

8 Integrating SCGE and I-O in Multiregional Modelling

Spatial distribution of demand for inputs and demand for transport services.
The intermediate inputs used in a region are in turn composed of an import aggregate, consisting of goods from different regions, the unit cost of this aggregate is given by

$$q_s^i = \left\{ \sum_r \varphi_r^i (\tilde{q}_{rs}^i)^{1-\sigma^i} \right\}^{\frac{1}{1-\sigma^i}}$$

the price $\tilde{q}_{rs}^i$ which, in turn, describes an assembly of all varieties which the monopolists in the supplying region r are offering along with the transportation costs of transporting the supply to the region s. We have

$$\tilde{q}_{rs}^i = \begin{cases} \left\{ \int_0^{n_r^i} (p_r^{Mi} + \xi^i(p^F + \tau^i)T_{rs}^{F,i})^{1-\varepsilon^i} dv \right\}^{\frac{1}{1-\varepsilon^i}} & r = 1,...,R-1 \\ (p_r^{Mi} + \xi^i(p^F + \tau^i)T_{rs}^{F,i}) & r = R \end{cases}$$

We have that the cost of receiving produce from another region is dependent on the producer price in the other region and the additional transport cost. Region R reflects the rest of the world and T reflects "distance". The corresponding demand functions for produced varieties from r demanded in s of sector i's supply per unit of import aggregate is given by Sheppard's lemma

$$z_{rs}^i = \begin{cases} \phi_r^i \left(\frac{q_s^i}{\tilde{q}_{rs}^i}\right)^{\sigma^i} \left(\frac{\tilde{q}_{rs}^i}{p_r^{Mi}+\xi^i(p^F+\tau^i)T_{rs}^{F,i}}\right)^{\varepsilon^i} n_r^i & r \neq R \\ \phi_r^i \left(\frac{q_s^i}{p_r^{Mi}+\xi^i(p^F+\tau^i)T_{rs}^{F,i}}\right)^{\sigma^i} & r = R. \end{cases}$$

It may be noted that the spatial distribution od demand with respect to origin of production is dependent of the size of the industry at the origin, since z_{rs}^i is dependent on $\tilde{q}_{rs}^i$ which in turn depends on the size of the industry n_r^i. This spatial distribution of demand for goods and services also generate a demand for transport services per unit of import aggregate which is given by

$$z_s^{F,i} = \sum_r T_{rs}^{F,i} \xi^i z_{rs}^i.$$

Production of transport services. The demanded transport services are produced by an aggregate transport service sector, whose unit cost of production are described by a simple Leontief technology

$$p^F = \sum_{is} \varphi_s^{F,i} q_s^i + \sum_{ks} \varphi_s^{F,k} w_s^k$$

and its demands for inputs are

$$a_s^{iF} = \varphi_s^{F,i} \quad \text{och} \quad b_s^{kF} = \varphi_s^{F,k},$$

per unit of output.

Household decisions, Euler equations and budget constraints. Households maximize an inter-temporal utility subject to budget constraints

$$U(c_{ry},...,c_{r(y+T)}) = \sum_{t=y}^{y+T} \beta^t u(c_{rt}/l_{rt})$$

We have the first order optimality conditions

$$u'(c_{rt}/l_{rt}) = u'(c_{r(t+1)}/l_{r(t+1)})\beta \left(\frac{w_{rt}^K}{p_{rt}^C} + (1-\delta) \right),$$

t = year, ...,year + horizon, together with the budget constraints of the households

$$k_{rt}w_{rt}^K + l_{rt}w_{rt}^L + g_{rt} = p_{rt}^C(c_{rt} + \delta k_{rt}), \text{ t = year + horizon +1}$$

$$k_{rt}w_{rt}^K + l_{rt}w_{rt}^L + g_{rt} = p_{rt}^C(c_{rt} + (k_{r(t+1)} - (1-\delta)k_{rt})) \text{ otherwise.}$$

These equations describe how much the households will invest in capital stocks for the coming year, and how much they will consume in the current time period. The decisions are taken under adaptive static expectations, with a planning horizon, beyond which the households think that they have reached a new steady state. The aggregate consumer product in region r is composed according to the following CES preferences, where the per unit expenditure on the consumption good is given by

$$p_r^C = \left\{ \sum_i \gamma^i (q_r^i)^{1-\sigma_c} \right\}^{\frac{1}{1-\sigma_c}},$$

and demands over the different types of goods is given by Hotelling's lemma

$$d_r^i = \gamma_i \left(\frac{p_r^C}{q_r^i} \right)^{\sigma_c},$$

per unit of demand for the consumption good.

Transfers. Total tax revenues are given by

$$g = \sum_{is} \tau^i z_s^{F,i} D_s^i$$

and this revenue is distributed to regions according to size.

$$g_r = \frac{gl_r}{\sum_r l_r}.$$

In this application of STRAGO this part of the model has not been activated since the only tax that has so far been modeled is a kilometer tax for goods freight, which is not relevant to the scenarios under consideration, and thus is set to zero.

Export demand describes the demand from the rest of the world, which is the part of demand that is not attributed to domestic regions. This demand is given by

$$E^i = \zeta^i \left(\frac{p_R^{Mi}}{q_R^i}\right)^{\varepsilon_i}.$$

Supply and demand equilibrates. In each region, the monopolists in each sector, make use of the total supply inputs available to them, this gives a restriction on the number of monopolists who are active in a region and sector

$$n_r^i = \frac{x_r^i}{\varepsilon_r^i F}.$$

The total demand of inputs of type i from region s can be written as

$$D_s^i = \begin{cases} d_s^i(c_s + \Delta_s) + \sum_j a_s^{ij} x_s^j + a_s^{iF} f & s = 1, \ldots, R-1 \\ E^i & s = R \end{cases}$$

where $\Delta_s = \Delta_{st} = (k_{s(t+1)} - (1-\delta)k_{st})$. This describes the total demand for the form of goods that go into consumption or capital formation, to the inputs of production and inputs to transport production. The system of product/service production can be closed in terms of inputs to the monopolists in each region and sector

$$x_r^i = \frac{\varepsilon_r^i \chi}{\varepsilon_r^i - 1} n_r^{i\frac{1-\sigma^i}{1-\varepsilon^i}} \sum_s \phi_r^i T_{rs}^{ice,i} \left(\frac{q_s^i}{p_r^{Mi} T_{rs}^{ice,i} + \zeta^i (p^F + \tau^i) T_{rs}^{F,i}}\right)^{\sigma^i} D_s^i,$$

which can be derived using the relationship $n_r^i x_r^{Mi} = \sum z_{rs}^i D_s^i$, stating that the monopolists produce is requested by the economic agents in the various regions.

Supply and demand for both labour and capital is in equilibrium

$$l_r = \sum_i b_r^{Li} x_r^i + b_r^{LF} f,$$

$$k_r = \sum_i b_r^{Ki} x_r^i + b_r^{KF} f,$$

the same applies to supply and demand of transport services.

$$f = \sum_{i,s} z_s^{F,i} D_s^i,$$

The elasticity of demand as perceived by a monopolist in sector i and region r is given by

$$\varepsilon_r^i = \varepsilon^i p_r^{Mi} \frac{\sum_s z_{rs}^i D_s^i (p_r^{Mi} + \xi^i (p^F + \tau^i) T_{rs}^{F,i})^{-1}}{\sum_s z_{rs}^i D_s^i}.$$

In typical applications of monopolistic competition with "iceberg" transport costs, this elasticity is independent of location, and is exogeneously given. In our case, where we model additive transport costs, the elasticity will have a spatial dependence, and we have to solve for the elasticities as part of the equilibrium.

Variables with expressed time index		
	c_{rt}	Total consumption in region r at time t
	k_{rt}	Capital in region r at time t
	p_{rt}^C	Price for the consumption aggregate in region r at time t
	w_{rt}^K	Capital rent in region r at time t
	w_{rt}^L	Wage rate in region r at time t
	g_t	Tax revenues at time t
	g_{rt}	Transfers to region r at time t
	Δ_{rt}	Investments in region r at time t

Variables with supressed time index

q_s^i	Price of input i i region s
$\tilde{q}_{rs}^i$	Aggregate price of variety i from region r i region s
p_r^i	Price of input aggregate in sector i in region r
p_r^{Mi}	Price of variety i in region r
p^F	Price of transport sevice
a_s^{ij}	Demand for input of good i per unit of input-aggregate of sector j in region s
b_s^{kj}	Input of factor k per unit of input-aggregate in sector j in region s
d_s^j	Demand for good j i region s per unit of consumption in region s
z_{rs}^i	Demand for varieties from sector i and region r per unit of input i in region s
$z_s^{F,i}$	Demand for transport services for good i per unit of import-aggregate in region s
D_s^i	Demand for inputs from sektor i in region r
x_r^i	Quantity of input-aggregate of sector i in region r
x_r^{Mi}	Quantity of variety in sector i in region r
f	Quantity of transport service
n_r^i	Number of firms producing varieties of good i in region r
ε_r^i	Elasticity of demand for sector i in region r

8   Integrating SCGE and I-O in Multiregional Modelling

Variables

Endogenous variables in the model are:

Appendix B Definition of Industries in rAps

1. Products of agriculture, hunting and related services
2. Products of forestry, logging and related services
3. Fish and other fishing products; services incidental of fishing
4. Mining
5. Food products and beverages
6. Textiles
7. Sawmills
8. Wood and products of wood and cork (except furniture)
9. Pulp industry
10. Paper and paper products
11. Printed matter and recorded media
12. Coke, refined petroleum products and nuclear fuels
13. Pharmaceutical industry
14. Other chemical products
15. Rubber and plastics
16. Other mining and quarrying products
17. Basic metals
18. Fabricated metal
19. Machinery and equipment n.e.c.
20. Electrical machinery and apparatus n.e.c.
21. Medical, precision and optical instruments, watches and clocks
22. Transport equipment
23. Furniture; other manufactured goods n.e.c.
24. Electrical energy, gas, steam and hot water
25. Water supply
26. Construction
27. Trade, maintenance and repair services of motor vehicles and motorcycles; retail sale of automotive fuel
28. Wholesale trade and commission trade services, except of motor vehicles and motorcycles; retail sale
29. Hotel and restaurant services
30. Land transport; railways
31. Other land transport
32. Water transport
33. Air transport
34. Supporting and auxiliary transport activities; activities of travel agencies

35. Post and delivery services
36. Telecommunication services
37. Financial intermediation services, insurance and pension funding services
38. Real estate services
39. Business services
40. *Education services*
41. *Health and social work services*
42. Membership organisation services n.e.c.
43. Public administration and defence services; compulsory social security services
44. *Other services*

Note: Industries in italics are represented both by private and public sectors

References

Ardelean A (2006) How strong is the love of variety? CIBER working paper no. 2006–006, Purdue University
Armington PS (1969) A theory of demand for products distinguished by place of production. IMF Staff Pap 16:159–178
Beaudry P, van Wincoop E (1996) The intertemporal elasticity of substitution: an exploration using a US panel of state data. Economica 63(251):495–512
Bröcker J (1998) Operational spatial computable general equilibrium modeling. Ann Reg Sci 32:367–387
Dixit AK, Stiglitz JE (1977) Monopolistic competition and optimum product diversity. Am Econ Rev 6(3):297–308
Dixon PB, Rimmer MT (2004) Disaggregation of results from a detailed general equilibrium model of the US to the state level, COPS, General working paper no. G-145, Monash University (http://www.monash.edu.au/policy)
Donnelly WA, Ingersoll D, Johnson K, Tsigas M (2004) Revised Armington elasticities of substitution for the USITC model and the concordance for constructing a consistent set for the GTAP model. Office of Economics Research note no. 2004-01-A, U.S. International Trade Commission
Ethier WJ (1982) National and international returns to scale in the modern theory of international trade. Am Econ Rev 72(3):389–405
ITPS (2009) Regional utveckling i Sverige – flerregional integration mellan modellerna STRAGO och rAps. A2009:004, Östersund
Krugman P (1991) Increasing returns and economic geography. J Polit Econ 99(3):483–499
Lundqvist L, Karlström A, Sundberg M (2004) Godstransporter and SCGE – en alternativ väg. Kort PM, Transport- och lokaliseringsanalys, KTH
Sundberg M (2009) Essays on spatial economies and organization. Doctoral thesis in infrastructure. Royal Institute of Technology, Stockholm
Wilson A (2010) The general urban model: retrospect and prospect. Pap Reg Sci 89(1):27–48

Chapter 9
Interjurisdictional Competition and Land Development: A Micro-Level Analysis

Jae Hong Kim and Geoffrey J.D. Hewings

Abstract A considerable number of recent studies show that metropolitan areas having a more fragmented governance structure tend to show a sprawling pattern of development. This may suggest that a fragmented institutional setting can generate a higher level of interjurisdictional competition that often hinders systematic management of the development process, thus offsetting the benefits from disaggregated local governance, such as welfare and fiscal efficiency gains. While previous studies typically assess this issue through metropolitan-level analysis, this research examines how the institutional setting influences land development at a micro-scale (i.e., section: 1 mile × 1 mile). More specifically, the present study (1) quantifies the institutional conditions in each section, taking the jurisdictional boundaries into account and (2) measures its effect on land use conversion rate by employing a quasi-likelihood estimation method. An empirical assessment of the U.S. Midwest case suggests that interjurisdictional competition, particularly the race for specific small land areas, does accelerate land use conversion, although the analysis results vary to some extent by the measurement of the institutional factor.

9.1 Introduction

A great deal of scholarly attention has been paid to various implications of the institutional structure of local governance and planning. A typical example is provided by Tiebout (1956) and many subsequent studies (see e.g., Ross and Yinger 1999; Howell-Moroney 2008), which examine the effectiveness of disaggregated

---

J.H. Kim (✉)
Department of Planning, Policy, and Design, University of California, Irvine, CA, USA
e-mail: jaehk6@uci.edu

G.J.D. Hewings
Regional Economics Applications Laboratory, University of Illinois at Urbana-Champaign, Champaign, Urbana, IL, USA
e-mail: hewings@illinois.edu

governance in providing a broad range of tax and public service mixes for diverse residents who would be able to 'vote with their feet.' A large branch of the fiscal decentralization literature (see e.g., Oates 1999; Wilson and Wildasin 2004; Wildasin 2006) also suggests that efficiency gains can be attained when citizens are served by disaggregated local governments.

The disaggregated (or fragmented) governance structure, however, is often viewed as a significant challenge in other respects. For instance, numerous environmental studies (e.g., Yaffee 1997; Kunce and Shogren 2005) indicate that the fragmented institutional setting is more likely to generate destructive interjurisdictional competition, thereby causing failure in environmental planning and resource management, as indicated by 'the tragedy of the commons' (Hardin 1968). Furthermore, in recent years, it has been suggested that fragmentation in local governance can hinder smart management of development processes and can result in a sprawling pattern of development (Razin and Rosentraub 2000; Glaeser et al. 2001; Fulton et al. 2001; Carruthers and Ulfarsson 2002; Carruthers 2003; Ulfarsson and Carruthers 2006). The rise of regionalism (Foster 2001; McKinney and Essington 2006) and megaregion-based planning (Dewar and Epstein 2007; Innes et al. 2011; Yaro 2011) seem to be a response to the perceived challenges generated by political fragmentation.

This study examines how the institutional setting influences the land development process. While previous studies typically investigate the relationship between political fragmentation and development patterns through metropolitan-level analysis, the present research conducts a micro-scale analysis to better understand the detailed mechanism of the development process in relation to local governance structure and interjurisdictional competition. This is accomplished by employing spatially-explicit indicators of the institutional conditions, designed to represent the degree of potential interjurisdictional competition over land areas.

The remainder of this paper is structured as follows. Section 9.2 discusses the previous studies that explore and/or examine the interrelationship between governance structure and land development pattern. Section 9.3 presents the design of the micro-level empirical analysis used to look into this issue more closely. The empirical analysis outcomes, using the case of the Midwest region in the United States, are presented and discussed in Sect. 9.4, followed by a conclusion.

9.2 Governance Structure, Interjurisdictional Competition, and Land Development Pattern

There is a voluminous literature on interjurisdictional competition. In particular, a large number of studies have discussed the nature of the competition (i.e., why and for what they are competing with each other) and examined the effects of the competition on residents' welfare, fiscal efficiency, and so forth (i.e., what the consequence tends to be). In addition, recent years have witnessed a variety of research methods designed to aid the studies, ranging from dynamic game theory-oriented approaches to agent-based modeling and spatial econometric estimation (see e.g., Zellner et al. 2009; Zhang 2011).[1]

Although it has not drawn as much attention as other aspects have, land use patterns have recently been examined in conjunction with interjurisdictional competition that arise from political fragmentation in local governance. Some recent studies have attempted to empirically test the notion that a more fragmented local governance structure is responsible for sprawl, which is generally presumed to lead to many contemporary urban problems, such as environmental disruption, segregation, inefficient public service provision, and many more.[2,3] In other words, the research has scrutinized whether a fragmented institutional setting tends to generate destructive interjurisdictional competition, thereby hindering the systematic management of the growth process.

For instance, in his study on the causes of sprawl, Pendall (1999) tests a hypothesis that "land use will be more dense (less sprawling) if ... municipal fragmentation is limited" (p. 560). While he conducts a county-level regression analysis to test the hypothesis and other suppositions, the level of fragmentation is measured by using two metropolitan-level variables – (1) physical size of the 25th-percentile jurisdiction and (2) number of municipalities in each metropolitan area. According to the analysis results, the first fragmentation variable exhibits a significant positive coefficient which suggests that land development pattern tends to be more compact in less fragmented metropolitan areas.

Razin and Rosentraub (2000) also examine the association between sprawl and political fragmentation in local governance and find a positive correlation in the case of North American metropolitan areas. In their regression analysis, however, the fragmentation indicators, such as the number of local governments per 10,000 residents and the existence of multipurpose metropolitan-level government (p. 823), turn out to be statistically insignificant predictors of residential sprawl. Two Brookings Institution's reports (Glaeser et al. 2001 and Fulton et al. 2001) examine the sprawl issue in relation to many factors, including political fragmentation. They find a larger area of land consumption and employment suburbanization that indicate a more sprawling pattern of development, in the metropolitan regions with a higher level of fragmentation.

---

[1] Brueckner (2003) provides a comprehensive summary of the empirical research on the interjurisdictional competition, particularly the studies using spatial econometrics.

[2] The notion of the association between political fragmentation and sprawl has been suggested by Downs (1994) and Lewis (1996). Furthermore, in the growth management literature, coordinated planning is often regarded as a necessary condition for the successful control of sprawl.

[3] Whereas sprawl is generally regarded as a main cause of the urban problems, Gordon and Richardson (1997) and some others contend that such development patterns are the natural consequence of economic decentralization and the American preference. It also needs to be noted that the concept of sprawl is still somewhat elusive and often defined differently, even though a few studies (Ewing 1996; Tsai 2005) have attempted to identify and clarify the characteristics of sprawl.

Carruthers and Ulfarsson (2002) empirically test the connection in a more rigorous manner. They employ a structural equation system with an emphasis on the relationships among multiple key variables in urban environments (i.e., population density, urbanized land, property values, and government spending for infrastructure) and analyze the effects of political fragmentation on the interrelated variables. From the county-level data for three time points (1982, 1987, and 1992) in 14 selected states, they find that the fragmentation, measured in terms of per capita municipal government units and special districts, has a significant negative effect on density.[4] Later, after refining the model and expanding the sample coverage, they (Ulfarsson and Carruthers 2006) investigate the issue again, this time giving explicit consideration to the endogeneity problem. In other words, the fragmentation and sprawl are assumed to have a cyclical relationship in model formulation. Their empirical analysis reinforces the hypothesized relationship. More specifically, it is suggested that fragmentation induces a lower density (i.e., more sprawling pattern of development) and vice versa, over time.

Although the issue has not been widely investigated yet, recent studies show that the structure of local governance may have a substantial effect on the pattern of land development for urban purposes. Previous studies, however, typically examine the issue by conducting regional or county-level analysis. Moreover, in explaining the relationship, they focus on the level of density or the spatial extent of urbanized areas in a particular year rather than on the dynamic changes in the variables. In the land use change literature, the influence of institutional relationships has not been explicitly taken into account, although there are a number of spatially explicit micro-level studies. To fill this gap, the present study conducts a more disaggregated level analysis of the potential causal linkage from institutional environments to land development. In the next section, attention will be directed to the empirical analysis.

9.3 A Micro-Level Analysis

This study examines how the institutional setting influences the process of land development at a disaggregated geographical scale. The unit of analysis here is the section, 1 mile × 1 mile grid zones. More specifically, consideration is given to the sections located in the metropolitan statistical areas (MSA), which are completely within the 11 Mid-west states – Illinois, Indiana, Kansas, Michigan, Minnesota, Missouri, Nebraska, North Dakota, Ohio, South Dakota, and Wisconsin.[5] There are

---

[4] In this study, the dependent variable is not density change but the level of population density. Therefore, the outcome does not necessarily indicate that the fragmented governance induces a more rapid density decline, although it may suggest that the fragmentation is inversely associated with density level.

[5] This analysis employs the definition of the metropolitan statistical area boundaries, used for Census 1990, as it analyzes how institutional structure in early 1990s affects land development between 1992 and 2001.

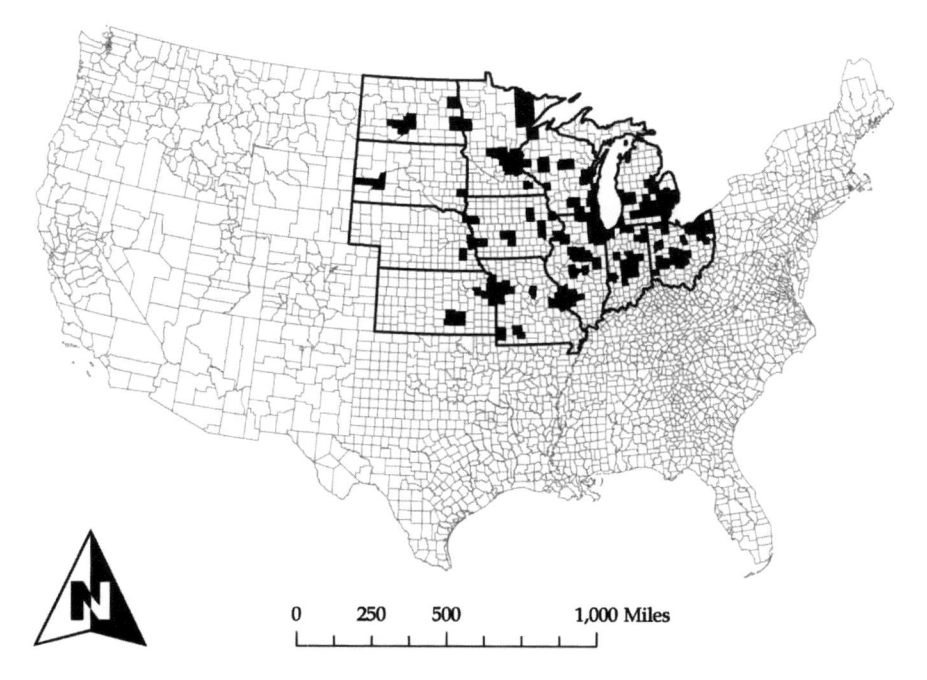

Fig. 9.1 Study areas: sections in 82 Midwest metropolitan areas

approximately 116,000 sections, and almost 80 % of the entire sections were not included in any incorporated places in 1990. Figure 9.1 shows the extent of the study area. Figure 9.2 provides a close view on the case of Kansas City metropolitan region, as an example.

Similar to previous research, the analytic framework is designed to estimate the effect of governance or institutional factors in $t-1$ (G_{t-1}) on the dependent variable of interest (Y) controlling for other potential determinants (X).[6]

$$Y = f(G_{t-1}, X) \tag{9.1}$$

In this study, the dependent variable (Y) is the development probability between year 1992 ($t-1$) and 2001 (t) in each section, which is defined as follows.

$$Y = \frac{Area\ of\ new\ development\ between\ 1992\ and\ 2001}{Undeveloped\ area\ in\ 1992} \tag{9.2}$$

One of the most important concerns here is how to effectively measure the institutional conditions for this micro-level analysis. In thinking about this issue, it

---

[6] The institutional factors in $t-1$ (G_{t-1}) is used to avoid the endogeneity problem (i.e., the influence of development pattern on G), although this may not be a perfect solution.

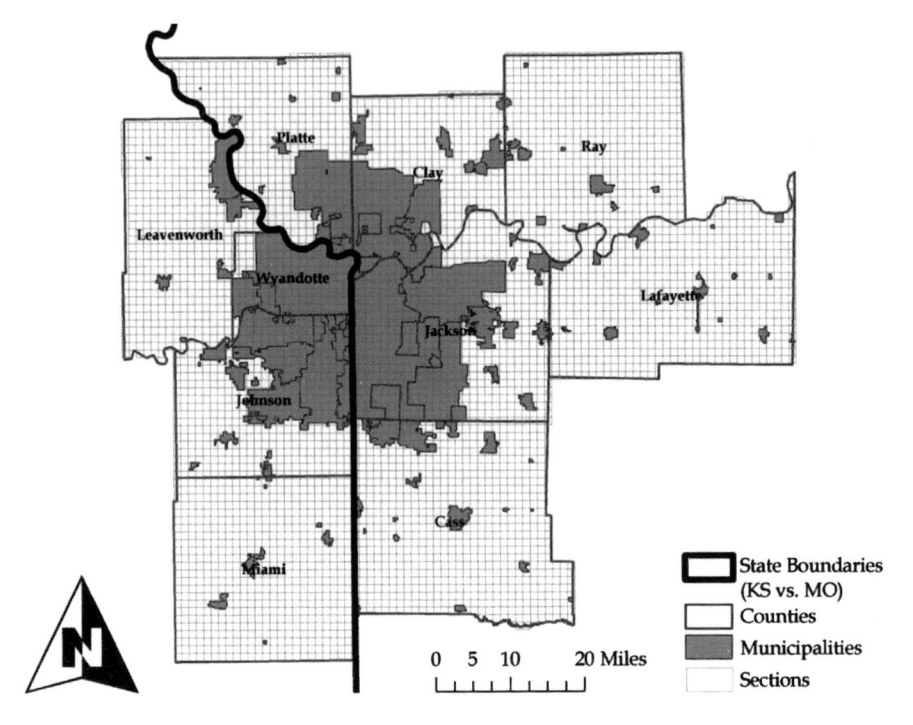

Fig. 9.2 Study areas – the case of Kansas City MSA

is essential to understand how land development can be affected by interjurisdictional competition. The first possibility is that land development at the micro scale is influenced by the overall institutional environment in the region as a whole. The localities within a highly fragmented region may be more likely to experience competition with others. In particular, as they maximize their tax bases while minimizing the amount of public service expenditures, the municipalities tend to compete over fiscally favorable types of development. For instance, to this end, suburban communities may employ fiscal zoning strategies (White 1975; Windsor 1979) and exclude multi-family housing unit development (Rolleston 1987). Such regulatory actions, arising from competition, can lead to a larger area of developed land consumption (i.e., lower density) and a more dispersed pattern of development.[7] This point is well recognized by previous studies that analyze the effect of political fragmentation on sprawl at regional or county level, such as Carruthers and Ulfarsson (2002) and Ulfarsson and Carruthers (2006). Following the approach used in these studies, the number of municipalities per thousand residents (i.e., *MSAFRAG*) is used as a component of G_{t-1} to capture this type of institutional influence.

---

[7] As synthesized in Kim (2011), exclusionary land use controls are found to induce a more sprawling pattern of development (see e.g., Shen 1996; Pendall 1999; Levine 2006).

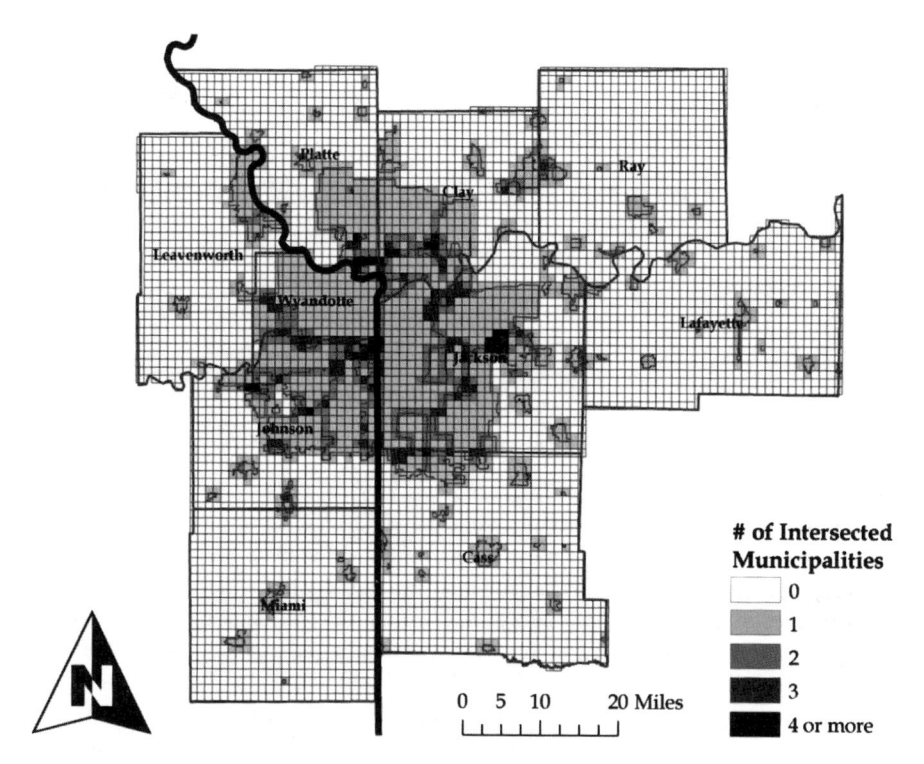

Fig. 9.3 Number of intersected municipalities – the case of Kansas City MSA

The other possibility is the effect of the competition among some municipalities at a particular location point within a region. A land area in the middle of two jurisdictions may be more likely to be developed not only because of its proximity to multiple localities but also due to the 'race to the bottom'. This potential effect is probably very critical in land development at a micro scale.

Spatially explicit competition can be quantified for each section in various ways. For instance, similar to the indicator for aggregated-level analysis, we can use the number of localities that share each section (*PLCOUNT*). By counting how many municipalities are involved in each section, the potential level of interjurisdictional competition at the micro level can be represented by this metric. However, the metric does not work properly, when a section is still surrounded by multiple localities, thus under a higher degree of potential competition, but it is not intersected with the jurisdictional boundaries, as shown in Fig. 9.3.

To deal with this issue, a complementary measurement is needed. This study employs the concept of entropy (Wilson 1970), which has been widely used in regional science and land use studies, particularly for the quantification of land use mix (see e.g., Cervero 1989; Frank and Pivo 1994; Krizek 2003; Frank et al. 2004; Brown et al. 2009). More specifically, considering the balance between the influences of three nearest places around the section, the entropy index for each section is calculated as follows.

$$Entropy = \sum_{j=1}^{3} \frac{-P_j \cdot \ln(P_j)}{\ln(3)} \qquad (9.3)$$

$$P_j = \frac{\left(\frac{1}{dist_j}\right)^2}{\sum_{j=1}^{3}\left(\frac{1}{dist_j}\right)^2} \qquad (9.4)$$

where $dist_j$ indicates the Euclidian distance between the section and j-th locality.

With the range of values between 0 and 1, by definition, the index can represent the different balance levels among three jurisdictions. A higher index value indicates a more balanced distribution of the powers.[8]

Although it may appropriately represent the power balance among nearby localities, the entropy index as it stands does not necessarily indicate a larger degree of potential competition over the section. In particular, a section that is far from urbanized areas would not be under competition, even if it is in the middle of multiple municipalities, and thus has a high entropy index (Fig. 9.4).

Therefore, a modified index was developed as follows:

$$EntropyD2 = Entropy \cdot \left(\frac{1}{1 + dist_1}\right)^2 \qquad (9.5)$$

where $dist_1$ is the Euclidian distance between the section and the nearest place.

Like the original form of the entropy index, by definition, the modified index has a value that is between 0 and 1. In addition, it will discern the sections by proximity to the localities as well as by power balance. Through this, the level of potential interjurisdictional competition can be better represented (Fig. 9.5).

Admittedly, many other factors need to be controlled in order to precisely estimate the effect of the governance factors. Among others, consideration needs to be given to the fact that the probability of development in a section will generally be higher, if the section is in a rapidly growing region. To control for such an effect of regional growth, an index that represents the expected probability of land development in individual MSA has been developed:

$$MSADPINDEX = \frac{(MSAPOP_{2001} - MSAPOP_{1992}) + \delta \cdot MSAPOP_{1992}}{MSAPD_{1992}} \cdot \frac{1}{MSAUDL_{1992}} \qquad (9.6)$$

---

[8] The index value 1 indicates that all three surrounding municipalities have the same distance to the section, whereas 0 means the section is included in a single place, so competition over the section may not occur.

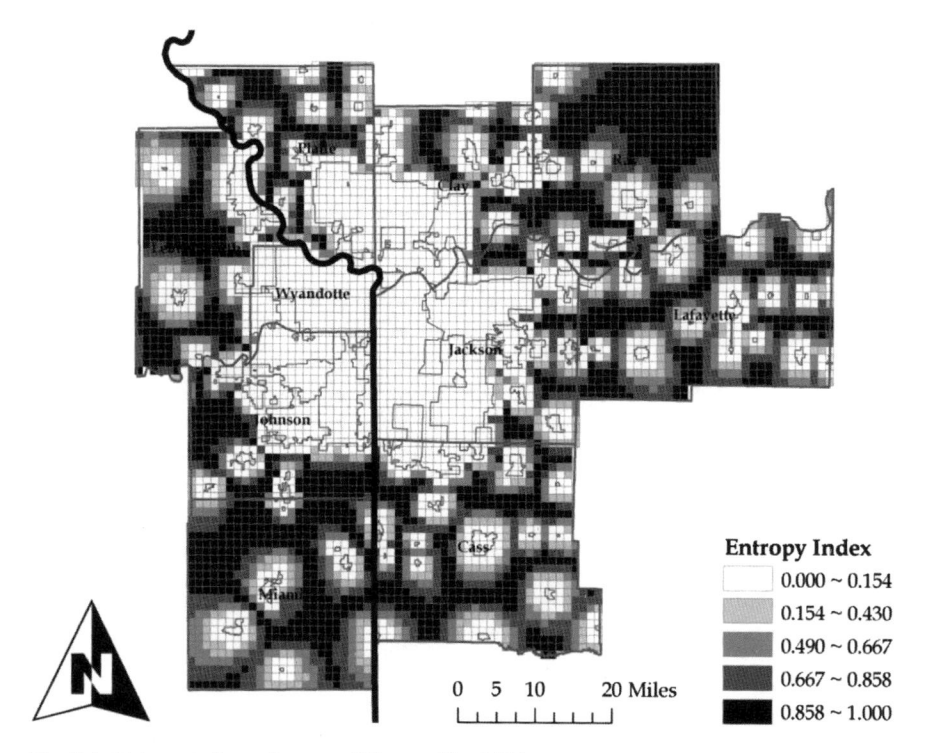

Fig. 9.4 Entropy index – the case of Kansas City MSA

where $MSAPOP_{1992}$ and $MSAPOP_{2001}$ represent the regional population in years 1992 and 2001, respectively; and $MSAPD_{1992}$ and $MSAUDL_{1992}$ indicate the region's population density and total area of undeveloped land in 1992. δ is a parameter introduced here to reflect the trend of density decline, as new development still occurs even if the MSA's population declines.[9]

Furthermore, some other region-wide determinants, such as the regional per capita income growth rate from 1992 to 2001 (*PCINCGR*) and population density in 1992 (*MSAPD92*), are included in the analysis as control variables. Additional explanatory variables are also used to consider the effects of ecological conditions and other characteristics of individual sections.

Table 9.1 summarizes all variables used in this analysis and their data sources. As presented in this table, a broad range of data sources are utilized in this empirical

---

[9] We use $\delta = 0.1$, which results in a positive value of the index (i.e., new development) even for the most rapidly declining MSA, between 1992 and 2001 in the Midwest.

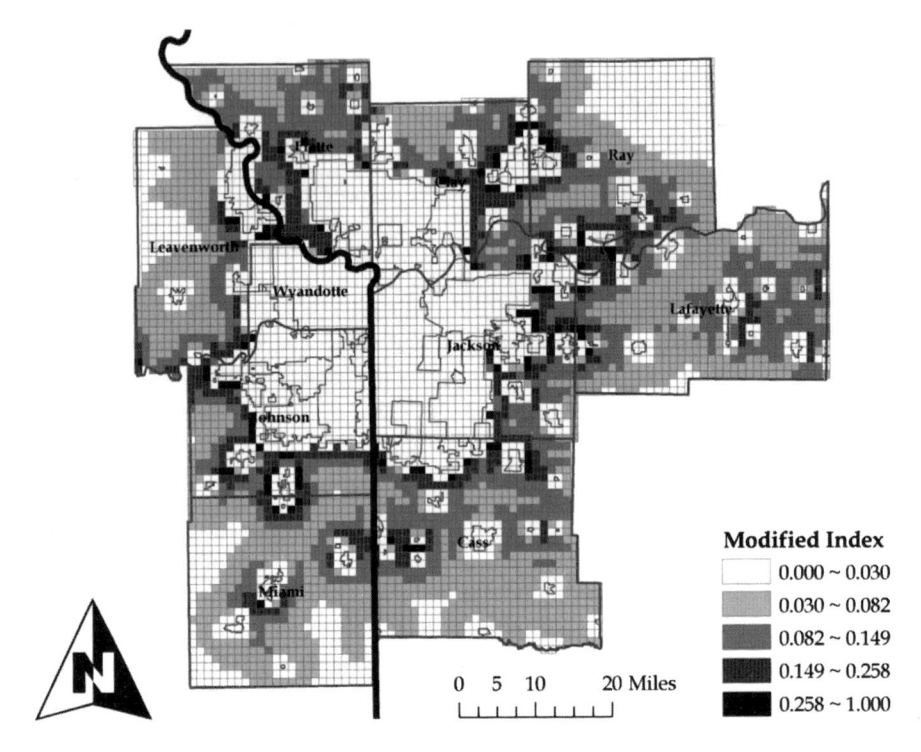

Fig. 9.5 Modified entropy index – the case of Kansas City MSA

study. For the dependent variable (i.e., Y: each section's development probability between 1992 and 2001), we employ the USGS NLCD 1992/2001 Retrofit Land Cover Change dataset, which identifies land use changes between 1992 and 2001 at the 30 m × 30 m cell level.[10] To compute the governance factors (i.e., G_{t-1}) for each section, Census 1990 as well as 1990 MSA and municipal boundary shapefiles are used.

The descriptive statistics of the variables are presented in Table 9.2. The sample size (i.e., the number of sections with valid data) totaled 113,738, after some sections with missing information were excluded.

---

[10] The two NLCD (National Land Cover Database) products (i.e., 1992 and 2001) use different classification schemes and methodologies, so that they could not be effectively used together for the research on land use changes. The NLCD 1992/2001 Retrofit Change Product addresses the compatibility issue between the two NLCD and identifies land use change between 1992 and 2001 (US Geological Survey 2008).

Table 9.1 Variables and data sources

Category	Variable	Description	Data sources
Dependent variable	Y	Development probability between 1992 and 2001	NLCD 1992/2001 retrofit change product
Governance factors (G_{t-1})	MSAFRAG	Regional fragmentation indicator (i.e., the number of municipalities per thousand residents in the MSA) in 1990	US Census 1990
	PLCOUNT	Number of municipalities intersected with the section's boundaries in 1990	Census boundary shapefiles
	ENTROPYD2	Modified entropy index in 1990	Census boundary shapefiles
Control variables (X)	MSADPINDEX	MSA's expected average of the development probability	REIS[a]; NLCD 1992/2001 retrofit change product
	PCINCGR	MSA's per capita income growth between 1992 and 2001	REIS
	MSAPD92	MSA's population density in 1992	REIS; NLCD 1992/2001 retrofit change product
	RC1	Dummy variable, indicating the presence of interstate-highway in the section	National highway planning network data
	RC2	Dummy variable, indicating the presence of other types of expressway in the section	National highway planning network data
	WATER	Dummy variable indicating the presence of water	NLCD 1992/2001 retrofit change product
	PERAGG	Percentage of agricultural and grasslands out of total undeveloped land in 1992	NLCD 1992/2001 retrofit change product
	PERDEV	Percentage of developed land out of total land surface in 1992	NLCD 1992/2001 Retrofit change product
	DISTCE	Distance between the section and the central city	Census boundary shapefiles

[a]REIS (Regional Economic Information System), US Bureau of Economic Analysis.

9.4 Estimation and Analysis Result

One of the most notable features of the empirical analysis in this study is that Y is a fractional variable (i.e., a proportion, having [0,1] range) in the model. Due to this special characteristic of the dependent variable, linear regression using ordinary least squares (OLS) method is not appropriate. Papke and Wooldridge (1996) provide a more robust estimation technique for this issue, a quasi-likelihood estimation method. With many respects, this method has advantages over traditional log-odds ratio conversion, as a treatment of the issue. Among others, it is more useful in handling the samples with $y = 0 \, or \, 1$. Such incidents (i.e., the

Table 9.2 Descriptive statistics

Category	Variable	Mean	Standard deviation	Min	Max
Dependent variable	Y	0.04165	0.15270	0.00	1.00
Governance factors (G_{t-1})	MSAFRAG	0.09219	0.05115	0.02	0.30
	PLCOUNT	0.32191	0.71548	0.00	11.00
	ENTROPYD2	0.05300	0.06473	0.00	0.83
Control variables (X)	MSADPINDEX	0.03555	0.04256	0.00	0.38
	PCINCGR	1.48725	0.05948	1.32	1.59
	MSAPD92	2.99492	1.15661	0.99	8.39
	RC1	0.05944	0.23645	0.00	1.00
	RC2	0.07157	0.25777	0.00	1.00
	WATER	0.71587	0.45100	0.00	1.00
	PERAGG	0.72284	0.30424	0.00	1.00
	PERDEV	0.15723	0.24179	0.00	0.99
	DISTCE	12.85138	13.67884	0.00	96.00

samples with 0 or 1 development probability) are numerous in this study, making the quasi-likelihood estimation method especially desirable here.[11]

Basically, Papke and Wooldridge's (1996, p.621) idea is to set the model in a nonlinear manner, as follows.

$$E(y_i|x_i) = H(x_i \cdot \beta) \qquad (9.7)$$

where $H(\bullet)$ is a function, satisfying $0 \leq H(z) \leq 1$ for all $z \in R$. For instance, $H(\bullet)$ can be a logistic function:

$$H(z) = \frac{\exp(z)}{1 + \exp(z)} \qquad (9.8)$$

As demonstrated by Papke and Wooldridge (1996), once the problem is framed in this way, the estimation can be accomplished by maximizing the following likelihood function.

$$Max \sum_{i=1}^{n} y_i \cdot \log\{H(x_i \cdot \beta)\} + (1 - y_i) \cdot \log\{1 - H(x_i \cdot \beta)\} \qquad (9.9)$$

Further, if the group size (m_i) is known, as it is in this analysis, the likelihood function can be conveniently modified to reflect the varying weights by sample.

---

[11] Approximately 65 % of the entire sections (73,879 out of 113,738) had attracted no new development between 1992 and 2001, thus have $y = 0$, whereas much less number of sections (about 1.2 %) show $y = 1$.

$$Max \sum_{i=1}^{n} m_i \cdot [y_i \cdot \log\{H(x_i \cdot \beta)\} + (1 - y_i) \cdot \log\{1 - H(x_i \cdot \beta)\}] \quad (9.10)$$

This weighted likelihood estimation is useful here, as some sections contain a large number of undeveloped land cells in 1992 (identified by the USGS NLCD 1992/2001 Retrofit Land Cover Change dataset), while some others have relatively fewer undeveloped cells. Once the model is estimated with consideration of such variation, the nature of land development in relation to institutional settings, rather than the section-based changes, can be revealed. This would be meaningful, since the section is an artificial unit of analysis meant to investigate the issue at a certain disaggregated level of geography.

Therefore, the model is estimated using three different methods – (1) Linear regression using OLS, (2) unweighted quasi-likelihood estimation method by solving Eq. 9.9 and (3) weighted quasi-likelihood estimation method by solving Eq. 9.10.[12] The outcomes are presented in Table 9.3. It needs to be noted that the estimated coefficients by linear regression and quasi-likelihood estimation are not directly comparable, as they are based upon completely different specifications. Furthermore, note that the R-squared for the quasi-likelihood estimation presented in the table is calculated from Eq. 9.11, for a comparison with linear regression result.

$$R^2 = 1 - \frac{\sum_{i=1}^{n} \left(y_i - H\left(x_i \cdot \hat{\beta}\right)\right)^2}{\sum_{i=1}^{n} (y_i - \bar{y})^2} \quad (9.11)$$

First of all, the explanatory power of the quasi-likelihood estimation is greater than that of linear regression. Although the weighted case shows a relatively poorer fit, in terms of the section level R-squared, compared with the unweighted likelihood estimation, it is still better than the linear regression result and meaningful in the sense that it reveals the effects of many of the considered factors on each land cell's development. For several control variables, the model estimation outcomes confirm significant effects with expected signs. For instance, the positive estimated coefficients are found for *MSADPINDEX*, which suggests that land development are more likely to occur as the regional population growth rate goes up. The positive coefficients of *PERAGG* also correspond to the expectation that development probability increases if the section has a large proportion of agricultural or grasslands, as opposed to forests and barren fields, out of the total undeveloped land area (i.e., the ecological conditions are more favorable to development). Furthermore, as anticipated, *PERDEV* shows a positive sign, indicating that development probability is higher in the section that has a larger area of existing development (i.e., agglomeration benefits).

---

[12] For the maximization of the likelihood function, BFGS (Broyden 1970 – Fletcher 1970 – Goldfarb 1970 – Shanno 1970), which is a quasi-Newton method, is employed in R.

Table 9.3 Analysis result (All sections with valid, complete data, n = 113,738)

Variable	Linear regression Estimated coefficient	Standard error	Quasi-likelihood estimation (unweighted) Estimated coefficient	Standard error	Quasi-likelihood estimation (weighted) Estimated coefficient	Standard error
C (Intercept)	−0.11930***	0.00820	−5.72691***	0.75334	−7.29477***	0.02549
MSADPINDEX	0.08869***	0.01026	2.26198***	0.56163	3.26104***	0.01825
PCINCGR	0.01938***	0.00558	−0.56565	0.51967	0.67702***	0.01746
MSAPD92	0.00355***	0.00045	−0.03346	0.03181	−0.02604***	0.00112
MSAFRAG	0.03850***	0.00830	−3.18747**	0.98472	−2.85348***	0.03141
RC1	−0.01132***	0.00139	0.13150	0.07800	0.10741***	0.00289
RC2	−0.02187***	0.00124	0.21346*	0.09199	0.30786***	0.00290
WATER	−0.02775***	0.00073	−0.44858***	0.06252	−0.09793***	0.00228
PERAGG	0.07200***	0.00114	2.16418***	0.10215	1.40407***	0.00408
PERDEV	0.45160***	0.00192	5.90655***	0.12453	6.03715***	0.00459
DISTCE	0.000094***	0.00003	−0.03740***	0.00600	−0.06129***	0.00019
PLCOUNT	0.00762***	0.00066	0.18870***	0.03359	0.25122***	0.00127
ENTROPYD2	−0.0227100***	0.0055650	0.5042437	0.5407871	0.0785906***	0.0162625
R-squared	0.5139		0.6956		0.6245	

*** 0.1 % level significant.
** 1 % level significant.
* 5 % level significant.

Although the outcomes are different in linear regression, according to the quasi-likelihood estimation, the positive coefficients are found for RC1 and RC2, which represent the positive effects of arterial road networks on land development. The negative sign of *DISTCE*'s coefficient suggests that new development is less likely to occur in areas that are very far from the central city of the MSA. Of interest are the estimated coefficients of the governance or institutional factors (G_{t-1}), such as *MSAFRAG*, *PLCOUNT*, and *ENTROPYD2*. From the quasi-likelihood estimation, negative estimated coefficients for *MSAFRAG* are found, while the variable exhibits a positive sign in the linear regression. This may suggest that region-wide political fragmentation is not a significant driver of micro-level land development, when other factors are controlled.[13] In contrast, *PLCOUNT*, and *ENTROPYD2* have positive effects on micro-level land development. All of the three estimation methods exhibit 0.1 %-level significant positive coefficients, particularly for *PLCOUNT*. This indicates that land areas shared by multiple municipalities tend to be developed more rapidly. This may be evidence that interjurisdictional competition does exist over small areas and that it has a significant effect on land development in the field.

The model is estimated again only with data for the unincorporated sections to explore the effect of interjurisdictional competition on land development in this specific context. The sample size is still large enough, as over 75 % of entire sections are out of any jurisdictional boundaries. *PLCOUNT* is not included in this estimation, since it is zero for the unincorporated sections. Table 9.4 presents the outcomes.

Again, the unweighted quasi-likelihood estimation shows the highest R-squared, much better than that of linear regression. The estimation outcomes are generally similar to those of the earlier case with an entire set of the sections. One notable finding here is that *ENTROPYD2* has a significant positive effect, which is consistently found in every model estimation with the unincorporated area samples. This suggests that unincorporated land areas are more likely to be developed if they are close to multiple jurisdictions; thus they can be considered as under a power balance that can induce interjurisdictional competition.

9.5 Summary and Discussion

This study attempts to quantify the institutional environments with emphasis on political fragmentation and resultant interjurisdictional competition, and then examine how the land development process is influenced by the institutional

---

[13] It needs to be noted that this finding may be attributable to the measurement of the fragmentation. In other words, the outcome can be different, if other metrics, such as Herfindahl index or other inequality indicators, are used.

Table 9.4 Analysis result (Unincorporated areas only. n = 87,674)

Variable	Linear regression Estimated coefficient	Standard error	Quasi-likelihood estimation (unweighted) Estimated coefficient	Standard error	Quasi-likelihood estimation (weighted) Estimated coefficient	Standard error
C (Intercept)	−0.01969***	0.00278	−6.95313***	1.94403	−6.54099***	0.04795
MSADPINDEX	0.00747	0.00498	6.23598**	1.99004	5.57988***	0.05020
PCINCGR	0.000941***	0.00187	0.24234	1.36633	−0.05401	0.03335
MSAPD92	−0.00223***	0.00017	−0.14224	0.11348	−0.12782***	0.00278
MSAFRAG	−0.01065***	0.00278	−3.79214	2.69186	−2.58958***	0.06094
RC1	−0.01060***	0.00059	0.15243	0.24889	0.20915***	0.00659
RC2	−0.00737***	0.00047	0.42068	0.23281	0.53782***	0.00556
WATER	−0.00262***	0.00024	−0.00760	0.19424	0.07780***	0.00475
PERAGG	0.00064	0.00039	0.73207*	0.31672	0.32170***	0.00834
PERDEV	0.23540***	0.00127	6.81645***	0.31509	7.50165***	0.010009
DISTCE	0.00017***	0.00001	−0.02446*	0.01213	−0.03019***	0.00030
ENTROPYD2	0.0215100***	0.0017970	2.6562184***	0.7384290	2.9571849***	0.0198648
R-squared	0.3165		0.4262		0.3019	

*** 0.1 % level significant.
** 1 % level significant.
* 5 % level significant.

environments. To this end, a micro-level analysis is conducted with the data for small land areas in 82 Midwest metropolitan statistical areas.

The outcomes of the empirical analysis using a quasi-likelihood estimation method suggest that institutional setting and interjurisdictional competition do matter. In particular, it is found that land areas that are shared or surrounded by multiple jurisdictions are more likely to be developed. This may imply that local municipalities often compete over a particular land resource to attain their fiscal or development goals and that such competition is a significant driver of micro-level land use conversion.

Although the present study does not explicitly test a certain policy action, it seems to provide some policy implications. Among others, given that land development is significantly influenced by interjurisdictional relations, states may want to ensure how to better manage development processes through appropriate and systematic coordination of local actions rather than leaving the 'race to the bottom' phenomenon that seems to characterize the present situation. This is particularly important, as land development is mostly irreversible and subsequent investment is path dependent. If the competition can result in an undesirable development pattern and can generate an inefficient spatial structure of the region, the benefits of the disaggregated governance and local autonomy will be offset by higher fiscal and environmental costs.

This study also suggests the need for further research on the subject. In the literature on land use or urban spatial structure, the influences of institutional conditions are often neglected, while attention has been paid to externalities, scale effects, and many other factors of development. This is also the case in urban system modeling. Careful consideration needs to be given to such critical institutional settings to better understand and describe the complexity and dynamics of spatial development processes.

References

Brown BB, Yamada I, Smith KR, Zick CD, Kowaleski-Jones L, Fan JX (2009) Mixed land use and walkability: variations in land use measures and relationships with BMI, overweight, and obesity. Health Place 15:1130–1141
Broyden CG (1970) The convergence of a class of double-rank minimization algorithms. IMA J Appl Math 6:76–90
Brueckner JK (2003) Strategic interaction among governments: an overview of empirical studies. Int Reg Sci Rev 26:175–188
Carruthers JI (2003) Growth at the Fringe: the influence of political fragmentation in United States metropolitan areas. Pap Reg Sci 82:475–499
Carruthers JI, Ulfarsson GF (2002) Fragmentation and sprawl: evidence from interregional analysis. Growth Change 33:312–340
Cervero R (1989) America's suburban centers: the land use – transportation link. Unwin Hyman, Boston
Dewar M, Epstein D (2007) Planning for "Megaregions" in the United States. J Plann Lit 22:108–124

Downs A (1994) New visions for metropolitan America. Lincoln Institute of Land Policy, Cambridge, MA

Ewing RH (1996) Characteristics, causes, and effects of sprawl: a literature review. Environ Urban Stud 21:1–15

Fletcher R (1970) A new approach to variable metric algorithms. Comput J 13:317–322

Foster KA (2001) Regionalism on purpose. Lincoln Institute of Land Policy, Cambridge, MA

Frank LD, Pivo G (1994) Impacts of mixed land use and density on utilization of three modes of travel: single-occupant vehicle, transit, and walking. Transport Res Rec 1466:44–52

Frank LD, Andresen MA, Schmid TL (2004) Obesity relationships with community design, physical activity, and time spent in cars. Am J Prev Med 27:87–96

Fulton W et al (2001) Who sprawls the most? How growth patterns differ across the U.S. Bookings Institution, Washington, DC

Glaeser EL, Kahn M, Chu C (2001) Job sprawl: employment location in U.S. Metropolitan areas. Bookings Institution, Washington, DC

Goldfarb D (1970) A family of variable metric updates derived by variational means. Math Comput 24:23–26

Gordon P, Richardson HW (1997) Are compact cities a desirable planning goal? J Am Plann Assoc 63:95–106

Hardin G (1968) The tragedy of the commons. Science 162:1243–1248

Howell-Moroney M (2008) The Tiebout hypothesis 50 years later: lessons and lingering challenges for metropolitan governance in the 21st century. Public Adm Rev 68:97–109

Innes JE, Booher DE, Di Vittorio S (2011) Strategies for megaregion governance. J Am Plann Assoc 77:55–67

Kim JH (2011) Linking land use planning and regulation to economic development: a literature review. J Plann Lit 26:35–47

Krizek KJ (2003) Operationalizing neighborhood accessibility for land use – travel behavior research and regional modeling. J Plann Educ Res 22:270–287

Kunce M, Shogren JF (2005) On interjurisdictional competition and environmental federalism. J Environ Econ Manag 50:212–224

Levine J (2006) Zoned out: regulation, markets, and choices in transportation and metropolitan land use. RFF Press, Washington, DC

Lewis P (1996) Shaping suburbia: how political institutions organize urban development. University of Pittsburgh Press, Pittsburgh

McKinney M, Essington K (2006) Learning to think and act like a region. Land Lines 18:8–13

Oates WE (1999) An essay on fiscal federalism. J Econ Lit 37:1120–1149

Papke LE, Wooldridge JM (1996) Econometric methods for fractional response variables with an application to 401 (k) plan participation rates. J Appl Econom 11:619–632

Pendall R (1999) Do land-use controls cause sprawl? Environ Plann B Plann Des 26:555–571

Razin E, Rosentraub M (2000) Are fragmentation and sprawl interlinked? North American evidence. Urban Aff Rev 35:821–836

Rolleston BS (1987) Determinants of restrictive suburban zoning: an empirical analysis. J Urban Econ 21:1–21

Ross S, Yinger J (1999) Sorting and voting: a review of the literature on urban public finance. In: Cheshire P, Mills ES (eds) Handbook of regional and urban economics, vol 3. North-Holland, New York, pp 2001–2060

Shanno DF (1970) Conditioning of quasi-Newton methods for function minimization. Math Comput 24:647–656

Shen Q (1996) Spatial impacts of locally enacted growth controls: the San Francisco Bay Region in the 1980s. Environ Plann B Plann Des 23:61–91

Tiebout CM (1956) A pure theory of local expenditures. J Polit Econ 64:416–424

Tsai YH (2005) Quantifying urban form: compactness versus 'sprawl'. Urban Stud 42:141–161

U.S. Geological Survey (2008) Completion of the National Land Cover Database (NLCD) 1992–2001 land cover change retrofit product. Open-File Report 2008–1379. U.S. Department of the Interior

Ulfarsson GF, Carruthers JI (2006) The cycle of fragmentation and sprawl: a conceptual framework and empirical model. Environ Plann B Plann Des 33:767–788

White MJ (1975) Fiscal zoning in fragmented metropolitan areas. In: Mills ES, Oates WE (eds) Fiscal zoning and land use controls: the economic issues. Lexington Books, Lexington, pp 31–100

Wildasin DE (2006) Fiscal competition. In: Weingast BR, Wittman DA (eds) Oxford handbook of political economy. Oxford University Press, Oxford, pp 502–520

Wilson AG (1970) Entropy in urban and regional modelling. Pion, London

Wilson JD, Wildasin DE (2004) Capital tax competition: bane or boon. J Public Econ 88:1065–1091

Windsor D (1979) Fiscal zoning in suburban communities. Lexington Books, Lexington

Yaffee SL (1997) Why environmental policy nightmares recur. Conserv Biol 11:328–337

Yaro RD (2011) America 2050: towards a twenty-first-century national infrastructure investment plan for the United States. In: Xu J, Yeh AGO (eds) Governance and planning of mega-city regions: an international comparative perspective. Routledge, New York, pp 127–147

Zellner ML et al (2009) The emergence of zoning games in exurban jurisdictions: informing collective action theory. Land Use Policy 26:356–367

Zhang J (2011) Interjurisdictional competition for FDI: the case of China's "development zone fever". Reg Sci Urban Econ 41:145–159

Part II
Micro-scale Approaches

Chapter 10
Occupation, Education and Social Inequalities: A Case Study Linking Survey Data Sources to an Urban Microsimulation Analysis

Paul Lambert and Mark Birkin

Abstract This chapter describes how a dynamic microsimulation model of the urban population, initiated from census data, can be productively linked with rich socio-economic survey data resources in order to give a more effective understanding of the development of occupations and educational qualifications through time and hence socio-economic inequalities. Small area modelling of these elements is a key feature of understanding labour markets and hence the distribution of employment. A selection of strategies for linking data resources are discussed, some analytical results presented, and discussion given on how these approaches can be facilitated by the 'NeISS' infrastructural provision.

In the urban simulation model, the city is represented as a complete but synthetic array of households and their constituent individuals. The model is capable of projections forward in time through the incorporation of demographic processes relating not only to the major events of fertility, mortality and migration, but also the formation and dissolution of households and changes in individual health status. The model is operational for every city or region in Great Britain. The chapter looks at how this model can be linked with survey data from the British Household Panel Survey, and the combined results used to explore the evolution of two major socio-economic variables – occupation and education – within the microsimulation model in a manner designed to engage with extended research traditions in the sociological study of social stratification.

P. Lambert
School of Applied Social Science, University of Stirling, Stirling, UK
e-mail: paul.lambert@stir.ac.uk

M. Birkin (✉)
School of Geography, University of Leeds, Leeds, UK

10.1 Introduction

10.1.1 Simulation Research Infrastructures

Traditionally, social simulation analyses have been developed in isolated environments using highly bespoke programming and development, but new infrastructural provisions in the field have the potential to support more widely accessible and replicable social simulation analyses which may also be more challenging and at a more comprehensive scale. Below, we explore how an extended infrastructure for handling data files and simulation models, the 'National e-Infrastructure for Social Simulation' (cf. Birkin et al. 2010), can help to develop a more comprehensive picture of the evolution of social inequalities through time through a case study of microsimulation. An existing urban microsimulation model is expanded in terms of its exploitation of measures of crucial socio-economic variables concerned with occupations and educational qualifications.

Whilst it is recognised that socio-economic measures are impactful for simulation model results – and these are often thoroughly evaluated in simulation applications (e.g. Zaidi et al. 2009) – in urban simulation models there has typically been less capacity for incorporating relatively complex measurements of socio-economic circumstances. This may arise because the constraints of modelling areal boundaries and geographical mobility set upper limits on the number of transition probabilities which might reasonably be calculated (e.g. Birkin et al. 2009), and in addition because the large scale data resources, such as census records, on which profiles are initiated, are conventionally restricted in the range of socio-economic details they measure (e.g. CCSR 2005). Accordingly, simulation models in general, and urban simulation models in particular, have often relied upon relatively crude measurement instruments to cover socio-economic circumstances such as in measures of occupations or of educational qualifications.

Elsewhere in social statistics, analysts routinely benefit from a wide array of detailed measured variables covering individuals' socio-economic and socio-demographic circumstances and, typically, numerous other detailed features of contemporary lifestyles. For instance, a large and complex social survey such as the British Household Panel Survey (University of Essex 2010), is able to sustain detailed longitudinal analysis of socio-economic differences and their relationships to numerous factors relevant to exploring social inequalities, such as social origins, attitudes, health and lifestyles (e.g. Jenkins 2011; Rose 2000). In this chapter we explore how an urban microsimulation initiated on UK census data can be enhanced through linkage with data from the BHPS, and a simulation based upon the BHPS, in order to explore simulated patterns of occupations, educational qualifications and social inequalities through time.

10.1.2 Variable Operationalisations in Social Stratification Research: Occupations

The domain of interest concerns the measurement and analysis of 'social stratification', a term commonly used in sociology to refer to analysis of the enduring structures of social inequalities and their reproduction through time (e.g. Platt 2011; Bottero 2005). Social stratification analyses feature an array of possible indicators of social circumstances (e.g. Shaw et al. 2007), but most typically make use of data about occupational circumstances, because occupational data is seen as easy to collect, relatively stable, and of particularly strong empirical importance to the social inequality structure (see esp. Treiman 1977). Occupational data is usually exploited by collecting detailed occupational descriptions then using those to construct occupation-based social classifications which attempt to indicate individuals' relative positions within the social structure. However, many different occupation-based classifications are available, and there is much debate about the qualities and contributions of alternative measures (cf. Rose and Harrison 2010; Chan 2010). Indeed, given such controversies, a good strategy when working with occupational data is to derive several alternative measures and undertake some degree of sensitivity analysis – the GEODE project (www.geode.stir.ac.uk) is an online resource which tries to bring together data on occupations to facilitate the operationalisation and comparison of different occupation-based schemes (see Lambert and Bihagen 2012).

Table 10.1 illustrates typical features of occupational data available in survey analysis, describing the relatively detailed occupational coding that is initially collected (based in this case on the 'Standard Occupational Classification' for the UK, see ONS 2000) and a small selection of occupation-based social classifications. We see great variety in the form of the occupational measures available to us, which arguably has two significant impacts. First, the empirical properties of the measures, in terms of their functional form and relationship with other factors, can change (see Table 10.1). Second, the theoretical explanations which may be linked to each measure may be different. We can highlight for instance recent interest in analysing detailed and disaggregated occupational units following an increasingly influential argument that fine-grained differences in occupations affect the mechanisms through which people experience social inequalities (e.g. Jonsson et al. 2009). The latter perspective is of particular interest to simulation analyses since most simulation studies have hitherto used highly aggregate measures of occupational positions and might therefore miss important differences within the social structure.

However, forward simulation through time of occupational positions cannot be seen as a straightforward endeavour. For one thing, it is well-established that measures of occupations are related to and influenced by other important measurable differences such as in age, gender, educational level, ethnicity, and in measured income or wealth (see also Table 10.1). Social stratification research usually proceeds by characterising the structure of social inequality through

Table 10.1 Occupational data from the BHPS

Occupational data	#Categories	Form	Correlation with Age	Correlation with Gender
Standard occupational classification 2000 (SOC2000), 3-digit unit groups (see ONS 2000)	392	Categorical	0.40	0.67
SOC2000 3-digit minor groups (ONS 2000)	126	Categorical	0.33	0.61
Goldthorpe class (Erikson and Goldthorpe 1992)	11	Categorical	0.17	0.40
Goldthorpe class 2-class version (I + II v's all others)	2	Categorical	0.07	ns
NS-SEC analytical version (Rose and Pevalin 2003)	8	Categorical	0.16	0.32
Registrar general social class (Szreter 1984)	7	Categorical	0.14	0.35
Four-class occupational ranking (used in Sect 10.4.2 below) (derived by recoding Goldthorpe scheme)	4	Categorical	0.13	0.30
HG scale (Goldthorpe and Hope 1974)	n/a	Metric	0.09	ns
Cambridge scale (Prandy 1990)	n/a	Metric	0.06	0.14

Source: BHPS, all adults in work sampled in 2008 excluding Northern Ireland, N ~ = 7,000. Correlations shown if statistically significant at 95 % threshold ('ns' = 'not significant'). Correlations are Pearson's R or square root of r2 from bivariate regression models predicting age (linear) and gender (dummy)

occupation-based measures whilst also describing how other social differences relate to the occupational structure, and accordingly it can be presumed that a simulation style analysis ought to incorporate a complex multivariate account in order to deal adequately with occupational measures. Second, major institutional changes through time have significant impacts upon the distribution of occupations, for instance in recent decades in the UK there have been major, often regionalised, shifts of 'de-industrialisation' and public sector growth which have led to changes in the profile of occupational positions, however measured, through time (e.g. Oesch 2006).

Lastly, the complex categorical form of occupations is particularly difficult to apply statistical predictive approaches to. Whilst some occupation-based social classifications have the appearance of hierarchical indicators of relative position, the longitudinal analysis or simulation of occupational positions is challenging because there are strong temporal and institutional dependencies involving occupations which means that individuals do not move seamlessly 'up' and 'down' the occupational structure over time. Indeed, on the contrary, many individuals stay in the same occupations through their entire lives (Blossfeld and Hofmeister 2005), often in the same, or if not in similar, occupations to those of their parents (Jonsson et al. 2009); and when individuals do change occupations it is very often within a restricted range of related career trajectories (e.g. Stewart et al. 1980). Moreover, locality effects are also significant influences of occupational positions since the available pool of occupations, defined by the local industrial distribution, shapes the options available to individuals. A simulation analysis of occupations therefore ought ideally to build strong temporal dependencies (and other institutional and locality considerations) into its approach. Figure 10.1a, b

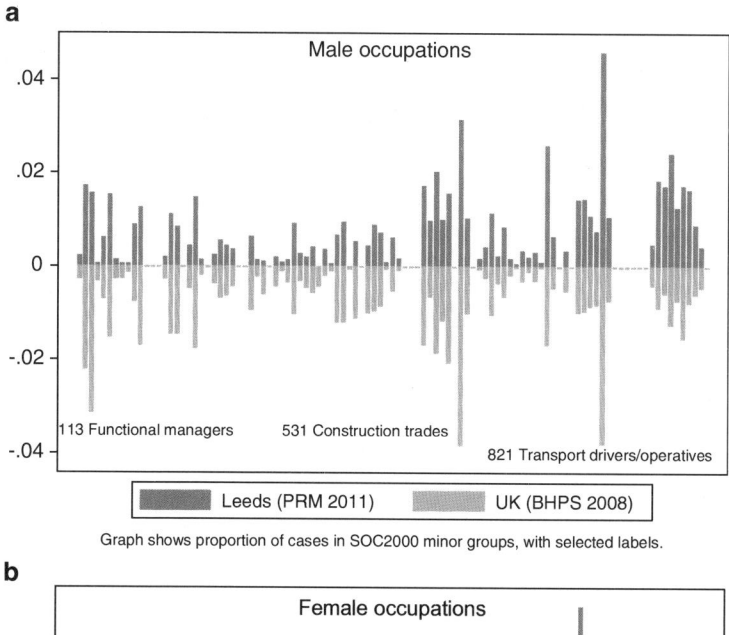

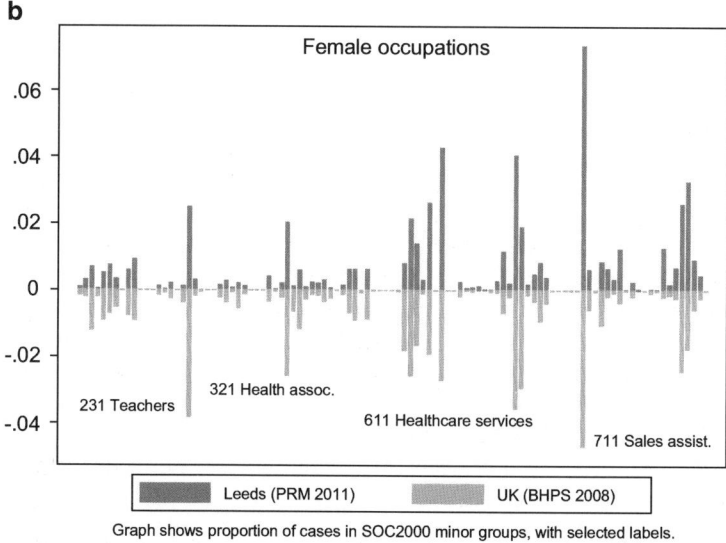

Fig. 10.1 (**a**) Detailed occupational distribution for males (**b**): Detailed occupational distribution for females

depict occupational distinctions amongst the adult population of a UK city (Leeds, as estimated through the 'PRM' as described below) and amongst the UK as a whole (as estimated from the BHPS). They show how the occupational profile of Leeds is quite typical of the country as a whole, although there are many other urban environments where this profile is much more distinctive. The figures also serve to reiterate the idea that if each particular minor group might be thought,

Table 10.2 Educational data from the BHPS

Educational qualifications data	#Categories	Form	Correlation with.. Age	Correlation with.. Gender
BHPS educational qualifications (University of Essex 2010)	12	Categorical	0.51	0.21
ISCED (Schneider 2011)	8	Categorical	0.49	0.08
BHPS 4-category educational code (see GEEDE, www.dames.org.uk/geede)	4	Categorical	0.33	0.06
Graduate indicator (dummy variable)	2	Categorical	0.10	0.02
Low qualifications indicator (dummy variable)	2	Categorical	0.32	0.04
Scaling of BHPS educational codes by Cambridge scale for employed	n/a	Metric	0.20	ns

Source: BHPS, all adults sampled in 2008 excluding Northern Ireland, N ~ = 14,000. Correlations shown if statistically significant at 95 % threshold ('ns' = 'not significant'). Correlations are Pearson's R or square root of r2 from bivariate regression models predicting age (linear) and gender (dummy)

theoretically, to be a distinctive long-term social position, then a longitudinal analysis ought ordinarily to assume stability within occupations over time at quite a precise level of detail.

10.1.3 Variable Operationalisations in Social Stratification Research: Educational Qualifications

Alongside occupational information, social stratification research also makes central the use of socio-economic measures concerned with educational qualifications. Educational systems are portrayed as key social institutions which define an individual's long-term social experience both in socio-economic terms and in other respects (e.g. Bills 2004). Table 10.2 shows information on educational qualifications measures on a typical UK dataset in a style similar to Table 10.1. As with occupational data, educational measures are usually recorded in the first instance in a more detailed categorical format, then recoded into simpler qualifications-based schemes of educational levels (the GEEDE service, www.dames.org.uk/geede, in a similar way to the GEODE service, seeks to collate information resources on ways of coding and exploiting information on educational qualifications). On the one hand, educational profiles of individuals are highly stable through time (at least after age 25) so forward microsimulation can often reasonably be reduced to the assumption of stability (e.g. Zaidi et al. 2009). Nevertheless educational data remain difficult to include in wider simulations through time, due to their complex categorical form, their very strong age–cohort dependencies (arising from educational expansion and transformations in educational provision through time – e.g. Muller and Wolbers 2003), and the high level of correlation between educational outcomes and other measurable factors (including high within-household correlations – e.g. Smits (2003)).

10.2 A Microsimulation Model of Small Area Populations

Microsimulation models (MSM) have been popular with economists for a number of years as a means for understanding the distributional consequences of financial policies as well as individual economic attitudes and behaviours. MSM rest on the idea that grouped populations are inadequate for this purpose, and that the representation of individual household units and their constituent individuals is more powerful and more flexible. For example, if it is possible to estimate the composition of a household (number of adults, number of children), the economic activity, occupation and income of individual adult members, and key behaviours, such as consumption patterns and vehicle ownership, then the impact of alternative tax-benefit regimes can be calculated from the bottom up. One of the most important aspects of distributional inequalities are inter-area variations, and for this reason spatial MSM have also been of enduring interest for geographers and regional scientists. Good examples are provided by Ballas et al. (2005) and Smith et al. (2009), while an excellent summary of the benefits of the MSM approach is presented by van Imhoff and Post (1998).

In view of the apparent attraction of synthetic micro-data about small area populations, Birkin et al. (2009) have outlined a Population Reconstruction Model (PRM) which creates spatially disaggregate individual and household data within urban areas. With the addition of dynamic simulation tools (Wu et al. 2011) it is also possible to generate realistic projections of changing demographic patterns over time.

Whilst the PRM provides persuasive simulated data on socio-demographic and socio-economic profiles through time to a high level of geographical detail, it does not naturally engage with the measures and approaches of social stratification research. This is unfortunate since the ongoing development of social inequality structures is of great interest. Accordingly, as part of work for the ongoing 'National e-Infrastructure for Social Simulation' project, an application area has sought to investigate the enhancement of the PRM outputs with measures and approaches from studying social inequalities. Research questions of interest include how a combination of ageing and widening educational participation may affect social inequalities; what is the impact of cohort-specific immigrant influxes on the distribution of skills in a region; and how do local concentrations of employment and fine-grained industrial sector transformations impact different age cohorts and their social stratification outcomes. The potential of individual-based microsimulation frameworks to model the accumulation of wealth through the combination of work, benefits, unearned income and inheritance is also evident.

10.3 BHPS Simulation of Occupational and Educational Measures

How might detailed sociological variables be added to high-quality simulation models such as the Population Reconstruction Model, and what would be the impact of doing so? As an initial exercise, we began with a detailed socio-economic dataset and tried to build a forward simulation model from it in order to assess the impact of occupation-based and education-based measures on simulation model results. Accordingly, using data from the nationally representative British Household Panel Survey (University of Essex 2010), we constructed a simple longitudinal predictive simulation model and made arrangements for it to be supported in the NeISS infrastructure in a way which would allow researchers to compare across different occupation-based and education-based measures (see also Sect. 10.5 below). However, our simulation model was of an unsophisticated variety and we are convinced that a more complex approach is required, particularly for analysis involving areal disaggregations.

The BHPS is a panel survey with repeated measures on individuals for up to 18 different yearly panels (there are around 4,000 adults who have been interviewed every year since the study began in 1991, in addition to around 28,000 other adults who have been included in the BHPS at some stage in its history but who do not have such regular records linked to them). Our analysis involved estimating a number of regression models to characterise the relationships involving socio-economic and socio-demographic measures from recorded contacts. Predicted values were then calculated using those regression model parameters for future points in time, and these were taken as the simulated values for the next year, and forward in time simulations were then continued on the basis of previous observed data and predicted values. Regression models were prepared in a relatively generic manner through which alternative measurement representations of key variables, and alternative starting sample structures, could be specified and the models re-run. Additionally, to accommodate the typical interest in simulation analysis of testing hypothetical simulations under different starting conditions, a number of model terms and data structures were used which were designed to represent major changes in structural relations such as could be influenced by social policies – for example, major educational expansion was represented by increasing qualifications levels for the younger birth cohorts only, and different profiles of population ageing were depicted through changing sample weights for different age groups.

At first sight, results suggest major influences associated with the strategy for measuring occupational and educational qualifications. Across a variety of applications, we frequently found a substantial gap between the results of different analyses depending upon the measure used; moreover, this gap was often bigger than the differences between different simulation scenarios themselves. For instance, Fig. 10.2 shows simulated trajectories under different measurement schemes for educational qualifications for the outcome of a 'social mobility' profile of the entire BHPS sample (i.e. the size of the correlation between the occupational advantage score for respondents and their parents). The differences between the full and dashed lines represents the difference between a population with and without additional educational expansion, whereas the difference between different shadings of lines, which is often more substantial, shows the impact of measuring educational attainment in one of four different ways (as in rows 2–5 of Table 10.2). The key point is that the lines from different measures are substantially different, indicating that operationalisation strategy can be enormously important.

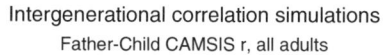

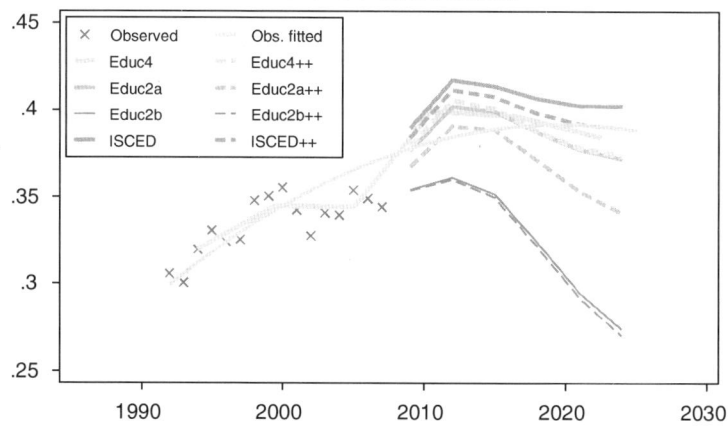

Fig. 10.2 Results from the BHPS simulation model: Substantial effects of educational coding

Unfortunately our BHPS-based simulation analysis itself must be considered as rather limited in its current form. Our own analysis of robustness of the model results, summarized in Fig. 10.3, highlights the fact that the simulated results seem not to match observed results in an overlapping period, and that simulated results are often dramatically distinct from observed results. The major problem with the current results is likely to lie with the homogenising nature of the calculation of simulations from predicted values of regression models – this will tend to push all cases towards average patterns and rapidly eliminate heterogeneity over time. More generally, there are numerous complications to the simulation scenario that might reasonably be added were we to take the model further (cf. Zaidi et al. 2009). Nevertheless, we suspect that the contribution of indicating the likely impact of social classification and measurement options will prove to be robust.

10.4 Imputations of Complex Measures to Existing Simulated Datasets

A second approach to using more detailed socio-economic measures in simulation analysis is to merge the detailed information from another source onto what could be considered a more reliable simulation result, such as the PRM outputs from a particular locality and point in time. This strategy is effectively one of imputation, since the distribution of measured variables on simulated datasets will be used in order to approximate likely socio-economic measures from more detailed data.

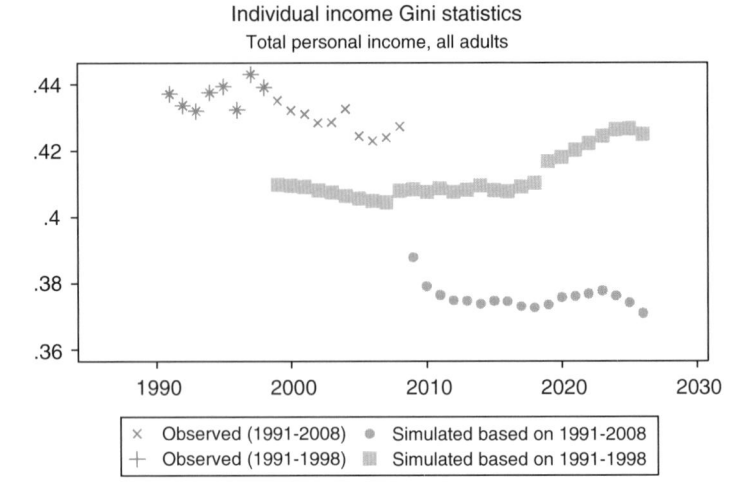

Individual income Gini statistics
Total personal income, all adults

| × Observed (1991-2008) | • Simulated based on 1991-2008 |
| + Observed (1991-1998) | ▪ Simulated based on 1991-1998 |

Source: Simulations on the BHPS balanced panel subsample (weighted by xrwght).

Fig. 10.3 Results from the BHPS simulation model: regression model probably serves to reduce heterogeneity within sample

We have undertaken such an analysis linking the results of the Population Reconstruction Model with plausible detailed measures imputed from the BHPS.

10.4.1 A 'Nearest Case' Methodology

Imputation of values from one dataset to another can be supported through many different approaches but we focussed upon finding exact matches on key measured variables, and an assessment of 'nearness' of cases based on combined profiles as revealed through regression analysis. In summary, a series of measures that were collected on both the PRM and the BHPS were identified and coded into a standardised format (measures of gender, age, occupation, household size, ethnicity, economic activity, car ownership and subjective health). Three outcome factors measured on the BHPS were then identified: own occupational CAMSIS scale score; educational level; and cigarette smoking. These three were selected to indicate factors which relate to socio-economic inequality but in each case with some different emphasis. A multivariate outcome regression model was then carried out to summarise how the explanatory variables predict the outcomes for the BHPS cases (the multivariate outcomes approach advocated by Goldstein 2003, c6, was applied). Our initial models are parameterised at the individual level only, but could be adapted to include dependence on data from household sharers within the BHPS. Subsequently, for all cases (from both the BHPS and PRM) it was then possible to calculate predicted values for these regressions for each case, then those predicted values could be used as a measure of 'nearness' of cases between the

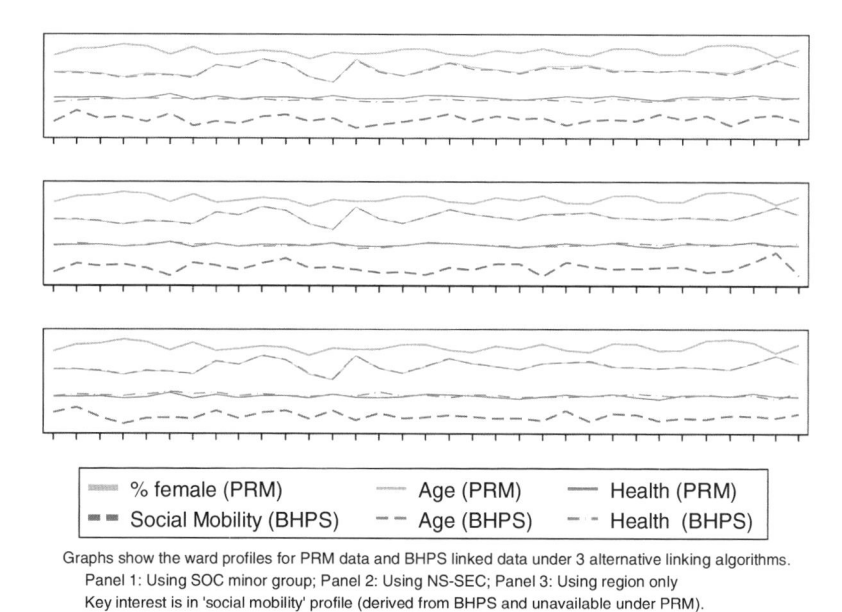

10 Occupation, Education and Social Inequalities: A Case Study Linking Survey...

| ▬▬▬ % female (PRM) | ——— Age (PRM) | ——— Health (PRM) |
| ▬ ▬ Social Mobility (BHPS) | – – Age (BHPS) | – – Health (BHPS) |

Graphs show the ward profiles for PRM data and BHPS linked data under 3 alternative linking algorithms.
Panel 1: Using SOC minor group; Panel 2: Using NS-SEC; Panel 3: Using region only
Key interest is in 'social mobility' profile (derived from BHPS and unavailable under PRM).

Fig. 10.4 PRM 2011 data fused with BHPS records – indicative results

samples. The allocation of imputed values was implemented by sorting the dataset with combined PRM and BHPS data, prioritising any key variables (gender and occupational position were used), then, within those categories, sorting by predicted values of the regressions. At this point, PRM cases could be matched with the nearest available BHPS record. From this, it was possible to attach imputed data from the BHPS respondents to appropriately near PRM cases, and conduct analyses upon those PRM cases with supplemented records from the BHPS.

This approach leaves open many options in the use of different socio-economic measurement instruments. Analysis in the initial calculation of regression models, and in later calculation of summary results from imputed data, can be undertaken using a variety of different approaches. We were particularly interested in the way in which the socio-economic measures are used in the imputation algorithm, results from an evaluation of which are summarised in Fig. 10.4. At issue here is which criteria are prioritised in identifying cases to match. In each of the three panels the first criteria was gender, then a second categorical criteria was used, then lastly 'nearness' as defined through the regression model predicted values was cited. The three panels of Fig. 10.4 show the analytical outcomes that emerge from applying three different second criteria (detailed occupations in panel 1; more aggregate occupational groups in panel 2; regional similarity only in panel 3): the figure shows estimated statistics based upon the PRM data with imputations, across a series of different wards in Leeds, and shows a number of instances whereby different estimates emerge from using different approaches to imputation.

10.4.2 Scenario-Based Models of a Local Labour Market

In this section, we consider how a simulation modelling approach combined with detailed survey data imputations might be deployed in order to assess the impacts of changing labour market conditions in a geographical region. Attention is confined to the metropolitan area of Leeds with a focus on intergenerational social mobility and spatial inequalities. The analysis comprises three elements:

1. In relation to a model baseline, we would like to address the following specific questions:

    (a) How does the composition of the labour market vary between small areas?
    (b) Is there a spatial pattern to social mobility?
    (c) How substantial are the spatial variations in social mobility?

2. With respect to a 'business-as-usual' scenario, we would like to know:

    (a) How will the labour market change over time?
    (b) Will these changes lead to increased social mobility?
    (c) Will there be an increase or a reduction in the level of spatial inequality between areas?

3. In the context of alternative economic development scenarios, then our indicative questions are of the following nature:

    (a) Which local areas are the biggest winners and losers under each scenario?
    (b) Which scenario supports the greatest level of social mobility in the region?
    (c) Which scenarios have a positive or a negative effect on spatial inequalities in the local labour market?

The baseline will be taken as the current year 2011 and the projection window will be 10 years 2011–2021. Three scenarios are considered over the decade of the projection:

1. Rebound: renewed economic growth in the Leeds area is powered by increased employment in the core domains of professional and financial services, which have both been strong drivers of prosperity in recent years;
2. Slump: there is a continuation of the current downturn in the economic fortunes of the city resulting in higher unemployment and a stagnation in productivity levels across all sectors
3. Infrastructure investment: in line with current government policy, major investments in construction, transport and communications are used as a tool to stimulate the local economy.

To enact the microsimulation, an initial population was created with individual attributes (age, gender, marital status, health status, ethnicity, headship) and household characteristics (household size, housing type, tenure and location) matching the 2001 census profiles of each census output area. These populations are then 'updated' to 2011 and projected on to 2021 using the dynamic simulation techniques described in Sect. 10.2 above. Occupation codes are imputed to each individual in the synthetic population using the methods introduced in Sect. 4.1. In this case occupation was imputed based upon age, gender, marital status and headship, and these probabilities are also scaled by individual census wards to reproduce the features of the local labour market (Leeds can be characterised as having a deprived urban core with high unemployment and disadvantaged occupations, enveloped by a much more affluent periphery). Similarly, parental occupations are imputed as a basis for the assessment of social mobility using predictor variables of age, gender and occupation. This approach adopts a much simpler data structure than the regression techniques described in Sect. 4.1 (in which household size, ethnicity, economic activity, car ownership and health were also introduced) but this more streamlined approach sidesteps the averaging problems which can occur with highly disaggregate models, as we noted at the end of Sect. 10.3.

Microsimulation will be adopted as the basis of the analytical approach, but even then – in the spirit of our earlier discussion – there are many decisions to be made in the model design and implementation. In reconstructing the baseline population, methods have varied from synthetic regeneration and deterministic reweighting to a variety of mathematical optimisation models including simulated annealing and genetic algorithms (see Harland et al. (2012) for a review). The incorporation of new data from social surveys can be attained by reweighting the profiles of panel data such as the BHPS to match small area characteristics, either existing or projected (as in the SimBritain model, for example – Ballas et al. (2005)), by regression modelling (as discussed above), or through contingency tables (e.g. Birkin and Clarke 1988) or other data merging procedures. Projections of the future population may be based around comparative statistics, in which the existing population is rescaled to aggregate estimates of population change, or using a truer model of individual and household dynamics based on either events (e.g. Gampe et al 2007) or transitions (Birkin et al. 2009). Finally, the projection of trends in our target variables is in itself a non-trivial matter, as indicated by the variations in trends and extrapolations of Figs. 10.2 and 10.3 above. In the simulations outlined below, we use a synthetic regeneration process to construct the baseline population, and add extra information from the BHPS using a contingency table analysis of the interrelationship between characteristics. Demographics are projected into the future using a model of transitions between states, and patterns of occupation and social mobility amongst demographic groups are assumed to be consistent at their current levels.

In the baseline model, six individual characteristics (age, gender, marital status, health status, occupation and ethnicity) and four household characteristics (household size, housing type, tenure and location) are estimated synthetically for the population of Leeds in 2001. The procedure starts by creating an 'empty' household for each identified unit in the 2001 census, which are then populated with individuals to match observed distributions (for example, a lot of single person households in the city centre, families in suburban areas, and bigger groups in student or multi-ethnic areas). The age of the head of household (or 'household representative person') and other family members is profiled, and other attributes are added in sequence using a combination of small area data and anonymised individual records (cf. Birkin et al. 2009).

Individuals and households are aged dynamically from 2001 to 2011 using transition estimates derived from analysis of the BHPS alongside other sources of population data such as the International Passenger Survey and ONS Vital Statistics. This combination of sources allows for the simulation of processes such as migration, household formation and dissolution, as well as fertility and mortality (Wu et al. 2011). Intergenerational social mobility can be assessed by adding a further characteristic from the BHPS questions on the occupations of parents when respondents were aged 14 (the more advantaged of the father's and mother's occupation is used in the analysis below, as relevant). In many sociological domains, the degree of 'upward' mobility (the extent to which the second generation improves upon their position compared to the first) is of greatest interest, so we construct summary indicators of upward mobility based upon comparing the occupations of respondents and their parents.

Table 10.3 shows the working population of Leeds broken down in to four occupation groups – professional and managerial; service occupations; skilled workers; and unskilled workers.[1] The occupation of the parent is broken down into four groups in the same way. This table illustrates that the distribution of occupations between generations is far from stationary (for example, only 29 % of skilled workers are the children of skilled parents) and also that although the dominant trend in social mobility is upwards, there is significant movement in both directions (for example 16,407 professional workers are the upwardly mobile children of unskilled parents, but also 8,544 unskilled workers are the children of professional parents). From this matrix, we construct two indicators of social mobility as follows:

Net upward mobility: the number of people working in higher ranked occupations than their parents less the number of people working in lower ranked occupations than their parents

Scaled upward mobility: the number of people working in higher ranked occupations than their parents divided by the number of people working in lower ranked occupations than their parents

For these purposes, the ranking of occupations is as implied in Table 10.3, with professional as the highest and unskilled as the lowest. Thus the measures of social mobility for Leeds are a net upward mobility of 12,365 and scaled upward mobility of 1.106. This compares to national scaled upward mobility of 1.090 as revealed by the BHPS and again reflects the fact that Leeds has an occupational structure which is quite representative of the country as a whole.

---

[1] These are a convenient classification which were calculated using the Goldthorpe classes as follows: Professional and Managerial – Service class higher; Small Business with Employees; Small business without employees; Service occupations – service class lower, routine non-manual, personal service, farmers, foremen; skilled workers – skilled manual workers; unskilled workers – semi-skilled and non-manual workers, agricultural workers. See also Table 10.1 above.

10 Occupation, Education and Social Inequalities: A Case Study Linking Survey... 217

Table 10.3 Net upward mobility for Leeds in the base year 2001

Occupation	Parental occupation				
	Professional	Service	Skilled manual	Unskilled	Total
Professional	27,226	20,045	7,581	16,407	71,259
	(38 %)	(28 %)	(11 %)	(23 %)	(100 %)
Service	37,467	50,118	22,204	36,294	146,083
	(26 %)	(34 %)	(15 %)	(25 %)	(100 %)
Skilled Manual	18,536	22,993	28,568	26,152	96,249
	(19 %)	(24 %)	(29 %)	(27 %)	(100 %)
Unskilled	8,544	16,123	12,655	30,073	67,395
	(13 %)	(24 %)	(19 %)	(44 %)	(100 %)
Total	91,773	109,279	71,008	10,892	380,986
	(24 %)	(29 %)	(19 %)	(29 %)	(100 %)

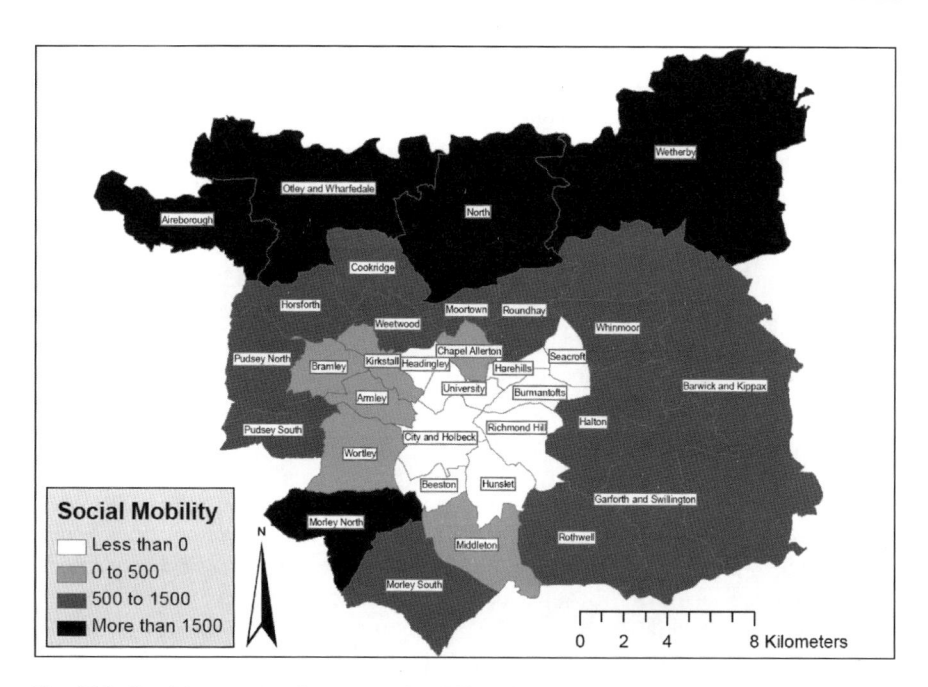

Fig. 10.5 Spatial summary of net upward mobility, simulation of Leeds for 2011

The spatial analysis of these data reveals some interesting patterns (see Fig. 10.5). Although net upward mobility is positive in the city as a whole, many central areas such as the deprived central and eastern neighbourhoods of Burmantofts, Harehills, Seacroft and Richmond Hill actually show negative net upward mobility. There are also negative levels in the wards of Headingley and University, but this could be an artefact of the occupation-based codes being less accurate indicators of social position for many of the students that reside in this area.

Upward social mobility by contrast is most strongly indicated in the suburban and rural communities such as Wetherby, Otley, Aireborough and Morley (see Fig. 10.5). This pattern is intuitively plausible, since these outer areas are also typically seen as the most desirable – in terms of the quality of their schools, housing, physical environment, social cohesion and quality of services – and would therefore naturally be seen as attractive to individuals and families in advantaged social positions. Indeed, these measures of upward social mobility are strongly related to occupational advantage – net upward social mobility correlates with using the proportion of the workforce in professional and service occupations with an r-squared of 0.85; while for scaled upward mobility the correlation yields an r-squared of 0.95.

Turning now to a series of 'what if?' analyses, we begin with a 'business as usual' (BAU) scenario in which the population of the city is projected dynamically for 10 years. In line with government estimates which form the basis for national planning, overall demographic growth of 10.7 % is envisaged (ONS 2010). In the spirit of the BAU concept, we assume that new jobs are made available to match the existing employment profile. In the first 'rebound' scenario we allow the number of new jobs created to be doubled over the BAU and furthermore all of these new jobs are in professional and service occupations. In the second scenario, 'stagnation', however, there is no overall change in the number of jobs provided, and demographic increases can only be accommodated through rising unemployment and reduced labour force participation rates. Finally, Infrastructure Investment requires that sectors such as transport, construction and health care are all prioritised. Here we assume that additional jobs are provided as in BAU but that their focus is in service occupations and manual employment (i.e. social groups 2 and 3).

Some results are presented in Table 10.4. Since the BAU scenario by definition involves little change over the baseline, then patterns of mobility can be expected to be relatively stable. Some increase in aggregate levels of net upward mobility can be observed here simply because of rising population and employment levels. In the rebound scenario, upward mobility increases as the expanding number of opportunities in high level occupations facilitates renewed upward movements. However once again, the benefits of this process are revealed unevenly across space. Here the 'Spatial Mobility Index' is simply the standard deviation in scaled upward mobility across the 33 wards of Leeds and so measures the spatial imbalance in upward mobility. In this scenario it shows a distinct positive tilt, reflecting higher levels in the prosperous suburbs and lower levels in the deprived inner city neighbourhoods. In the stagnation scenario, opportunities for social improvement are distinctly limited, and overall patterns of upward mobility are forced into reverse. Now we see that in most areas it is normal for downward movements to prevail, and that only the most prosperous of the suburbs and commuter towns can buck this trend. Although the scenario is completely different, the resulting spatial inequalities are qualitatively the same as for rebound – in fact, there is an even

Table 10.4 Upward mobility patterns in Leeds under various forward scenarios

		Scenario				
		Baseline	BAU	Rebound	Stagnate	Invest
Population		799,885	890,939	890,939	890,939	890,939
Total employment		380,986	418,440	455,894	418,440	418,440
Emp – 1		71,259	77,664	115,118	71,259	71,259
Emp – 2		146,083	160,551	160,551	146,083	168,657
Emp – 3		96,249	106,548	106,548	96,249	111,129
Emp – 4		67,395	73,677	73,677	104,849	67,395
Net upward mobility		12,365	13,292	36,465	−8,424	13,248
Scaled upward mobility		1.106	1.104	1.285	0.939	1.103
Spatial mobility index		0.263	0.261	0.339	0.242	0.246
Positive spatial outcomes	Wetherby	2,280	2,488	3,818	1,829	2,367
(net upward mobility)	Otley	1,844	2,009	3,250	1,333	1,909
	North	1,623	1,775	2,889	1,037	1,712
Negative spatial outcomes	Seacroft	−629	−691	−364	−1,289	−638
(net upward mobility)	Harehills	−625	−689	−260	−1,519	−600
	Richmond Hill	−540	−596	−272	−1,189	−538

greater increase in spatial inequality in this instance. The third and most interventionist of these scenarios seems to come closest to maintaining the status quo. There is a very slight reduction here in upward mobility versus BAU, but also a slight reduction of spatial inequalities as measured by the spatial mobility index.

10.5 Conclusions

The work which we have reported here has been undertaken with support from projects under the banner of 'e-Research' or 'e-Social Science' (ESRC) and 'Information Environments' (JISC). These programmes seek to explore ideas of data intensive research (e.g. Bell et al. 2009) which to some degree is a value-adding agenda – what kinds of services do we need to provide and combine together in order to extract the value from vast reservoirs of data which are increasingly abundant, but not necessarily of the very highest quality?

Our examples have illustrated the value of conducting multiple simulation models on comparable data using slightly different measurements, and of combining data sources. The NeISS project's e-research infrastructure has been designed to support these investigations. Firstly, it uses a virtual organisation model for data storage. This model has been proposed by computer scientists and offer a solution to the increasingly difficultly of collect all the relevant data resources together at a single point, such as a desk top machine or even a high performance computational device. In the virtual organisation, different data sets are maintained in a form in which they can be accessed and reconfigured from a variety of physical sources.

Leaving aside for the moment specific modelling and analysis tools, the social simulation research infrastructures which we are supported in NeISS should also be capable of providing visualisations of trends in space in time; and they should be capable of archiving, regeneration and perturbation so that particular numerical experiments can be queried, refined or extended. The power of the infrastructure comes from two basic components – a service and a workflow. The service is a means of representing any of the individual building blocks in such a way that it can be called by or connected to any other service/ building block. This means it is then possible to connect together individual services into more complex workflows. Combining the services in different ways allows an enormous variety of applications to be generated from a core of data resources and analysis tools.

In this work, we have begun to link micro-simulation models with a set of services which are specifically concerned with individual progression through the education system and particularly through the labour market. These models are written in software code (Stata is used here) which involves a specification of programme arguments which define the scope of the analysis (e.g. measurement approach, sample coverage and simulation condition), and can then be carried forward in a largely automated way, generating outputs subsequent to their application on suitable datasets. Such generic features are attractive since they provide us with options for sensitivity analysis which can be undertaken through an infrastructural resource which supports data and the execution of the software commands. In this case, therefore, the NeISS infrastructure, supports secure storage of data files, routines for running command files, and front end mechanisms for collecting parameters for command files and submitting jobs. The combination of microsimulation with sociological investigations of education and occupational mobility provides both an application of social simulation to a substantive problem and an exemplar of the way in which e-infrastructures have the potential to link together all kinds of socio-economic data and modelling approaches.

A major attraction of simulation infrastructure is the possibility of enhancing large scale but sparse datasets with richer information from other sources. Whilst other socio-economic analyses have used detailed survey resources such as the BHPS within a microsimulation at a national level (cf. Zaidi et al. 2009), in the context of wider social structural and demographic changes through time, the relatively small BHPS sample is not well suited to forward projections about the national level population on the basis of analysis of its own records. Moreover since the BHPS does not represent a complete or bounded population, crucial processes such as family formation and dissolution are not realistically modelled since wider structural constraints cannot be known (for example, whilst we could predict a transition probability for the dissolution of an existing household, we have no sensible information to use to impute the formation of a new household involving drawing in a new respondent). On the other hand, urban simulations focus upon within-areal change which is itself responsive to the wider areal profile they are much better placed to achieve an accurate simulation of population evolutions involving individuals within households through time. Forward projections can be shown to provide feasible parameters using a variety of model specifications (e.g.

Birkin et al. 2010). For data such as involved in an urban simulation, however, it may seem likely that a substantial level of imputation may be required to undertake an appropriate analysis of social stratification. In the examples above we have shown ways that resources from the two traditions might be productively linked.

Indeed, it is clearly desirable for the analysis of social stratification to incorporate measures of occupations and educational qualifications within simulation models, but there are many complex options and strategies to achieve this. Simulation analysis of occupations and educational qualifications is also substantively appealing since it is readily possible to imagine policy led changes in either distribution through time (e.g. expansion or contraction of educational provision; investment in certain occupational sectors). Additionally, one particularly promising avenue of research in simulation analysis involves undertaking sensitivity analysis of the properties of different occupation-based schemes and alternative educational qualifications taxonomies. The resources required to do so (multiple runs of similar simulations on modified data, and complex information on alternative measures) have not hitherto been easy to access, but this paper has shown examples of how the NeISS infrastructure facilitates such activities.

Acknowledgement This research has been supported by the JISC funded project 'National e-Infrastructure for Social Simulation' (NeISS, www.neiss.org.uk) and as part of the GeoSpatial Analysis node (TALISMAN) of the ESRC National Centre for Research Methods (www.ncrm.ac.uk).

References

Ballas D, Clarke G, Dorling D, Eyre H, Thomas B, Rossiter D (2005) SimBritain: A spatial microsimulation approach to population dynamics. Popul Space Place 11(1):13–34

Bell G, Hey T, Szalay A (2009) Beyond the data deluge. Science 323:1297–1298

Bills DB (2004) The sociology of education and work. Blackwell, London

Birkin M, Clarke M (1988) SYNTHESIS: a SYNTHetic Spatial Information System for urban modelling and spatial planning. Environ Plann A 20:1645–1671

Birkin M, Wu B, Rees P (2009) Moses: dynamic spatial microsimulation with demographic interactions. In: Zaidi A, Harding A, Williamson P (eds) New frontiers in microsimulation modelling. Ashgate, Farnham

Birkin M, Procter R, Allan R, Bechhofer S, Buchan I, Goble C, Hudson-Smith A, Lambert PS, de Roure D, Sinnott RO (2010) The elements of a computational infrastructure for social simulation. Philos Trans R Soc A 368(1925):3797–3812

Blossfeld HP, Hofmeister H (2005) GLOBALIFE – life courses in the globalization process: final report, Faculty of Social and Economic Science, Otto Friedrich University of Bamberg, Bamberg, and http://oldsite.soziologie-blossfeld.de/globalife/. Accessed 31 Aug 2012

Bottero W (2005) Stratification: social division and inequality. Routledge, London

CCSR (2005) The 2001 household CAMS codebook, Version 1.2. Cathie Marsh Centre for Census and Survey Research, University of Manchester, Manchester

Chan TW (2010) The social status scale: its construction and properties. In: Chan TW (ed) Social status and cultural consumption. Cambridge University Press, Cambridge

Erikson R, Goldthorpe JH (1992) The constant flux: a study of class mobility in industrial societies. Clarendon, Oxford

Gampe J, Zinn S, Willekens F, van den Gaag N (2007) Population forecasting via microsimulation: the software design of the MicMac project. Eurostat: methodologies and working papers; theme: population and social conditions. Office for Official Publications of the European Communities, Luxembourg, pp 229–233

Goldstein H (2003) Multilevel statistical models, 3rd edn. Arnold, London

Goldthorpe JH, Hope K (1974) The social grading of occupations. Clarendon, Oxford

Harland K, Heppenstall A, Smith D, Birkin M (2012) Creating realistic synthetic populations at varying spatial scales: a comparative critique of population synthesis techniques. J Artif Soc Social Simulation 15(1):1

Jenkins SP (2011) Changing fortunes: income mobility and poverty dynamics in Britain. Oxford University Press, Oxford

Jonsson JO, Grusky DB, Di Carlo M, Pollak R, Brinton MC (2009) Microclass mobility: social reproduction in four countries. Am J Sociol 114(4):977–1036

Lambert PS, Bihagen E (2012) Stratification research and occupation-based classifications. In: Connelly R, Lambert PS, Blackburn RM, Gayle V (eds) Social stratification: trends and processes. Ashgate, Aldershot forthcoming. ISBN 9781409430964

Muller W, Wolbers MHJ (2003) Educational attainment in the European Union: recent trends in qualification patterns. In: Muller W, Gangl M (eds) Transitions from education to work in Europe. Oxford University Press, Oxford

Oesch D (2006) Redrawing the class map: stratification and institutions in Britain, German, Sweden and Switzerland. Palgrave, Basingstoke

ONS (2000) Standard occupational classification 2000, vol. 1: Structure and description of unit groups. Office for National Statistics, London

ONS (2010) Sub-national population projections – 2008-based projections. Office for National Statistics, London

Platt L (2011) Understanding inequalities: stratification and difference. Polity, Cambridge

Prandy K (1990) The revised Cambridge scale of occupations. Social - J Brit Social Assoc 24(4):629–655

Rose D (ed) (2000) Researching social and economic change: the uses of household panel studies. Routledge, London

Rose D, Harrison E (eds) (2010) Social class in Europe: an introduction to the European socio-economic classification. Routledge, London

Rose D, Pevalin DJ (eds) (2003) A researcher's guide to the national statistics socio-economic classification. Sage, London

Shaw M, Galobardes B, Lawlor DA, Lynch J, Wheeler B, Davey Smith G (2007) The handbook of inequality and socioeconomic position: concepts and measures. Policy Press, Bristol

Smits J (2003) Social closure among the higher educated: trends in educational homogamy in 55 countries. Soc Sci Res 32(2):251–277

Smith DM, Clarke GP, Harland K (2009) Improving the synthetic data generation process in spatial microsimulation models. Environ Plann A 41(5):1251–1268

Stewart A, Prandy K, Blackburn RM (1980) Social stratification and occupations. Macmillan, London

Szreter SRS (1984) The genesis of the registrar-general: social classification of occupations. Brit J Sociol 35(4):522–546

Treiman DJ (1977) Occupational prestige in comparative perspective. Academic, New York

University of Essex and Institute for Social and Economic Research (2010) British household panel survey: waves 1–18, 1991–2009 [computer file], 7th edn. UK Data Archive [distributor], Colchester, SN: 5151, July 2010

van Imhoff E, Post W (1998) Microsimulation methods for population projection. Popul Engl Sel 10:97–138

Wu B, Birkin M, Rees P (2011) A dynamic microsimulation model with agent elements for spatial demographic forecasting. Social Sci Comput Review 29(1):145–160

Zaidi A, Evandrou M, Falkingham J, Johnson P, Scott A (2009) Employment transitions and earnings dynamics in the SAGE model. In: Zaidi A, Harding A, Williamson P (eds) New frontiers in microsimulation modelling. Ashgate, Vienna

Chapter 11
Firm Location Choice Versus Job Location Choice in Microscopic Simulation Models

Rolf Moeckel

Abstract Traditionally, land-use models simulate employment at the aggregate level. More recently, some models microsimulate single employees. Such models allow capturing the interaction between location decisions of different employees, such as a new agglomeration of office employment may attract additional restaurant or office supply employment. Within the last decade, a few models were developed that simulate location choice of firms rather than single employees. These models acknowledge that location decisions are not made at the level of individual employees, but rather the firm as a whole is the entity that relocates. From an academic point of view, the progress in simulating location choice of economic activity has been remarkable. However, stochastic variability between different model runs has questioned the validity of such models, at least at a detailed geographic scale. This chapter provides a brief history of employment modelling and describes the ILUMASS approach, which microsimulates firms, in more detail. Following, a synthetic city is described and two models are implemented and contrasted: One that simulates businesses and one that simulates employees as the decision-making unit. Recommendations are given for a reasonable level of disaggregation.

11.1 Introduction

An analysis of businesses in the United States suggests that a region with stable employment gains 2 % of its employment due to new establishments of firms and loses 2 % due to closures in an average year. Eight percent of employment is lost and another 8 % of employment is gained because some firms hire and other firms lay off employment (Birch 1984). This demography of firms, also called firmography, results in a sizeable number of firms looking for a new location every year.

R. Moeckel (✉)
Parsons Brinckerhoff, 6100 NE Uptown Blvd, Suite 700, Albuquerque, NM 87110, USA
e-mail: rolf.moeckel@udo.edu

There are three ways to simulate such location search in simulation models. The traditional approach is aggregate and simulates employment by zones. The second option is to microsimulate employment by treating single employees individually. The principal advantage of such approaches is the ability to models explicitly the interaction between employees, particularly agglomeration effects. The third option microsimulates firms instead of single jobs. This approach accepts that the decision-making unit for employment relocation is not a single employee but rather the firm as a whole. Furthermore, firmography can be simulated explicitly. It is very common that a firmographic change, such as growth or decline, triggers business relocations. Thus, it appears behaviourally richer to simulate firms instead of jobs. Results, however, are presented by number of jobs only, because microsimulation models are stochastic in nature and may only be analyzed in the aggregate.

Microsimulating businesses may be a benefit and a burden. While it sounds intuitively right to model firms instead of individual jobs, modelling firms may be computationally more challenging. Stochastic variation between different model runs is another concern if differences are larger than an acceptable standard deviation. If stochastic variations in a single scenario are greater than the change simulated in two different scenarios, the model becomes unusable for such analysis.

11.2 A Short History of Employment Simulation

There is a distinguished history of modelling employment. Figure 11.1 shows employment simulation models in chronological order. The graphic distinguishes aggregate approaches from disaggregate models that microsimulate employment. The majority of models are spatially explicit, i.e. the location of firms within the study area is defined either by zones, raster cells or parcels. The few models that are not spatially explicit simulate employment of an entire region without allocating them to zones (Shubik 1960; Forrester 1969, p. 133ff; Bergmann 1974; Eliasson 1978 [MOSES Model]; Kay 1979; van Tongeren 1995). Bold names show employment simulations that are integrated with models simulating households, and model names that are underlined simulate firms instead of jobs.

The model developed by Lowry (1964) attracted much attention as one of the very first spatial employment simulations. It was developed as a gravity model and first applied to Pittsburgh, Pennsylvania. The Lowry modelling concept has been applied in many other models, and until today derivatives of this model are developed if a simple employment model is needed. As this section focuses on microsimulation, models on the left half of Fig. 11.1 are not described with further detail. Documentation of the first aggregate models may be found in Hill (1965), Putnam (1967) for INIMP, and Goldner et al. (1972) for PLUM, which later was further developed by Putnam to the frequently applied DRAM/EMPAL land-use model. Marcial Echenique & Partners developed the MEPLAN model (Echenique et al. 1990), which has been applied to more than 25 regions worldwide (Hunt et al. 2005, p. 332). The three models TRANUS (de la Barra 1989; de la Barra and

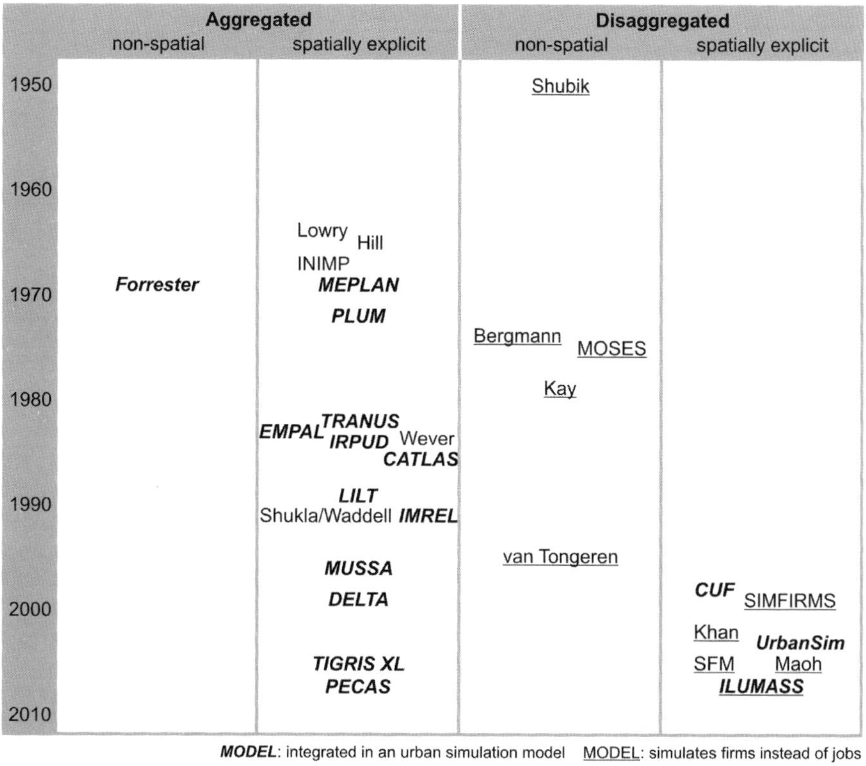

11 Firm Location Choice Versus Job Location Choice in Microscopic ... 225

	Aggregated		Disaggregated	
	non-spatial	spatially explicit	non-spatial	spatially explicit
1950			Shubik	
1960				
1970	Forrester	Lowry Hill INIMP MEPLAN PLUM		
1980			Bergmann MOSES Kay	
1990		EMPAL TRANUS IRPUD Wever CATLAS LILT Shukla/Waddell **IMREL**		
2000		MUSSA DELTA	van Tongeren	CUF SIMFIRMS Khan UrbanSim SFM Maoh
2010		TIGRIS XL PECAS		ILUMASS

MODEL: integrated in an urban simulation model   MODEL: simulates firms instead of jobs

Fig. 11.1 History of employment modelling

Rickaby 1982), MUSSA (Martínez 1996) and PECAS (Hunt and Abraham 2003) follow the MEPLAN methodology, with each further developing certain aspects of employment allocation. Recent additions to the PECAS model allow microsimulating the supply of floorspace (Hunt and Abraham 2009, pp. 52–109).

The IRPUD model has been developed by Wegener (1982) and is described in Chap. 2 of this book. This was one of the very first employment location choice model that uses logit models based on Domencich's and McFadden's work (1975). Further aggregate models shown in Fig. 11.1 include Wever (1983), CATLAS (Anas and Duann 1985), LILT (Mackett 1990), Shukla and Waddell (1991), and IMREL (Anderstig and Mattsson 1991, 1992). The DELTA model developed by Simmonds (2001, 1999) is described in more detail in Chap. 3 of this book, and TIGRIS XL (Zondag et al. 2005) is described in Chap. 4.

In the last decade, several approaches modelling businesses or jobs by microsimulation have been developed. Based on the understanding that microsimulation is better suited to simulate location behaviour and, in particular, the interactions of different agents, the number of highly disaggregate approaches is growing continuously.

The first spatially explicit microsimulation of jobs was developed for the California Urban Futures (CUF) model (Landis and Zhang 1998a, b). All potential uses, such as single-family, apartments, commercial, or industrial uses, bid for developable land. The forecasted job growth is allocated to raster cells of 100 by 100 m based on the highest bid. Logit models are used to select the sites.

The first complete microsimulation model for modelling business relocation and firmography instead of jobs was developed by van Wissen (2000, 1999). The model SIMFIRMS simulates businesses in the Netherlands from 1991 to 1998. Every firm is described by business type, age, size and location. Within the demographic cohort model age, size and location of firms may change. Location is defined by 40 regions (European Commission classification NUTS 2) in the Netherlands. To simulate business relocation, both push (drawbacks at current location) and pull (advantages at alternative sites) factors are considered. The decision to move or not is simulated by a binomial logit model mostly depending on the growth rate and the age of the firm. If a business decides to move, the location is chosen by a multinomial logit model, taking into account the distance from the current location and the vacant carrying capacity for this industry at the alternative locations. Though location choice of SIMFIRMS is not as sophisticated as in later models, van Wissen's work is a milestone in microscopic business simulation.

Khan et al. (2002) developed a microsimulation model for small and medium-sized businesses in a synthetic study area of ten by ten raster cells. Discrete choice models are used to simulate location decisions determined by the price of floorspace. Businesses buy and sell commodities from other business location depending on good's prices. The success of a business in reaching lower prices to buy and higher prices to sell determines the probability of growing, shrinking, going out of businesses, leaving the study area or splitting off. Business relocation is simulated by nested logit models. The access to input, such as raw materials, intermediate goods or services, and to customers is used to calculate the location utility.

Waddell and Ulfarsson (2003) developed another job microsimulation model that is embedded in UrbanSim (Waddell 2002, 2011). At the time of development, the level of spatial detail was unprecedented for employment simulation with raster cells of 150 by 150 m. Today, UrbanSim may be applied at the parcel level. Location choice is simulated by multinomial logit models aiming at utility maximization. To select a location, a uniform distribution is used to randomly sample a set of nine alternatives additionally to the site with the highest utility. The final location is selected out of these 10 alternatives by Monte Carlo sampling. Within UrbanSim the simulation of firms instead of jobs has been tested (Waddell et al. 2003, p. 52). While the results were similar, the simulation of jobs turned out to be computationally less complex than firms.

Maoh and Kanaroglou (2009) developed a model simulating small and medium-sized businesses with less than 200 employees for Hamilton, Canada. Decisions to stay, move within the study area, or out-migrate are simulated by multinomial logit models aiming at maximizing the utility. A second multinomial logit model selects a new location within a raster system of 200 by 200 m.

The Spatial Firm demographic Microsimulation (SFM) was developed by De Bok (2009) and is one of the few models that simulate firms instead of jobs. Special attention was given to calibrating the model with longitudinal micro data. Closeness to train stations and highways, accessibility of labour, customers, or suppliers are used as location factors. The model further includes urban economic location factors for specialization and diversity in location choice. Businesses that decide to move are offered 20 sites and a selection is made by a competing destinations model. For a new business, a location is selected randomly. The firmography model simulates firm growth, firm dissolution, and start-ups.

11.3 The ILUMASS Approach

At the Institute of Spatial Planning at the University of Dortmund, a microscopic integrated land-use transport model was developed. In cooperation with institutes at the German Universities of Aachen, Bamberg, Cologne and Wuppertal under coordination of the German Aerospace Centre, the project ILUMASS (Integrated Land-Use Transportation Simulation System) developed a model to microsimulate land-use, travel demand, traffic assignment and the environmental impacts (Wagner and Wegener 2007; Strauch et al. 2005). The model was implemented for the urban region of Dortmund in Germany with a population of 2.6 million. The ILUMASS model was the first approach that embedded the simulation of firms into a microscopic land-use/transport model.

One module of ILUMASS simulates firms, which is described in more detail in Moeckel (2009, 2007). A synthetic population of 978,029 jobs that are allocated to 80,291 firms was created based on publicly available business census data. 44 business types were distinguished, and the number of full-time employees, part-time employees, telework employees and open positions was defined for each firm. The micro location is described by 100 by 100 m raster cells. The four floorspace types farmland, industrial, retail and office floorspace were created to house businesses. As data on floorspace are very scarce, floorspace necessary for the base year employment was generated based on average floorspace-per-employee data, and a fixed rate of vacant floorspace was added systemwide. The floorspace rent was estimated by interpolating between 3,800 observed land-value prices that were geo-referenced for the study area.

Business relocation is simulated by logit models. Figure 11.2 shows a flowchart of the business relocation model. First, a binomial logit model simulates whether a firm decides to relocate or not based on the utility of the current location. If the current site is too small and the firm can expand into neighbouring parcels, this option is preferred to relocating the firm. Furthermore, some firms may establish a branch, keeping most of their employment at the current site and finding a new site for a subset of its employment.

Alternative sites that are analysed by the firm are selected with two multinomial logit models. The first one selects a neighbourhood and second one selects a vacant site within that neighbourhood. Location factors that are evaluated differently by each business type include, among others, size of the site, price, accessibilities of

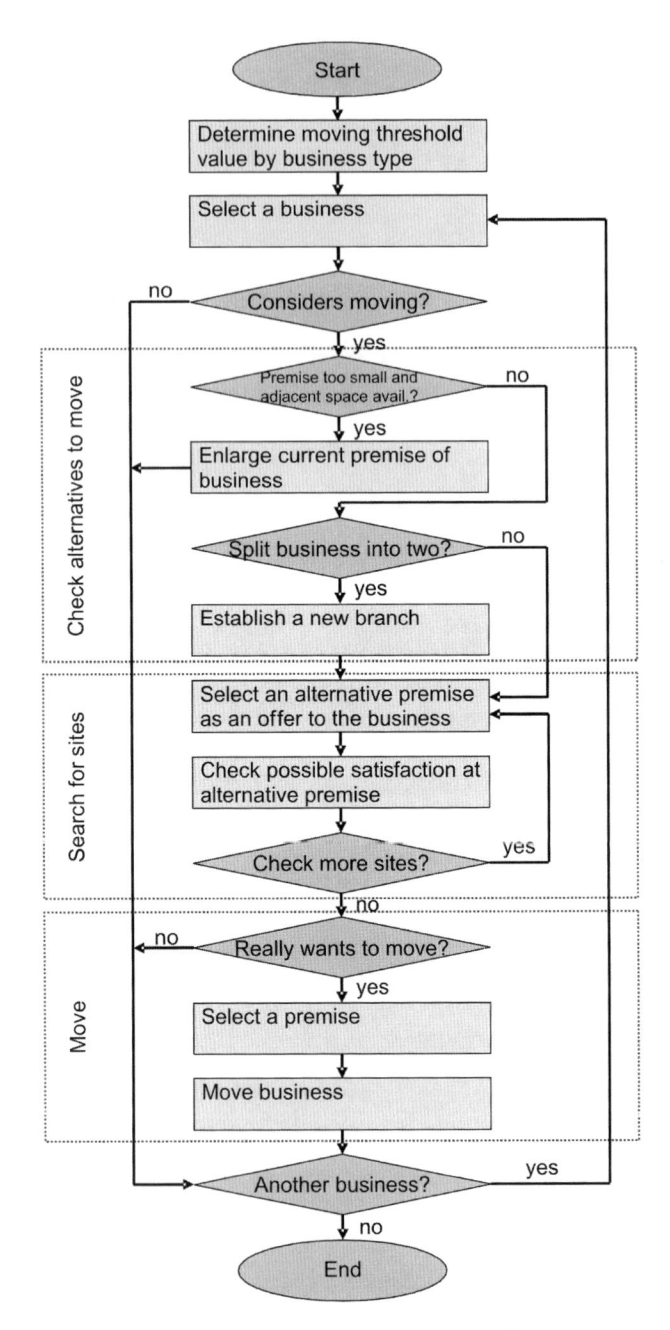

Fig. 11.2 Flowchart of the business relocation model in ILUMASS

other employment types and of transport infrastructure, purchasing power or local taxes. A firm may evaluate up to 10 different sites. The number of sites evaluated is simulated dynamically, depending on the quality of the sites found. Finally, a binomial logit model is used to simulated whether – given the sites found by firm – it really wants to move. The firm has the option to decline moving at this point, it may postpone relocating in expectation of better alternatives in the future. If the firm decides to move, a multinomial logit model is used to select one of the analyzed sites and the firm relocates.

While models simulating population need to update demographic features, models simulating firms need to adjust firmographic attributes. Changes in firmography, particularly growth, are the most important drivers for relocation. In ILUMASS, firmographic changes include establishment of a firm, growth, decline and closure. These events are simulated by Markov models using transition probabilities. Firmographic events are affected by the overall economic growth and by the economic restructuring process, both of which are given exogenously. Business establishments and business closures are limited to firms of 100 employees or less. Larger firms can be established or closed exogenously. The establishment or closure of very large firms happens relatively rarely and should be considered a historic event. Simulating such major changes endogenously may result in very different model outcomes from run to run purely based on the random numbers chosen by the model. Likewise, excessive growth or decline of business is not modelled endogenously. While the smallest firms are assumed to develop more dynamically and may grow or decline by up to 200%, larger firms are expected to change size less vigorously, and firms with more than 200 employees may change by 5% only.

Floorspace is updated in response to actual demand. If the vacancy rate of a certain floorspace type in a zone drops below an exogenously given threshold, new floorspace of that type is developed. Developers are searching for sites in this zone to develop that floorspace type. The newly generated floorspace is released to the market with 1 year delay to reflect time needed to recognize the demand, to plan the site, to request the required building permits and to actually construct the floorspace. A 1-year delay had been chosen as an average duration: Construction of some sites may take years from planning to completion, while other floorspace may be converted from one use to another and added to the market within a few weeks. Floorspace may only be built in areas with the appropriate zoning, which may limit the development of floorspace even if demand is high. The demand frequently exceeds the available land particularly in city centres.

The model has the base year 2000 and runs in yearly intervals till 2030. A series of scenarios were analysed, including scenarios with alternative zoning, rise in fuel prices, reduction of business taxes in selected cities and establishment of major new firms. For demonstration purpose, two scenario results are shown in Fig. 11.3. The left graphic shows the change in employment from 2000 to 2030 in the base scenario. The city centre of Dortmund in the centre of the graph loses employment, while several suburban centres grow substantially. The right graphic shows the change of employment in the compact city scenario, a scenario that only allowed additional floorspace development within the city of Dortmund in the centre of the study area. Only few areas outside of Dortmund are able to gain employment, mostly due to filling in vacancy that existed in the base year. The entire city of Dortmund gains substantially, mostly due to lack of competition from surrounding cities.

Being a microsimulation, the model works stochastically and generates slightly different results in every model run. Figure 11.4 shows the range of model results

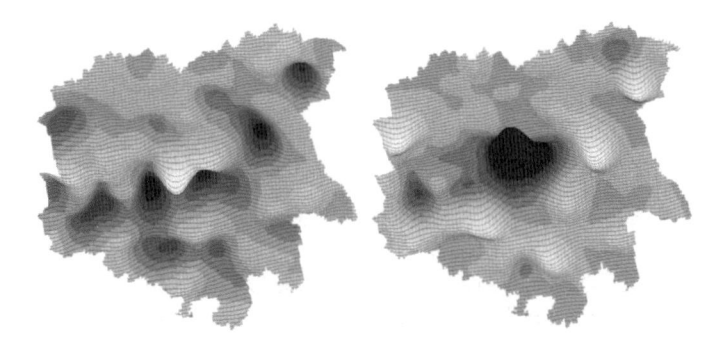

Fig. 11.3 Employment change from 2000 to 2030 in the base scenario (*left*) and a compact city scenario (*right*)

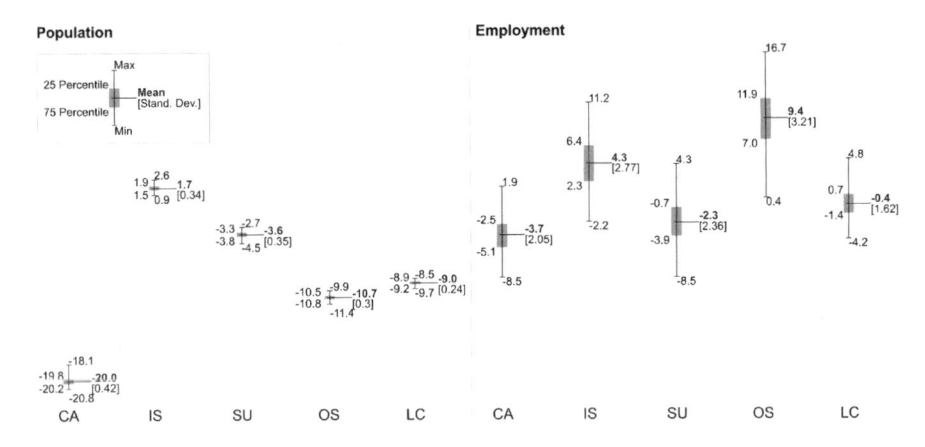

Fig. 11.4 Stochastic variation across 150 model runs for percent change of population and employment from 2000 to 2030

across 150 model runs. The left side shows percent change of population from 2000 to 2030, and the right side shows percent change of employment for the same time period. For employment, different model results are triggered by applying random numbers both in location choice of businesses and in simulating firmographic events. The range of results for employment is much larger than for population. This is caused by having fewer firms (80,291) than households (2,618,379) in the study area. Having 32-times more households than firms results in many more relocation decisions made by households than by firms. The larger number of household moves smoothes out the differences between model runs. Furthermore, the average relocation of a household changes the distribution of population by 2.2 persons, which is the average household size in the area. The average size of firms is 13.9 employees. Given that only firms of 100 employees or less are considered for relocation, the effective average business size for relocation is 7.4 employees. Still, the average relocation of a firm affects the socio-economic data more than three times as much as the average move of a household.

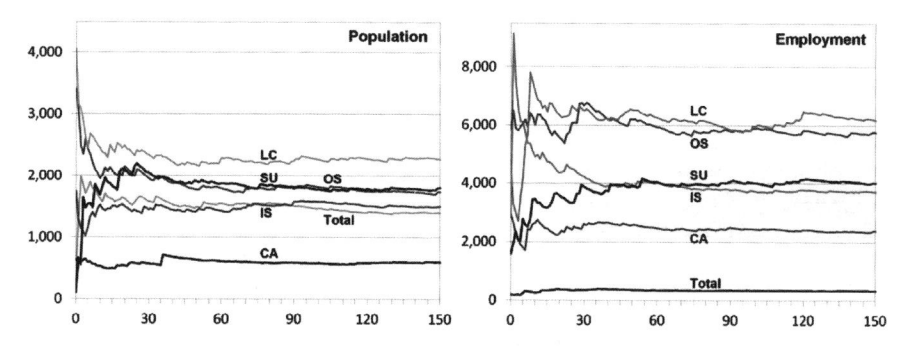

Fig. 11.5 Change of standard deviation over 150 model runs for population and employment

Different subareas were defined in Fig. 11.4: CA = Central Area (or Central Business District), IS = Inner Suburbs, SU = Suburbs, OS = Outer Suburbs and LC = Other Larger Cities. The large range of results found on the employment side suggests that the model has to be run many times and a mean should be calculated to get a representative result that is not biased due to random numbers chosen. It is worth noting that the standard deviation is the smallest for LC, which has the largest number of firms and households. The standard deviation is inversely proportional to the size of the zone.

Figure 11.5 shows the development of the standard deviation with an increasing number of model runs. Shown is the standard deviation of the change in population and employment between 2000 and 2030; i.e. population in the region LC has a standard deviation of 2,090 after 150 model runs, which means that the change in population (which is an average loss of 870,945 people or 9 %) ranges ± 0.24 %. The standard deviations for employment are about twice as large as for population, reinforcing the difference between population and employment shown in Fig. 11.4. After 40 model runs, fairly stable standard deviations were reached for both population and employment.

However, calculating the average change of population and employment shows that after 14 model runs the average across all completed model runs oscillates by 1 % point. In other words, after 14 model runs, an average has been reached that is within +/− 0.5 % of a true average. Running the model 14-times and calculating an average generates a representative model result. The number of model runs necessary to reach a stable average most likely can only be determined experimentally, as shown here for the ILUMASS model. The required number of model runs is dependent on three variables:

- Number of choices: With an increasing number of decision makers (here: firms), fewer model runs are necessary.
- Number of alternatives: With an increasing number of locations (or zones), possible outcomes spread further and more model runs are necessary.
- Selectivity: If the decision makers always seek for one of the best alternatives available, fewer runs are necessary; if decision makers are less focused on the alternatives with the highest utilities, the result becomes "more random" and more model runs are necessary.

The impact of number of choices and number of alternatives is discussed further by Wegener (2011). Different results in every model run are a significant problem in modelling. One option to avoid misinterpretations is to run the model many times and to take an average of the results, as suggested above for ILUMASS. As this would effectively increase the number of events modelled, the average of several model should be the most likely outcome. If a policy scenario is compared to a base scenario, both scenarios need to be run several times and the averages of each scenario can be compared against each other. Given the long runtimes of some models, many model runs may be computationally prohibitive. Even though the ILUMASS model runs with 25 min comparatively fast, running the model 14-times equals 6 h for every scenario.

Even though microsimulation has been applied for urban modelling since decades, only in the last few years researcher started to systematically analyze the impact of stochastic variations. MOSAIC is a microsimulation simulation of tours of commercial vehicles for the Portland Metro area (Donnelly 2007). Every model run leads to slightly different results. Sixty model runs were done, and the average value of vehicle-kilometers of travel, tons of freight transported and tons handled by trans-shipment facilities were calculated. Out of the 60 model runs, the one run was chosen as the final result that resembled the closest the average in these categories. This run is assumed to resemble the average result in all relevant dimensions of the model sufficiently well. Scenario analyses were only performed at a highly aggregate level, such as total tons shipped, total vehicle-kilometer of travel or total share of empty vehicle miles of travel. Hunt and Stefan (2007, p. 984) run a tour-based commercial vehicle model for the City of Calgary 10-times and average the results to avoid misinterpretations due to stochastic variation. Another possible approach to reduce the impact of the random number generator occasionally applied in travel demand modeling is to increase the number of households, employees and number of lanes by a factor (such as 10), run the model with the increased inputs and divide the simulation results by the same factor. Given the larger number of objects simulated, the stochastic variation becomes smaller.

The microsimulation land-use model LUSDR (Gregor 2006) acknowledges the impact of the random number generator on the results. As the model runs fast, many runs are done and the land-use cover of every model run is stored. The final result reported is not a single land-use cover outcome for every zone, but rather a range of different outcomes that occurred in different model runs. If, for example, one zone changes from agricultural land into industrial land in most runs, it can be concluded that this shift of land-use is likely to occur for this zone.

11.4 A City In Vitro

The previous section suggested that there might be a problem with microsimulating businesses, as different model runs may lead to quite different results. If single jobs were simulated instead of businesses, the number of events simulated would be

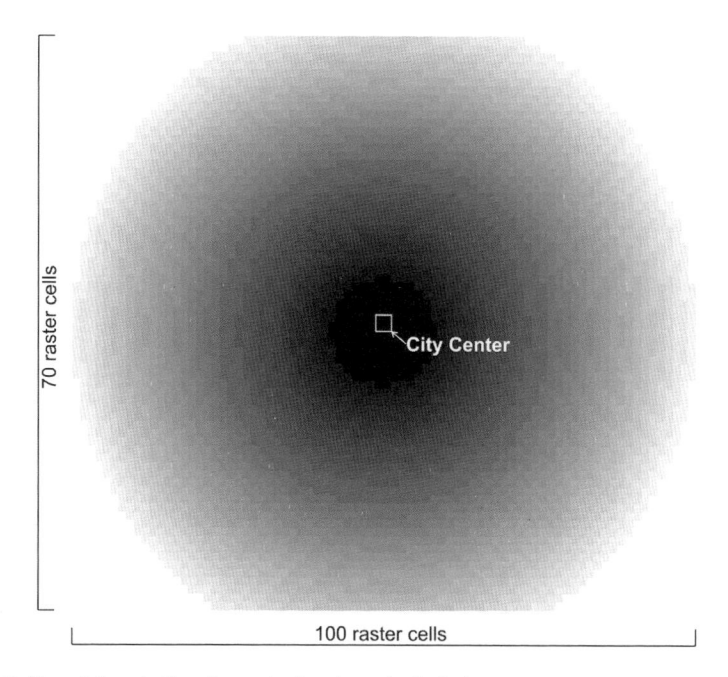

Fig. 11.6 Travel time to the city centre in a hypothetical city

larger. If in ILUMASS 100 businesses were simulated to relocate, the equivalent of 740 employees would be simulated in a job simulation model to achieve the same number of jobs being relocated. The larger number of events would be more likely to smooth out differences between each model run that are due to the random number generator.

To test the behaviour of simulating relocation of firms versus relocation of jobs, a synthetic city was built in a computer model. To single out the actual behaviour of firms or jobs, the artificial city has been simplified as much as possible. The goal in this context was not to build a usable employment relocation model, but rather to understand the differences of modelling firms and jobs. The city developed has a near elliptical shape and is subdivided into 4,760 cells, with a maximum width of 100 cells and a maximum height of 70 cells. Figure 11.6 shows the layout of this city, where the grey shades represent the travel time to the city centre. Travel time is assumed to be proportional to distance and uniform across the study area. To keep the behaviour of firms and jobs unobstructed by the model setup, travel time is the only location factor used to simulate employment relocation. Shorter travel times to the city centre are assumed to be more favourable, i.e. every firm or job would prefer to be located in the city centre.

As floorspace supply is limited, not every firm or job finds space to locate in the city centre. Floorspace is measured in employee spaces, and all employees require one employee space. Floorspace is generated based on travel time to the city centre.

More floorspace is generated close the city centre than in the outskirts, using the equation

$$fs_i = c \cdot \exp(\beta \cdot t_{i,cc}) \tag{11.1}$$

fs_i Floorspace in zone i
c Constant, set to 280
β Parameter, set to -0.1
$t_{i,cc}$ Travel time from zone i to the city centre

A total of 158,820 employee spaces were created. To analyze the effects of simulating firms versus simulating jobs, two models were implemented in this synthetic city:

(a) Business relocation: entire firms move from one location to another and relocate all their employees at the same time
(b) Job relocation: single employees move from one location to another independently

On the outset of a model run, businesses are generated and located in raster cells. As the model is set up as a microsimulation, single businesses need to be generated. Even in the case of simulating job relocation, businesses are generated at the beginning and not just jobs. This way, the initial setting is identical for both models. According to U.S. Census (2011) data, the average business size of firms with 1–100 employees is 9.77 employees. An exponential function was used to calculate probabilities for the size of a generated business:

$$p_i = \exp(\beta \cdot i) \tag{11.2}$$

p_i Probability of size i
β Parameter, set to -0.108

The parameter β was calibrated to resemble the distribution of business sizes as found in the census data. The number of firms is not predefined, but the model continues to create additional firms until the threshold of 100,000 employees is reached or exceeded. Using the same start value for the random number generator in each model run, the same number of 100,014 jobs in 10,248 firms is created every time the model runs. A location is selected for each firm using a multinomial logit model (Domencich and McFadden 1975, p. 53 ff). It should be noted that in a true model application, it would be inconsistent with multinomial logit theory to apply this model across all 4,760 cells at the same time. The multinomial model has been developed to make a selection from a reasonably small sample, replicating the behaviour of humans who are unable to compare thousands of options at the same time. In this particular case, however, it seems acceptable to violate that condition

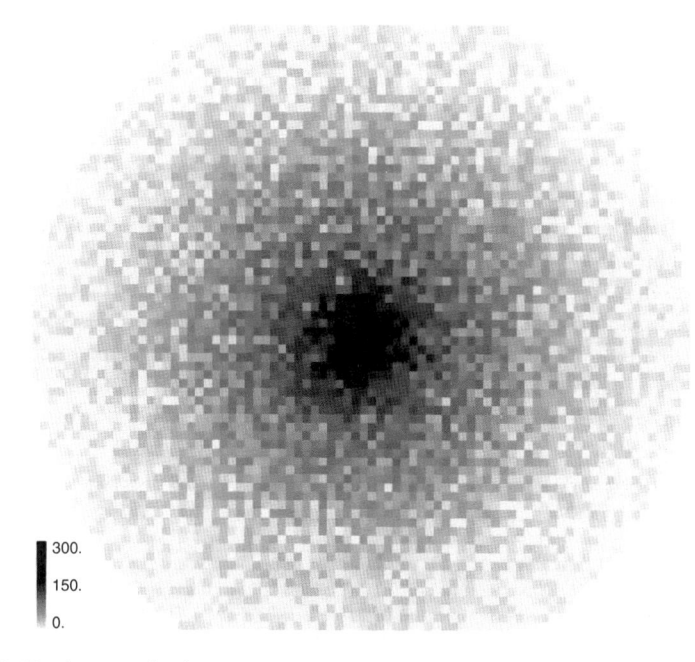

Fig. 11.7 Employment density per raster cell at time t = 0

of the logit model as the selection process itself is not under scrutiny, but only the different behaviour in firm and job location search.

To ensure that even larger firms find sufficient floorspace to locate, the synthetically created firms are sorted by size and the larger firms are allocated first. Only travel time to the city centre is used as a location factor. The inverse of the travel time is assumed to be the utility of a raster cell. Only those raster cells are considered that have at least the number of vacant floorspace required by the business.

$$p_i = \frac{vf_i \cdot \exp(\beta \cdot u(t_{i,cc}))}{\sum_J vf_j \cdot \exp(\beta \cdot u(t_{j,cc}))}, \quad \text{where} \quad u(t_{i,cc}) = \frac{1}{t_{i,cc}} \quad (11.3)$$

p_i Probability of zone i
β Parameter, set to 1
$t_{i,cc}$ Travel time from zone i to city centre
vf_i Vacant floorspace in zone i

The initially generated employment density is shown in Fig. 11.7. The city centre has the highest employment density, with a gradually declining density towards the outskirts. The pattern is not perfectly gradual, as the stochastic element in the location choice as well as the size distribution of business sizes added some randomness to the pattern.

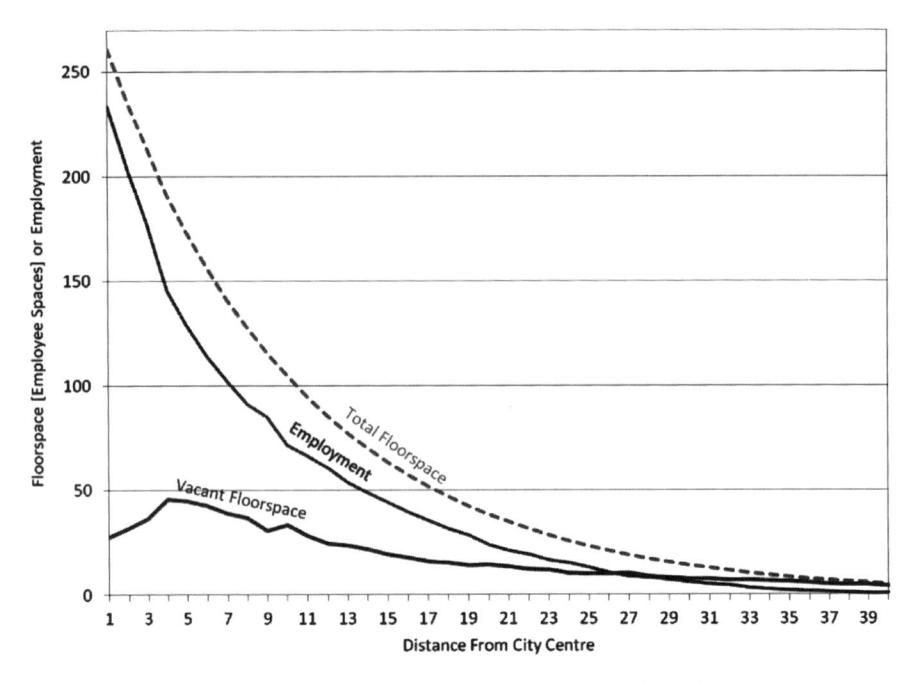

Fig. 11.8 Average employment and floorspace density by distance from city centre

The summary of employment and floorspace density by distance from the city centre is shown in Fig. 11.8. Total floorspace shows the gradual curve generated by Eq. 11.1. Total employment more or less follows the shape of the floorspace curve, leaving a fairly even distribution of vacant floorspace.

The model runs over 30 simulation periods. In every simulation period, 5,000 employees (or roughly 5 % of all employees) are relocating. In model (a), the relocation model randomly chooses firms and relocates them until the threshold of 5,000 relocated employees is reached or exceeded. In model (b), exactly 5,000 single employees are chosen randomly to relocate in each simulation period.

Floorspace is hold constant and not updated over time. This simplification was chosen to better expose the different relocation behaviour in the two models, without obscuring the results by unexpected chain reactions between relocation behaviour and the behaviour of developers building additional floorspace.

No employment is added or removed in this experiment. In reality, some firms grow and others decline, some emigrate from and others immigrate to the study area, and some firms go out of business and other firms are newly established. All these firmographic changes have been omitted in this simulation to focus on the main differences between the relocation of jobs and businesses, unobstructed by the effect of different firmographic model designs. In reality, firmographic events are the most important trigger for firm relocation. For example, a firm that has grown is likely to require a larger site to locate. In this simplified model, the reason for relocating is not further analyzed. A constant number of 5,000 employees is assumed to relocate every simulation period.

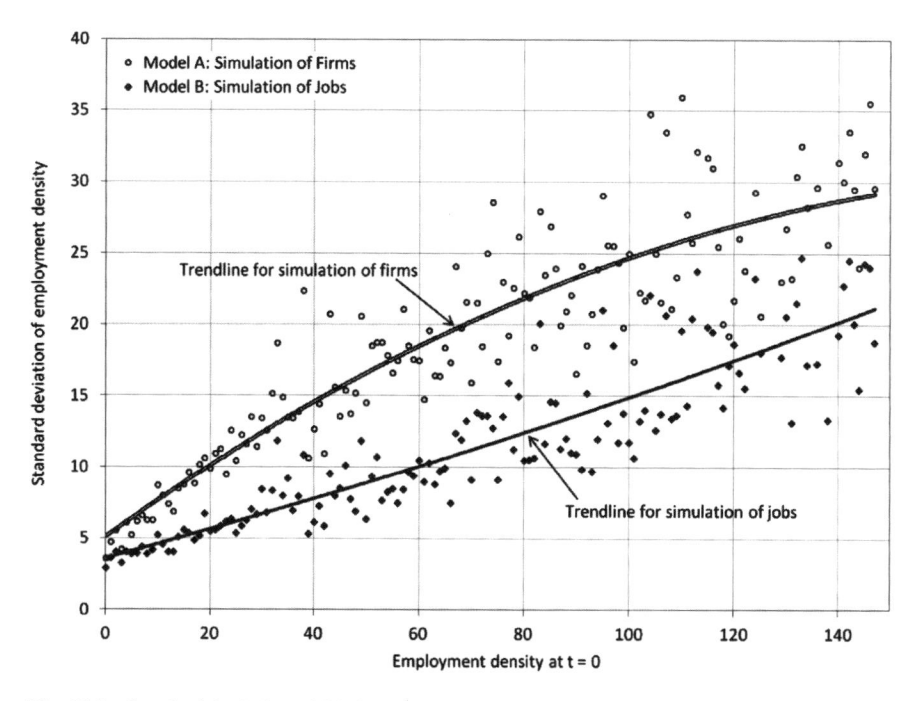

Fig. 11.9 Standard deviation of 30 time periods for firms and job relocation

Figure 11.9 shows the standard deviation of employment density between these 30 simulated time periods. The abscissa shows the employment density in the base year, and the dots represent the standard deviation of all raster cells that had this density in the base year. Standard deviations are calculated across all 30 simulation periods, thus the standard deviation measures the variation of employment density that zones experience over 30 periods. Naturally, the standard deviation has to be smaller on the left side of the diagram, as these raster cells have few employees in the base year and do not offer the floorspace for larger shifts in employment. To the right side, the standard deviation is higher as more floorspace allows for larger growth and decline.

The circled dots show the standard deviation of model A, which simulates the location choice of firms. The grey diamonds show the standard deviation of the simulation of jobs. As the trendlines (shown as polynomial trendlines at the order of 2) suggest, the standard deviation for the simulation of firms is significantly larger than for the simulation of jobs. On the average, it is 64 % larger for firm than for jobs.

Even though the same utility function was used for simulating firms and for simulating businesses, the results after 30 simulation periods are quite different. Figure 11.10 shows employment and floorspace densities relative to the distance to the city centre. The employment density in model A, which simulated firms as the decision-making unit, is higher close to the city centre than in model B, which simulated relocation of single jobs. Accordingly, there is more vacant floorspace closer to the city centre when simulating jobs. The reason for this behaviour is that there are

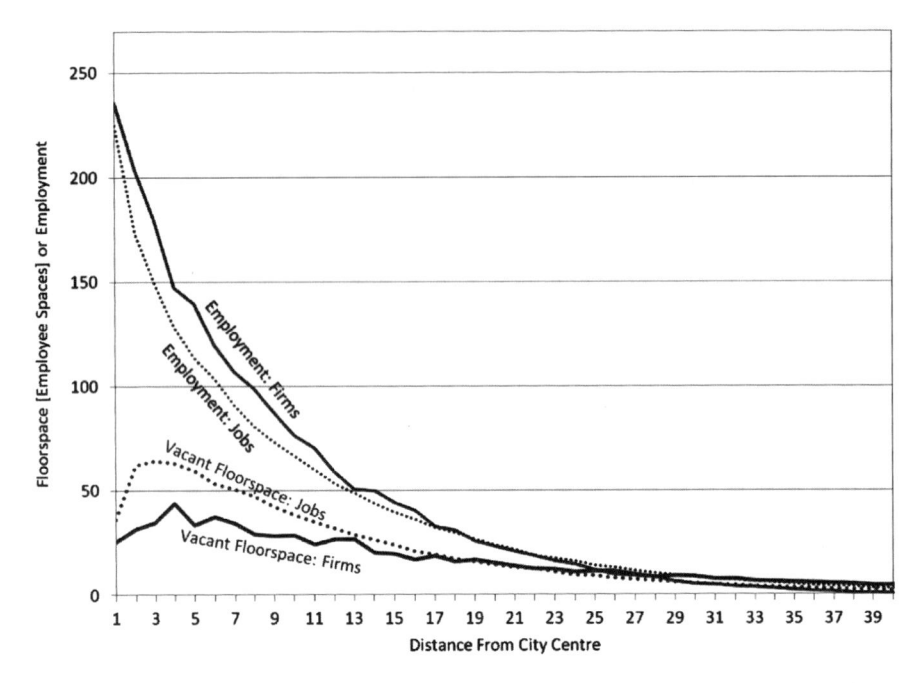

Fig. 11.10 Employment and floorspace densities after 30 simulation periods in model A (Firms) and model B (Jobs)

many more zones on the outskirts, making it more likely to be chosen by single employees as their new location, even though the utility of these zones is comparatively low. In model A, in contrast, entire businesses are moved and the new location has to provide space for all employees of the firm. As there are only few vacant employee spaces available in the outskirts, model A is unable to allocate much employment to those zones. Given the higher floorspace density close to the city centre, model A tends to allocate larger firms to more central locations. Model B, on the other hand, may allocate a single employee wherever a single employee space is vacant.

The curves of model B are much smoother than the curves of the firm's model. This reflects the higher number of transactions simulated in model B, where every transaction relocated only a single employee, allowing to smooth out the curves.

11.5 Conclusions

If behavioural replication of real-world business relocations was the only objective of urban modelling, the choice would be simple. The simulation of firms offers the most realistic patterns. When relocating, firms need to find space for the entire firm. When simulating jobs instead of firms, every last vacant employee space in the most desirable locations is filled in, leading to an unrealistically low vacancy rate in parts of the study area.

However, replicating behaviour cannot be the only purpose of urban modelling. Stochastic variability between different model runs has become a serious concern of microscopic urban modelling. If the differences between several model runs of the same scenario are larger than the differences between the base and a policy scenario, conclusions about that scenario are invalid as it is unclear how much of the change is due to the alternative scenario settings and how much is due to random differences between two model runs. Within the synthetic city described above, the simulation of jobs resulted in significantly less stochastic variability. This increases the validity of single model runs, making it more reasonable to compare a base scenario with policy scenarios. Possible disadvantages include that interactions between firms, such as agglomeration effects or crowding out of economically less successful businesses, have to be reduced to interactions between jobs. Firmographic processes are reduced to disappearance or emergence of single jobs. The simulation of jobs leads to much smoother distributions of employment, a side effect that seems to be unrealistic. The relocation of entire firms, as happening in reality, is likely to lead to rather discontinuous distributions.

In terms of stochastic variability, aggregate approaches that do not need a random number generator at all are the preferred alternative. Those models produce exactly the same result between different model runs. If two scenarios are compared with each other, differences in the model results doubtlessly can be attributed to the changes in the scenario setup, there is no random noise added to the outcome. Unfortunately, aggregate models limit the depth of attributes assigned to employment and make it challenging to reflect agglomeration effects.

In actual applications of urban models, the behavioural difference between simulating jobs and simulating firms often becomes concealed. In every simulation period, firmographic events change the employment data, developers add and remove floorspace, changing accessibilities affect utilities, etc. This additional noise to the system hides that the model design significantly impacts the results. The artificial city described in the previous chapter allows highlighting some of the benefits of simulating firms instead of jobs. The differences in model results are worrisome. Even though model A and model B see the same utility function to choose a location, the model design affects the simulation outcome. The simulation of jobs leads to highly smooth employment distributions. If a large enough number of simulation periods is simulated, an almost perfect equilibrium is found, which would never be reached in reality. As single employees in model B need only single employee spaces to locate, an employment distribution is found that almost perfectly reflects the utility function.

As the initial setup of employment in the base year was generated by allocating firms, the pattern does not change much over 30 simulation periods in model A simulating business relocation. In model B, in contrast, employment that was allocated as firms at the beginning is relocated as single jobs, changing the pattern substantially. As real-world employment is allocated in firms, the base year data for any model are likely to reflect the heterogeneous employment distribution of employment in firms. Hence, the simulation of firms is likely to update these employment data more realistically than the simulation of jobs.

Whether the differences between simulating jobs and firms actually matter can only be answered on a case-by-case basis. If only larger aggregations of zones shall be analyzed, the stochastic variability added by the simulation of firms is likely to be negligible. If, however, smaller areas are analyzed and the runtime makes multiple runs prohibitive, simulating jobs should be preferred over simulating firms. In complex land-use/transport interaction models, the uncertainty in the model system may be much larger than the differences between simulating jobs or firms. If employment is mostly simulated to provide workplaces in a transport-dominated model system, the simulation of jobs instead of firms is most likely sufficient. If, however, scenarios are analyzed that focus on changes of employment, it is likely to be too much of a simplification to simulate jobs instead of firms.

References

Anas A, Duann LS (1985) Dynamic forecasting of travel demand, residential location and land development. J Reg Sci 56(1):37–58

Anderstig C, Mattsson L-G (1991) An integrated model of residential and employment location in a metropolitan region. Pap Reg Sci 70(2):167–184

Anderstig C, Mattsson L-G (1992) Appraising large-scale investments in a metropolitan transportation system. Transportation 19:267–283

Bergmann BR (1974) A microsimulation of the macroeconomy with explicitly represented money flows. Ann Econ Soc Meas 3(3):475–489

Birch DL (1984) The contribution of small enterprises to growth and employment. In: Giersch H (ed) New opportunities for entrepreneurship. J.C.B. Mohr, Tübingen, pp 1–17

U.S. Census (2011) Statistics about business size (including small business) from the U.S. Census Bureau. http://www.census.gov/econ/smallbus.html. Accessed 22 Sept 2011

De Bok M (2009) Estimation and validation of a microscopic model for spatial economic effects of transport infrastructure. Transport Res A Policy Pract 43:44–59

de la Barra T (1989) Integrated land use and transport modelling: decision chains and hierarchies, vol 12, Cambridge urban and architectural studies. Cambridge University Press, Cambridge

de la Barra T, Rickaby PA (1982) Modelling regional energy-use: a land-use, transport, and energy-evaluation model. Environ Plann B Plann Des 9(4):429–443

Domencich TA, McFadden D (1975) Urban travel demand: a behavioural analysis, vol 93, Contributions to economic analysis. North-Holland, Amsterdam/Oxford

Donnelly R (2007) A hybrid microsimulation model of freight flows. In: Taniguchi E, Thompson R (eds) City logistics V. Institute for City Logistics, Kyoto, pp 235–246

Echenique MH, Flowerdew ADJ, Hunt JD, Mayo TR, Skidmore IJ, Simmonds DC (1990) The MEPLAN models of Bilbao, Leeds and Dortmund. Transport Rev 10:309–322

Eliasson G (1978) A micro simulation model of a national economy. In: Eliasson G (ed) A micro-to-macro model of the swedish economy. Papers on the Swedish model from the symposium on micro simulation methods in Stockholm, 19–22 Sept 1977. Almqvist & Wiksell International, Stockholm, pp 3–83

Forrester JW (1969) Urban dynamics. The MIT Press, Cambridge, MA

Goldner W, Rosenthal SR, Meredith JR (1972) Theory and application: projective land use model, vol 2. Institute of Transportation and Traffic Engineering, Berkeley

Gregor B (2006) The Land Use Scenario DevelopeR (LUSDR): a practical land use model using a stochastic microsimulation framework. In: 86th annual meeting of the Transportation Research Board, Washington, DC

Hill DM (1965) A growth allocation model for the Boston region. J Am Inst Plann XXXI (2):111–120

Hunt JD, Abraham JE (2003) Design and application of the PECAS land use modelling system. Paper presented at the 8th international conference on computers in urban planning and urban management, Sendai 27–29 May 2003

Hunt JD, Abraham JE (2009) PECAS – for spatial economic modelling. PECAS – for spatial economic modelling. HBA Specto Incorporated, Calgary

Hunt JD, Stefan KJ (2007) Tour-based microsimulation of urban commercial movements. Transport Res 41A(9):981–1013

Hunt JD, Kriger DS, Miller EJ (2005) Current operational urban land-use-transport modelling frameworks: a review. Transport Rev 25(3):329–376

Kay NM (1979) The innovating firm: a behavioural theory of corporate R & D. Macmillan, London

Khan AS, Abraham JE, Hunt JD (2002) Agent-based micro-simulation of business establishments. In: Congress of the European Regional Science Association (ERSA), Dortmund

Landis J, Zhang M (1998a) The second generation of the California urban futures model. Part 1: Model logic and theory. Environ Plann B Plann Des 25:657–666

Landis J, Zhang M (1998b) The second generation of the California urban futures model. Part 2: Specification and calibration results of the land-use change submodel. Environ Plann B Plann Des 25:795–824

Lowry IS (1964) A model of metropolis. Memorandum RM-4035-RC. Rand Corporation, Santa Monica

Mackett RL (1990) The systematic application of the LILT model to Dortmund, Leeds and Tokyo. Transport Rev 10(4):323–338

Maoh H, Kanaroglou P (2009) Intrametropolitan location of business establishments. Microanalytical model for Hamilton, Ontario, Canada. Trans Res Rec: J Trans Res Board 2133:33–45

Martínez FJ (1996) MUSSA: land use model for Santiago city. Transport Res Rec 1552:126–134

Moeckel R (2007) Business location decisions and urban sprawl: a microsimulation of business relocation and firmography, vol 126, Blaue Reihe. Institut für Raumplanung, Dortmund

Moeckel R (2009) Simulation of firms as a planning support system to limit urban sprawl of jobs. Environ Plann B Plann Des 36:883–905

Putman SH (1967) Intraurban industrial location model design and implementation. Reg Sci Assoc Pap 19:199–214

Shubik M (1960) Simulation of the industry and the firm. Am Econ Rev 50(5):908–919

Shukla V, Waddell P (1991) Firm location and land use in discrete urban space: a study of the spatial structure of Dallas-Fort Worth. Reg Sci Urban Econ 21:225–253

Simmonds DC (1999) The design of the DELTA land-use modelling package. Environ Plann B Plann Des 26(5):665–684

Simmonds DC (2001) The objectives and design of a new land-use modelling package: DELTA. In: Clarke G, Madden M (eds) Regional science in business, Advances in spatial sciences. Springer, Berlin, pp 159–188

Strauch D, Moeckel R, Wegener M, Gräfe J, Mühlhans H, Rindsfüser G, Beckmann KJ (2005) Linking transport and land use planning: the microscopic dynamic simulation model ILUMASS. In: Atkinson PM, Foody GM, Darby SE, Wu F (eds) GeoDynamics. CRC Press, Boca Raton, pp 295–311

van Tongeren FW (1995) Microsimulation modelling of the corporate firm: exploring micro–macro economic elations, vol 427, Lecture notes in economics and mathematical systems. Springer, Berlin

van Wissen LJG (1999) A micro-simulation model of firms. Applications of concepts of the demography of the firm. In: van Dijk J, Pellenbarg PH (eds) Demography of firms: spatial dynamics of firm behaviour, vol 262. Nederlandse Geografische Studies. Faculteit der Ruimtelijke Wetenschappen, Rijksuniversiteit Groningen, Utrecht, Groningen, pp 15–48

van Wissen LJG (2000) A micro-simulation model of firms: applications of concepts of the demography of the firm. Pap Reg Sci 79:111–134

Waddell P (2002) UrbanSim. Modeling urban development for land use, transportation, and environmental planning. J Am Plann Assoc 68(3):297–314

Waddell P (2011) Integrated land use and transportation planning and modelling: addressing challenges in research and practice. Trans Rev 31(2):209–229

Waddell P, Ulfarsson GF (2003) Accessibility and agglomeration: discrete-choice models of employment location by industry sector. In: 82nd annual meeting of the Transportation Research Board, Washington, DC, 12–16 Jan 2003

Waddell P, Borning A, Noth M, Freier N, Becke M, Ulfarsson GF (2003) Microsimulation of urban development and location choice: design and implementation of UrbanSim. Netw Spat Econ 3:43–67

Wagner P, Wegener M (2007) Urban land use, transport and environment models. Experiences with an integrated microscopic approach. disP 170(3):45–56

Wegener M (1982) Modeling urban decline: a multilevel economic-demographic model for the Dortmund region. Int Reg Sci Rev 7(2):217–241

Wegener M (2011) From macro to micro – how much micro is too much? Trans Rev 31 (2):161–177

Wever E (1983) Cohort analysis in economic geography. Tijdschrift voor Economische en Sociale Geografie 74(3):217–223

Zondag B, Schoemakers A, Pieters M (2005) Structuring impacts of transport on the spatial distribution of residents and jobs. In: 8th Nectar conference, Las Palmas, Gran Canaria, 2–4 June 2005

Chapter 12
Modelling Firm Failure: Towards Building a Firmographic Microsimulation Model

H. Maoh and P. Kanaroglou

Abstract This paper analyzes the survival and failure of small and medium (SME) size business establishments in the city of Hamilton, Canada. The objective is to develop a business failure module that will be used in the construction of an agent-based firmographic simulation model. Such model will be utilized within a microsimulation land use and transportation planning support system (PSS). A longitudinal database is extracted from Statistics Canada Business Register (BR) to determine the duration of survival for small and medium (SME) size business establishments. We follow the annual life trajectory of each establishment in the 1996 population till 2002 and record the year when the establishment failed. Business establishment failure is modeled by specifying and estimating discrete-time hazard duration models for 12 economic sectors. The estimation results suggest that the business failure process can be explained by the establishment characteristics, the industry it belongs to, the geographic location it is at and the macro-economic conditions in the city. The results also indicate a variation in the significance of those factors among the different economic sectors due to establishment heterogeneity.

12.1 Introduction

Firm demography or firmography is the study of processes that relate to the establishment of new businesses or to the growth, decline, failure, or migration of existing business establishments. In recent years, there has been an increasing

---

H. Maoh (✉)
Department of Civil and Environmental Engineering, University of Windsor, 401 Sunset Ave., N9B 3P4 Windsor, ON, Canada
e-mail: maohhf@uwindsor.ca

P. Kanaroglou
School of Geography and Earth Sciences, McMaster University, 1280 Main Street West, L8S 4K1 Hamilton, ON, Canada

interest to use firmography in the planning of regions and cities. Urban modelers consider firmography as an ideal approach for the development of the land use component of agent-based *Integrated Urban Models* (IUMs) (Maoh and Kanaroglou 2009). The latter are computer simulation programs used to assess the ramifications of urban planning policies (Miller et al. 2004). Research to develop firmographic microsimulation models within IUMs is still in its early stages as discerned from the limited but growing number of studies on the topic (see, for example, the work of Maoh and Kanaroglou 2007b; Hunt et al. 2003; Bliemer and De Bok 2006; Elgar and Miller 2006; Moeckel 2005, and Kumar and Kockelman 2008; Maoh and Kanaroglou 2009; DeBok 2009).

The work presented here contributes to this growing literature by designing a prototype agent-based firmographic model. The proposed model consists of a number of submodels that operate on a base year population to drive its evolution over time (See Maoh and Kanaroglou 2007b, 2009). This paper reports on the results achieved from formulating, specifying and estimating the firm failure component of this firmographic model.

The model in this paper will be used to predict the annual failure probabilities for individual business establishments. We make use of a longitudinal business establishment database that is extracted from the Business Register (BR) of Statistics Canada. We focus our attention on small and medium size establishments (SMEs) in the City of Hamilton in Ontario, Canada. SMEs are more dynamic in their firmographic behavior and are described as *"agent of change"* (Audretsch 2001). Conversely, large business establishments exhibit a more predictable firmographic behavior. Also, the population size of large firms is relatively small whereas SMEs comprise the majority of the firm population.

The current work makes two novel contributions. First, it provides an assessment of the factors influencing the intra-urban failure process of SMEs using longitudinal data for the city of Hamilton in Ontario, Canada. To our knowledge, the use of longitudinal data to study the intra-urban failure process of SMEs has not been done before. Second, it highlights cross-industry differences in terms of the determining factors that lead to the failure of establishments. Earlier efforts to model the hazard of failure for establishments were mainly concerned with the manufacturing sector and were conducted at the regional level (Audretsch and Talat 1995; Caves 1998; Audretsch et al. 1999; Strotmann 2007; Schwartz 2009). Studies to explain the factors affecting firm survival and failure within the urban context are scarce and underrepresented (Lerman and Liu 1984; Miron 2002).

The remainder of the paper is organized as follows. Section 12.2 reviews some of the recent mainstream research on firm survival and failure to highlight the factors and determinants that drive this firmographic process. Section 12.3 discusses the formulation of the discrete-time hazard duration model, specified to handle the failure process of business establishments. Section 12.4 presents and discusses the estimated models. The final section provides a conclusion to our study.

12.2 Literature Review

Most of the existing empirical studies on firm survival and failure are conducted at the national as oppose to the urban level. According to the reviewed literature, variables normally used to explain firm failure can be summarized into four general groups: (1) firm specific, (2) industry specific, (3) macro-economic and (4) spatial or geographic factors.

Firm specific factors reflect the individual characteristics of the business establishment. Examples of such establishment-based characteristics may include age, employment size, growth, organizational structure and adaptability to new technologies. According to the literature, young and small establishments are vulnerable and their exit rates are higher compared to large and more mature establishments (Fotopoulos and Louri 2000; Holmes et al. 2000; Caves 1998; Baldwin et al. 1997). The relation between establishment age and failure is referred to as the "liability of newness" hypothesis, whereas the relation between establishment size and failure is known as the "liability of size" hypothesis (Van Wissen 2000).

In examining the "liability of newness", Baldwin et al. (2000) found that 23 % of new Canadian firms are likely to exit during the first year of entry. A new firm has a 0.42 probability of surviving past its fourth year and a 0.20 likelihood of completing its first decade (Baldwin et al. 2000). As for their tests on the liability of size hypothesis, the Canadian study found that small firms in their first years of entry have a higher hazard of exit when compared with medium and large size firms. Similar results are reported for manufacturing firm in Portugal (Mata and Portugal 1994). The evidence suggests pronounced failure during the first year of entry and a 50 % failure rate within the first 4 years of operation.

Organizational structure of the firm refers to its ownership structure. Studies show that multi establishment firms are less likely to exit as compared to single establishment firms (Holmes et al. 2000). Such a result is expected since multi establishment firms have access to large capital resources. Change in technology can also affect the survival of business establishments (Mata and Portugal 1994).

Other firm specific factors that are less highlighted in the literature but appear to have a significant influence on the survival of firms relate to internal factors of the firm. Baldwin et al. (1997) conducted a survey to study the causes of firm bankruptcy in Canada. They found that almost half of the firms in Canada go bankrupt due to internal problems caused by managerial inexperience and lack of knowledge and vision. Their findings suggest that even firms that failed because of external events, such as economic downturn or increased competition, suffered from internal managerial deficiencies that did not allow them to react to the external shocks.

Industry specific factors refer to variables measured at the industry level. According to the literature, high industry growth can lead to firm exit. Baldwin et al. (2000) use the average firm size of an industry to measure the performance of new entrants. They argue that establishments usually enter at a suboptimal level and their likelihood to survive will decrease unless if they are able to increase their size to match the average firm size of the industry. Studies have also shown a variation among the survival rates of business establishments by industry (Agarwal and Audretsch 2001; Baldwin et al. 2000; Holmes et al. 2000).

Macro-economic conditions have been addressed in the firm survival and exit literature (Berglund and Brannas 2001; Baldwin et al. 2000; Holmes et al. 2000; Audretsch and Talat 1995). The most commonly used indicators of macro-economic conditions are unemployment rate, interest rate, exchange rate, average total income and growth rate in real GDP. High unemployment rates due to economic downturn and high levels of interest rates or exchange rates are expected to increase the likelihood of exit from the market (Holmes et al. 2000; Audretsch and Talat 1995). On the other hand, high growth in real GDP can prolong survival as suggested by Baldwin et al. (2000). Average total income in a given region would suggest that higher income areas are likely to generate higher demand for goods and services (Berglund and Brannas 2001).

Some studies made use of geographic factors to examine if survival varies by location (Baldwin et al. 2000; Fotopoulos and Louri 2000; Lerman and Liu 1984). Baldwin et al. (2000) show that exit of Canadian firms vary by location of province and this variation is dependent on the age of the firm. Their results suggest that Ontario hosts the most successful young firms when compared to the other provinces. Fotopoulos and Louri (2000) show that young Greek firms locating in Athens exhibit higher survival rates when compared to their counterparts elsewhere in Greece. Lerman and Liu (1984) devise a number of categorical variables including distance to the CBD and proximity to main routes to examine if the failure of retail trade stores varies across the urban space in Boston. Their findings suggest that business failure indeed is variable over space.

The levels of local competition and agglomeration economies are spatial measures at the industry level for a business establishment. These measures reflect the market power and market share effects, respectively. Typically, when certain establishments become more distinct than their rivals, they enjoy a higher degree of market power since the demand for their products becomes much higher than their rivals. Lerman and Liu (1984) devised a ratio measure of the total number of employees in similar types of retail store in a given zone to the total number of employees at the store of interest. Miron (2002), on the other hand, adopted a different approach to handle local competition in his survival model of retail establishments in Toronto. He defines local competition by counting the number of rival establishments within a 2 km search radius from a given establishment. The approach proposed by Lerman and Liu (1984) is perhaps more appropriate to capture the effect of local competition since it also accounts for the size of the rival firms in terms of employment. It should be noted, however, that competition and agglomeration are not easy to disentangle. One could argue that a lot of competition may result in higher failure rates (i.e. loss of market share). On the other hand, competition could also be a catalyst for lower failure as a result of knowledge spillover effects.

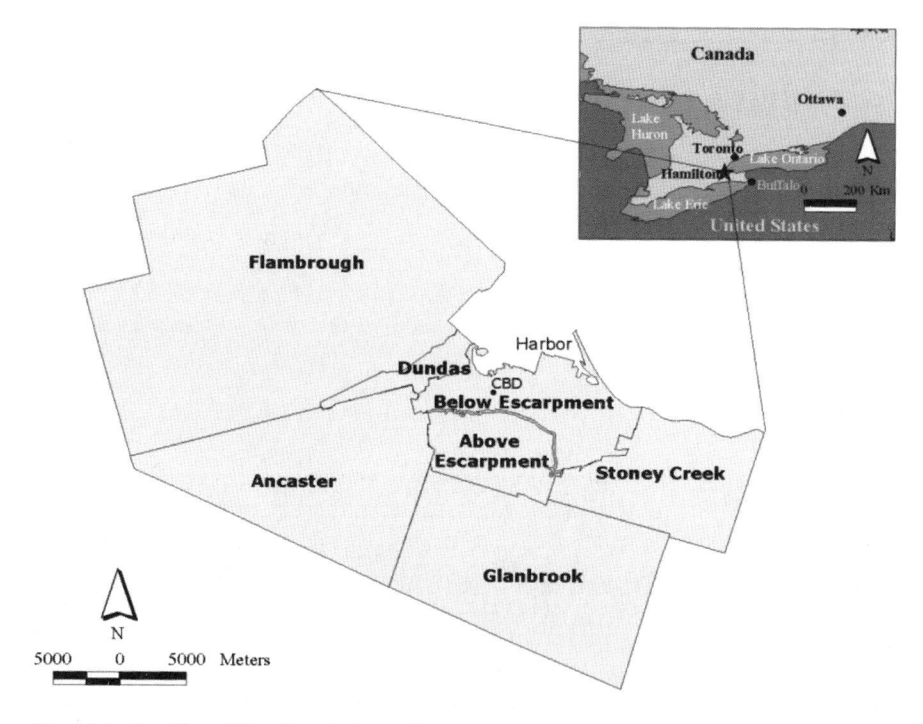

Fig. 12.1 The City of Hamilton in the regional context

12.3 Data and Model Formulation

12.3.1 Firmographic Database

The focus of our study is the City of Hamilton, Ontario, which is located south west of Toronto (See Fig. 12.1). The empirical analysis is based on individual business establishment data extracted from Statistics Canada Business Register (BR). The BR is a confidential source of information that includes an annual census of the universe of Canadian business establishments since 1990. Each establishment is attributed a unique Establishment Number (EN), created at the time the establishment is registered in the BR. The EN was utilized to create a longitudinal firmographic database of SMEs in the city of Hamilton for the time period 1990–2002. SMEs account for the majority of business establishments in the city of Hamilton. More than 94 % of the Hamilton's business establishment population was SME in the years 1990, 1996 and 2002. These accounted for 36 %, 37 % and 35 % of the total jobs in the 3 years, respectively.

Annual firmographic events (survivorship, birth, failure, in-migration and out-migration) in the longitudinal database were derived based on the establishment number and Standard Geographical Classification (SGC) code. An establishment is deemed to fail if it existed in year t in the Hamilton SGC but are no longer in any SGC in Canada in year $t + 1$. Once the annual firmographic events are measured, the annual records were joined using the EN in order to trace the life trajectory of each business establishment over time. Individual business establishment attributes that were derived from the BR and formed the basis of the empirical analysis include: (1) establishment number (2) geographic address in year t (3) paid worker jobs in year t (4) 1980 Standard Industrial Classification (SIC) in year t (5) age of the establishment and (6) annual firmographic event.

12.3.2 Model Formulation

We seek to estimate a model that will follow the evolution of a base year business establishment population to determine the factors that lead to failure and exit from the market. The base year population is a composite of n individual business establishments ($i = 1, 2, 3, \ldots, n$) and the establishment failure event is observed at time values that are positive and discrete ($t = 1, 2, 3, \ldots$). In our context, the base year is 1996 and the failure event is observed every year between 1996 and 2002. An observation for an individual establishment i will continue until time t_i, at which point either a failure occurs or the observation is right censored.

Right censoring in our data occurs if the individual establishment is still in operation by the year 2002. It might also occur if the establishment changes status from being non-incorporated to incorporate. Unfortunately, the latter type of right censoring cannot be observed or controlled and is recorded as failure. However, non-parametric survival analysis from Maoh (2005) suggests that this right censoring is more pronounced among large establishments with size greater than 50 employees. Consequently, we decided to model the survival of small and medium (SME) establishments with size less than or equal to 50 employees. A total of approximately 8,600 establishments make up the 1996 SME population.

The model will determine the hazard rate of a randomly selected business establishment i to shutdown at time t, given it was in operation between the base year and year ($T_i = t$). Following Allison (1995), the hazard rate can be expressed as the following conditional probability:

$$P_{it} = \Pr(T_i = t | T_i \geq t_i, \mathbf{x}_{it}) \tag{12.1}$$

The probability in Eq. 12.1 is dependent on a vector of observed explanatory variables $\mathbf{x}_{it}$ that may vary over time. Each explanatory variable is associated with a parameter β that infers how the covariate contributes to the explanation of the failure events. As noted in Maoh (2005), the hazard probability can be formulated as a binomial logit model:

$$P_{it} = \frac{1}{1 + \exp[-(\alpha_t + \boldsymbol{\beta}' \mathbf{x}_{it})]} \tag{12.2}$$

where α_t is a set of constants. A consistent and efficient estimator for any of the β's or α's that is asymptotically normal can be achieved via the maximum likelihood estimation method. Since the modeled population includes business establishments that existed prior to 1996, left truncation is present in the data. Since the age of the establishment is known to us, this type of left truncation was accounted for following the treatment suggested by Guo (1993) and Allison (1995). It should be noted that unlike a cross-sectional binomial logit model, the hazard model in Eq. 12.2 utilized discrete time observation per establishment to account for temporal effects. As such, if an establishment failed at time k between 1996 and 2002, k observations are used in the model to represent the establishment. The dependent variable takes on a value of 1 for time $t = k$ observation and zero for the other $k - 1$ observations. This treatment is referred to in the literature as *individual-year* (Allison 1995).

12.3.3 Model Specification

We follow the same scheme highlighted in Sect. 12.2 to classify our covariates into four types of factors as shown in Table 12.1. For the firm specific factors, we use the $Age(Y)_i$ and $Size_i$ to examine *liability of newness* and *liability of size* hypotheses as highlighted in Sect. 12.2 (Van Wissen 2000; Baldwin and Gellaty 2004). The $Size_i$ squared is also introduced to account for any non-linear effect that the size variable might exhibit (Baldwin et al. 1997).

$Growth_i$ is introduced as a time covariate to control for the relation between the growth in employment size and the survival of the establishment (Van Wissen 2000; Fotopoulos and Louri 2000). Other things being equal, establishments that grow over time reflect superiority in performance and are less likely to fail. $Relocate_i$ is another time covariate to examine if a change in business location has an influence on survival. Dijk and Pellenbarg (2000) note that intrametropolitan firm location occurs when the business is able to react to ongoing stress and loss in profit. Failure to do so will lead to a sharp decline in profit, resulting in failure.

Local competition impacting establishment i is measured as the ratio of employment in rival establishments to the employment of establishment i. Rivals are establishments with the same two-digit industry as establishment i that are located within a radius of 1,500 m from establishment i. This radius provided the most statistically significant results when compared with radius values of 3,000, 2,500, 2,000, 1,000 and 500 m. According to Lerman and Liu (1984), the devised ratio represents competitors with the same size as establishment i in the vicinity of i.

The *Agglomeration* variable is used to examine the positive effects of agglomeration economies on the success of businesses. Previous studies for the city of Hamilton suggested a strong presence of agglomeration economies in the form of geographic clustering among firms (Maoh and Kanaroglou 2007a). As such, agglomeration is measured as the number of establishments with the same two-digit industrial sector within 1,500 m from business establishment i. The radius of

Table 12.1 Description of covariates used in the failure hazard duration models

Covariate name	Covariate definition	Expected sign
Firm specific factors		
$Age(Y)_i$	1 if business establishment i is Y years old, 0 otherwise. $Y = 1, 2, \ldots, 6$ and $7+$	+
$Size_i$	Employment size of business establishment i in the base year	–
$Size_i^2$	Employment size of business establishment i in the base year squared	+
$Growth_i$	1 if business establishment i experienced growth in employment size in a given individual year t between 1996 and 2002, 0 otherwise	–
$Relocation_i$	1 if business establishment i relocated within the city of Hamilton in any given year between 1996 and 2002, 0 otherwise	–
Geography specific factors		
$Competition^k_i$	Ratio of the total number of employees of establishments in the same k two-digit industry of establishment i to the number of employees of establishment i. Total number of employees is measures from all business establishments of sector k within a 1,500 m radius around establishment i	+
$Agglomeration^k_i$	Total number of establishments in the same k two-digit industry of establishment i within a 1,500 m radius around establishment i	–
Core	1 if business establishment i is located below the escarpment excluding the CBD, 0 otherwise	
Inner suburbs	1 if business establishment i is located above the escarpment, 0 otherwise	
Outer suburbs	1 if business establishment i is located in any of the outer municipalities that includes: Dundas, Ancaster, Stoney Creek, Flamborough or Glanbrook, 0 otherwise	
Macro-economic specific factors		
UEMR	Annual unemployment rate in the Hamilton region	+
ATINC	Annual average total income (Cnd $) in the Hamilton region	–
Industry specific factors		
$INDUSTRY(s)$	See definition in the Appendix A	
$RASE^k_i$	Ratio of the average employment size of industry k at the two-digit that establishment i belongs to the annual employment size of establishment i	+

1,500 m is again based on experimenting with different values as in the case of the local competition variable. It should be noted that the *Agglomeration* and *local competition* variables tend to be correlated for industries that are less dispersed over space. This is particularly the case for establishments that exhibit strong central locational preferences.

Results from non-parametric survival analysis as reported in Maoh (2005) suggest a variability of survival by geography in the city of Hamilton. Thus, a number of geographical dummies were devised to characterize the different municipalities of the city. Also, the urban core in the Hamilton municipality is divided into three geographic areas: (1) Central Business District (*CBD*) (2) *Core* area excluding the CBD and (3) *Inner Suburbs*. In some cases, the outer municipalities were grouped to devise the *Outer Suburbs* covariate as described in Table 12.1. These covariates, other things being equal, assess the relation between establishment failure and urban morphology.

Two time covariates are included to control for macro-economic conditions in the model: (1) unemployment rate *UEMR* and (2) average annual total income *ATINC*. *UEMR* is used to capture the direct effect of economic downturn (Holmes et al. 2000; Audretsch and Talat 1995). The variable also captures part of the decline in demand for goods and services in certain sectors of the economy such as retail trade and certain services sectors. On the other hand, *ATINC* is employed to account for the purchasing power and high level of market demand in the city. Higher income levels increase the demand for goods and services and thus contribute to the success of businesses supplying these goods and services.

Industry specific factors are used to examine the relation between industry type and the hazard of failure and to account for establishment heterogeneity. A number of categorical variables are used to classify establishments by industry type at the two-digit or three-digit 1980 SIC codes. Although initially we had a large number of industrial dummies, the final specification only included those that were statistically significant. Furthermore, the effect of industry size on the hazard of failure is tested through the $RASE^k_i$ ratio, which is a time covariate. The variable compares the average employment size of the industry at the two-digit SIC that establishment *i* belongs to the annual employment size of the establishment. Baldwin et al. (2000) argue that establishments have a better chance of survival if their employment size can reach an optimal level that matches the average size of the industry they belong to. Alternatively, the hazard of failure is more pronounced if the gap between the establishment size and the average size of the industry increases over time.

12.4 Results and Discussion

The estimation results of the hazard duration models are shown in Table 12.2. The findings indicate that the models are well behaved and most of the estimated parameters are significant and meet our a priori expectation with regard to the hypothesized sign. The estimated coefficients of the *Age* covariates are in line with the "liability of newness" hypothesis (Van Wissen 2000) and suggest that failure is more prominent among younger establishments. Here, the different age parameters tend to decline as we move from *Age1* to *Age6*. It should be noted that age covariates that were not significant were dropped from the final specification of the model.

Based on the age parameters, market penetration for SME establishments occurs at different points during the lifetime of the establishment and depends on the type of industry. Market penetration in our context refers to the ability to enter and remain in the market beyond the infancy years when the firm is susceptible to failure. For instance, market penetration for finance insurance and real estate (FIRE) business establishments is more common after the first 2 years of operation,

Table 12.2 Estimation results of failure hazard models by economic sector

Covariate	M	C	T	W	R	FI	RE	B	H&SS	E	AF&B	OS
Constant	0.8792 –	5.2390 **	9.0369 –	3.6324 –	5.2866 ***	−5.3791 –	7.6808 **	6.4470 ***	10.7244 ****	−1.1632 ****	0.3912 –	7.8168 ****
Age(1)	0.7412 ****	0.9727 ****	1.2526 ****	0.7973 ****	0.9173 ****	0.8757 ****	0.5023 ***	1.0800 ****	0.7950 ****		0.7391 ****	1.0397 ****
Age(2)	0.9350 ****	0.7734 ****	1.1304 ****	0.8349 ****	0.8131 ****	0.5894 ***	0.4261 **	0.6310 ****	0.5867 ****		0.7704 ****	0.6046 ****
Age(3)	0.5569 ***	0.7289 ****	1.0753 ****	0.6594 ***	0.8746 ****			0.8320 ****	0.5047 ***		0.3345 **	0.6309 ****
Age(4)	0.3960 *		0.9993 ***		0.5911 ****			0.3694 **	0.7301 ****			0.7720 ****
Age(5)					0.5036 ***							0.7487 ****
Age(6)					0.4591 ***							
Size	−0.0472 ***	−0.0940 ****	−0.0099 –	−0.0794 ***	−0.0593 ****	−0.1204 ***	−0.0347 **	−0.0292 –	−0.000012 –	−0.2078 ***	−0.0761 ****	−0.0974 ****
Size²	0.0011 ***	0.0034 ****	0.0001 –	0.0025 ****	0.0015 **	0.0037 ****		0.0008 –	0.0001 –	0.0092 ****	0.0018 ****	0.0027 ****
Growth	−1.5766 ****	−1.5574 ****	−1.1205 –	−1.2497 ****	−1.5643 ****	−1.5777 ****	−0.9536 ****	−1.5788 ****	−1.3677 ****	−1.2008 ***	−1.2787 ****	−1.3264 ****
Relocation	−0.7801 ****	−0.9651 ****	−1.3959 ****	−1.1447 ****	−0.3631 ****	−0.9232 ****	−0.6609 ****	−0.5884 ****	−0.7129 ****		−0.5927 ****	−0.6204 ****
Competition	0.0011 ***	−0.0008 –	0.0001 –	0.0004 –	0.0015[a] **	0.0008 **	0.0010 **	0.0000 –			0.0001 ***	0.0000 –
Agglomeration	−0.0102 –	0.0037 –	−0.0283 –	0.000124 –	−0.0083 ***					−0.0600 ***		−0.0001 –
UEMR	0.2868 ****	0.0673 –	0.0938 –	0.0903 –	0.0789 *	0.2163 **	0.1908 ***	0.0756 –			0.1282 **	0.1214 ***

12 Modelling Firm Failure: Towards Building a Firmographic Microsimulation Model

ATINC	−0.0001 −	−0.0002 ****	−0.0003 ***	−0.0002 *	−0.0002 ****	0.0001 −	−0.0003 ***	−0.0002 ****		−0.0001 −	−0.0003 ****
CBD		0.6503 ****	0.7386 **		0.2207 **		−0.4994 **	0.1552 −	0.4103 ****		
Core area		0.3399 ****	0.6161 ***			0.6989 ****			0.3980 ****		
Inner suburbs		0.3588 ****	0.4639 *	0.5994 ****						0.3084 ***	
Outer suburbs	−0.2054 −										−0.1851 *
Dundas		0.6198 ***									
Stoney Creek					0.3018 ***						
Ancaster										1.0483 ****	
RASE	0.0043 **	0.0031 −	0.0115 −	0.0308 **	0.0148 **	−0.0711 ***	−0.0553 −	−0.0052 −	0.0159 *	0.0050 −	−0.0046 −
INDUSTRY(1)	−0.7172 ****										
INDUSTRY(2)	−0.2996 **										
INDUSTRY(3)		−0.3813 −									
INDUSTRY(4)		0.3528 −									
INDUSTRY(5)			−0.2429 −								
INDUSTRY(6)				1.2062 ****							

(continued)

Table 12.2 (continued)

Covariate	M	C	T	W	R	FI	RE	B	H&SS	E	AF&B	OS
INDUSTRY(7)						0.6494 ***						
INDUSTRY(8)								−0.3700 ***				
INDUSTRY(9)								−0.3517 ***				
INDUSTRY(10)								−0.5616 ****				
INDUSTRY(11)									2.6128 ****			
INDUSTRY(12)									1.2138 ****			
INDUSTRY(13)									0.5429 ****			
INDUSTRY(14)									0.8732 **			
INDUSTRY(15)											−0.5365 *	
INDUSTRY(16)												−0.7011 ****
Pseudo-R²	0.19	0.16	0.21	0.17	0.14	0.16	0.08	0.12	0.10	0.20	0.13	0.14
No. Obs	2,861	5,841	1,049	2,377	6,345	768	1,978	4,465	5,680	308	2,936	6,059
−2 Log L	1,420.35	3,351.69	666.44	1,285.07	4,141.61	642.24	1,244.17	2,635.35	2,441.28	186.03	2,020.80	3,270.72

The significance level is provided below each parameter. Significance at the 1 %, 5 %, 10 % and 15 % level is indicated by ****, ***, ** and * respectively. Insignificant parameters are indicated with −

Note: *M* manufacturing; *C* construction; *T* transportation; *W* wholesale trade; *R* retail trade; *FI* finance insurance services; *RE* real estate services; *B* business services; *H&SS* health and social services; *E* educational services; *AF&B* accommodation, food and beverage services; *OS* other services.

[a] The parameter value is based on interacting the *Competition* covariate with *SIC63* industrial dummy variable.

other things being equal. That is, the first 2 years of operation tend to be the most difficult for FIRE establishments. Retail trade establishments, on the other hand, face more difficulty to remain in the market and while their failure diminishes with age, it is still common among older establishments. The same could be said about establishments within the "other services" industrial sector, where market penetration is more common after the fifth year of operation. By comparison, educational service establishments can penetrate the marker at the early stages in their lifetime.

Establishment size also seems to impact the hazard probability in all sectors expect for transportation, business service, and health and social service industries. The negative and significant parameter of the *Size* covariate suggests that smaller establishments are more susceptible to failure and as size increases, the propensity of failure decreases. However, this relation is not linear as indicated by the positive and significant parameter of the *Size*-squared in some sectors. The hyperbolic relation between size and failure suggests that, other things being equal, failure will decrease with establishment size up to a certain point after which it will start increasing (Maoh 2005).

The growth parameter suggests that establishments exhibiting an increase in their employment size in a given year are less likely to fail. While this result is intuitively appealing, the introduction of this variable in the model has a significant impact from a predictive point of view. The devised failure model is part of a larger firmographic microsimulation model, which includes a separate growth model. During simulations, establishments are subjected to the growth and the failure models to determine their transitional status. All else being equal, establishments that exhibit decline in employment size will have a higher propensity of failure compared to their growing counterparts and vice-versa. The failure status is then determined from the annual failure probabilities using Monte Carlo simulations. Therefore, the *Growth* covariate in the failure model works as a performance indicator.

The same could be said about the *Relocation* covariate. Typically, relocation occurs if the establishment requires additional floorspace to expand although growth does not always trigger relocation. It can also occur if the establishment is trying to overcome losses in profit due to the low attractiveness of its current location or if it is eager to increase its profit by relocating to a new site (Dijk and Pellenbarg 2000; Maoh 2005). This relocation action reflects, to a certain extent, superiority in performance. However, other things being equal, the relocation in our model does not suggest total immunization against failure. An examination of the 1996 cohort shows that 34 % of the relocating establishments between 1996 and 2002 have failed before 2002 after the relocation was observed. By comparison, 48 % of non-movers have failed between 1996 and 2002. As the results indicate, relocation decreases the hazard of establishment failure but this relation is variable by the type of economic sector. It can be seen that establishment from the wholesale trade and transportation sectors attain higher benefits by relocating.

Despite the temporal nature of the hazard failure model, the simulations will be conducted in 1 year time steps to determine the probability of an establishment failing or surviving. This is done because it is not feasible to determine the relocation of an establishment for more than 1 year at a time given the cross-sectional nature of the mobility sub-model location (see Maoh and Kanaroglou (2007b) for the details of modeling this process). Admittedly, the way the Relocation variable is specified is primarily done for explanation purposes in the hazard model. That is, we wanted to examine the degree by which relocation affects the hazard of failure among the different economic sectors during the observation period. However, since the hazard model is temporal in nature, it is possible to calculate the hazard of failure and survival at any given point in time t, ($t = 1, 2, \ldots, 6$). Therefore, we will assume that the recent relocation status of the establishments in the base year population is known before starting the simulations. This way the probability of failure in the hazard model can take into account the impact of relocation on failure since the failure model is the starting point of simulations in any given year. Relocation for all successive simulation years will be based on the mobility model shown in Fig. 12.2 and discussed in Maoh and Kanaroglou (2007b).

As expected, the competition and agglomeration parameters have a significant impact on the propensity of failure of certain establishments. However, due to multicolinearity, the specification of both variables simultaneously was not possible in some of the models. This is especially the case for establishments in the city core that belong to the finance insurance, real estate, business, education and AF&B industries. In such cases, one of the variables was included on the basis of statistical significance. The results suggest that local competition is significant and increase the hazard of failure in the manufacturing, retail trade, finance insurance, real estate and AF&B models. However, within the retail trade sector, competition appears to impact the hazard of failure of "automotive vehicles, parts and accessories, sales and service" establishments (SIC 63). On the other hand, the agglomeration parameter appears to be significant in the retail trade and education service models and tends to decrease the propensity of failure. This is expected in the case of retail trade since geographic clustering (one form of localization economies) is vital for the success of such establishments. This relation also holds to some extend in the case of educational establishments.

The categorical variables depicting urban morphology suggest that the probability of failure varies over space. However, the general pattern of results suggests that central establishments have a higher hazard compared to suburban establishments. These findings are in line with the observed urban sprawl in the city of Hamilton, which has led to the decay of the CBD in the last two decades (Maoh and Kanaroglou 2007a). Suburban establishments enjoy accessibility to clients with higher income. Also, suburban areas are experiencing a continuous development, which in turn is attracting more establishments away from the city center. By comparison, central areas are becoming less attractive to both newly formed establishments (Maoh and Kanaroglou 2009) and to high-income households. As a result, the hazard of failure for central establishments has been on the rise and this trend is expected to continue, at least in the near future.

In terms of the role of macro-economy, the *UEMR* parameter suggests that high failure rates are associated with high unemployment rates, due to economic

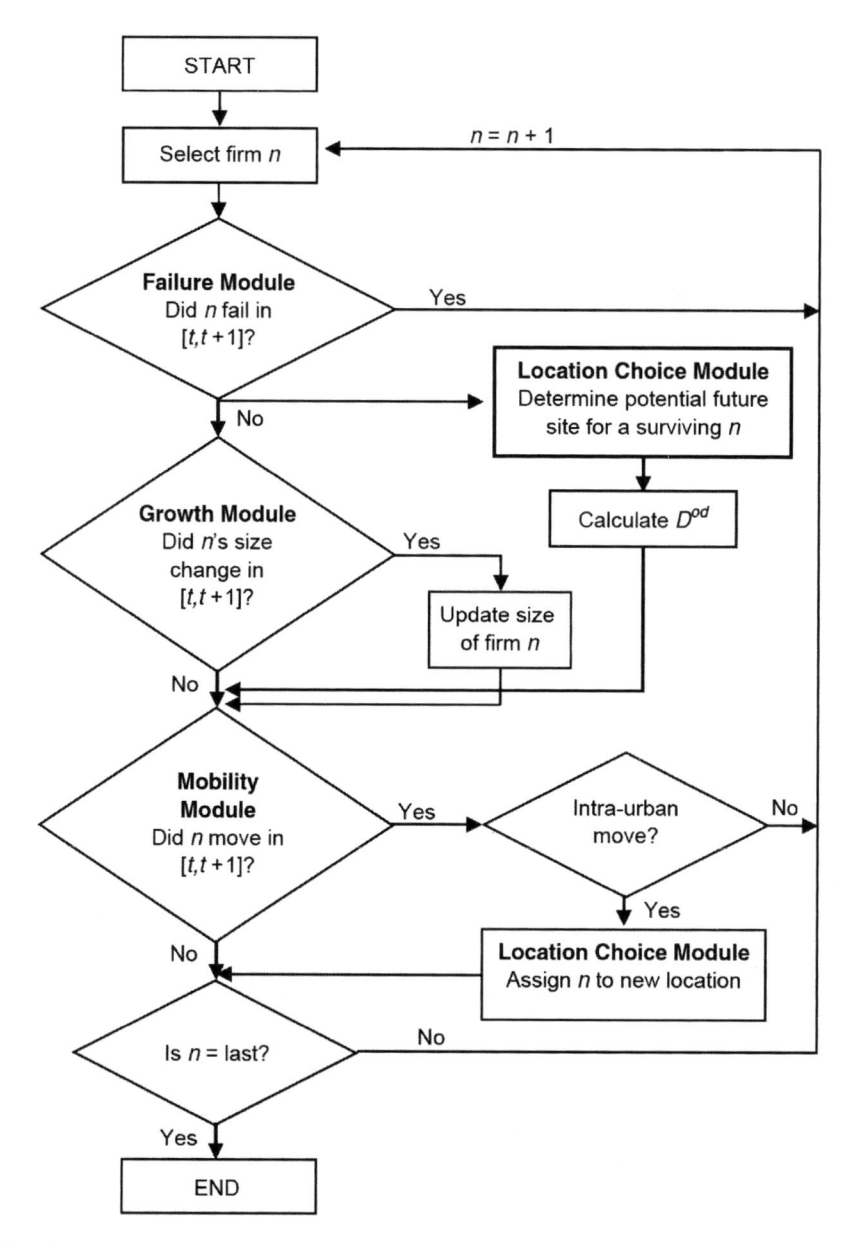

Fig. 12.2 General structure of Hamilton's firmographic model for one time step operation on a base year population (Source: Maoh and Kanaroglou 2009)

downturns, in Hamilton. This is evident in the case of manufacturing (Holmes et al. 2000; Audretsch and Talat 1995), finance insurance, real estate, AF&B and other service establishments. On the other hand, the *ATINC* parameter suggests that an increase in the levels of income in Hamilton will lead to a higher purchasing power, decreasing the likelihood of failure. This is more evident among retail trade, real estate, business service, other service, construction and transportation establishments. This is expected since a lot of these establishments tend to sell their products locally.

Industry specific variables suggest a variation in the propensity of failure among establishments from the different industries, as can be seen from the parameters of the industry covariates in Table 12.2. The results not only hint to the variation in survival by industry but also highlight which industries are more or less susceptible to failure, all else being equal. For example, the findings suggest that product-differentiating industries, as shown in the manufacturing model, are less likely to fail, other things being equal. We believe this is the case since many of those small manufacturing establishments have the advantage of holding a larger share of the local market. Given their size and nature, these establishments have the ability to produce more tailored products or change production rapidly to meet the demand of the local market. On the other hand, labour-intensive manufacturing firms have lower failure rates since labour constitutes the bulk of the cost in their production process and labour wages among these establishments are typically low.

Finally, the relation between the *RASE* covariate and the hazard probability suggests that the gap between the size of the establishment and its industry's average size could help explain the failure of establishments from particular industries. The results suggest that establishments from the manufacturing, wholesale trade and retail trade sectors are required to match their size to the average size of the industry in order to remain in business. Failure to do so will increase the hazard of failure (Baldwin et al. 1997), other things being equal. More interestingly, the results of the *RASE* parameter in the finance insurance model is counter intuitive. One possible explanation for this result is that the gap between the size of a failing finance insurance establishment and its industry's average size is not very large. In fact, Troske (1996) notes that finance insurance establishments enter the market at a much larger size and experience a much smaller change in that size the first 10 years relative to manufacturing establishments. According to the author, the latter establishments are only half the industry's average size 5 years prior to their failure. By comparison, finance insurance establishments that are destined to fail maintain the average size of the industry 5 years prior to their failure.

12.5 Conclusion

The objective of this paper was to develop an econometric model with two purposes in mind: first, to investigate the failure process of small and medium size establishments with an application to the city of Hamilton, Canada; second, to use the model as the basis for the development of a business failure module in a firmographic microsimulation system. Such a firmographic system will be developed to simulate the different firmographic processes that will drive the spatio-temporal evolution of Hamilton's business establishment population. Therefore, in addition to the business failure module as shown in Fig. 12.2., the microsimulation system to be developed includes: (1) an establishment mobility module that determines if a business establishment will move within the city or leave the city over time, (2) a location choice module that determines the location of intra-urban, in-migrating and newly formed establishments in the city, and (3) a growth-decline module that predicts the change in employment size of business establishments over time.

The significance of the empirical work presented in this paper is twofold. First, it provides an analytical framework for studying and modeling the business establishment failure process in the urban context. It does so by highlighting the determinants that explain the intra-urban failure process using real world data. Second, it adds to the growing efforts to model and potentially simulate the demography of firms (Van Wissen 2000).

Unlike earlier efforts on firm failure from the literature, our work also emphasizes the geographical element of the business establishment failure process. Similar efforts are rare and not very elaborate (Miron 2002; Lerman and Liu 1984). Our findings advance the current state of knowledge on how the failure process operates over space in the city. The results also show how certain geographical factors such as local competition and agglomeration economies influence the survival of SME businesses from the different economic sectors. Furthermore, the analysis establishes a relationship between urban form and business failure. These are important aspects that have not been addressed elaborately in the past. The estimation of separate models by economic sector also highlights the cross-industry differences with regard to the determinants that explain the failure of SME establishments over time and space. Such elaborate analysis is absent from the current literature.

While our attempts to utilize the Business Register (BR) data to explain the failure process of SME businesses were insightful, the specification of the models suffered some limitations and was constrained by the quality of the available Business Register (BR) data. For instance, firm specific internal factors were dismissed from the specification of the model despite their importance in explaining the failure process as shown by Baldwin et al. (1997). This is an area that needs to be addressed in future research.

Acknowledgments The authors thank Statistics Canada and the Social Sciences and Humanities Research Council of Canada for supporting this research.

Appendix A Definition of the *INDUSTRY(s)* Covariates in Table 12.1

INDUSTRY(s) $= 1$ if establishment i belongs to specific economic sectors s, 0 otherwise where $s = $ **1, 2, ..., 16**. The following is the definition of the different s values:

1 = production differentiating manufacturing industries (Vinodrai 2001); 2 = labour intensive manufacturing industries (Vinodrai 2001); 3 = industrial and heavy (engineering) construction industries (SIC 41); 4 = service industries incidental to construction (SIC 44); 5 = other transportation industries (SIC 458); 6 = apparel and dry goods wholesale industries (SIC 53); 7 = deposit accepting intermediary industries (SIC 70); 8 = accounting and bookkeeping service industries (SIC 773); 9 = architectural, engineering and other scientific and technical business service industries (SIC 775); 10 = offices of lawyers and notaries business service industries (SIC 776); 11 = hospitals service industries (SIC 861); 12 = non-institutional health service industries (SIC 863); 13 = non-institutional social service industries (SIC 864); 14 = offices of social services practitioners (SIC 867); 15 = lodging houses and residential clubs service industries (SIC 912); 16 = membership organization service industries (SIC 98)

References

Agarwal R, Audretsch D (2001) Does entry size matter? The impact of the life cycle and technology on firm survival. J Ind Econ XLIX:21–43

Allison P (1995) Survival analysis using the SAS system: a practical guide. SAS Institute Inc., Cary

Audretsch D (2001) Research issues relating to structure, competition, and performance of small technology-based firms. Small Bus Econ 16:37–51

Audretsch D, Talat M (1995) New firm survival: new results using a hazard function. Rev Econ Stat 77:97–103

Audretsch D, Santarelli E, Marco V (1999) Start-up size and industrial dynamics: some evidence from Italian manufacturing. Int J Ind Organ 17:965–983

Baldwin J, Gellaty G (2004) Innovation strategies and performance in small firms. Edward Elgar, Cheltenham

Baldwin J, Gray T, Johnson J, Proctor J (1997) Failing concerns: business bankruptcy in Canada. Statistics Canada, Ottawa

Baldwin J, Bian L, Dupuy R, Gellatly G (2000) Failure rates for new canadian firms: new perspectives on entry and exit. Statistics Canada, Ottawa

Berglund E, Brannas K (2001) Plants' entry and exit in Swedish municipalities. Ann Reg Sci 35:431–448

Bliemer M, De Bok M (2006) Infrastructure and firm dynamics: calibration of micosimulation model for firms in the Netherlands. Trans Res Rec: J Trans Res Board 1977:132–144

Bok M (2009) Estimation and validation of a microscopic model for spatial economic effects of transport infrastructure. Trans Res A Policy Pract 43(1):44–59

Caves R (1998) Industrial organization and new findings on the turnover and mobility of firms. J Econ Lit 36(4):1947–1982

Dijk J, Pellenbarg P (2000) Firm relocation decisions in the Netherlands: an ordered logit approach. Pap Reg Sci 79:191–219

Elgar I, Miller E (2006) Conceptual model of location of small office firms. Trans Res Rec: J Trans Res Board 1977:190–196

Fotopoulos G, Louri H (2000) Location and survival and new entry. Small Bus Econ 14:311–321

Guo G (1993) Event-history analysis for left-truncated data. Sociol Methodol 23:217–243

Holmes P, Stone I, Braidford P (2000) New firm survival: an analysis of UK firms using a hazard function. In: Proceedings: The Royal Economic Society 2000 conference, University of St. Andrews, Scotland
Hunt J, Khan J, Abraham J (2003) Microsimulating firm spatial behavior. In: Proceedings of 8th international conference on computers in urban planning and urban management, Sendai
Kumar S, Kockelman K (2008) Tracking size, location, and interactions of businesses: microsimulation of firm behavior in Austin, Texas. Transport Res Rec 2077:113–121, Journal of the Transportation Research Board
Lerman S, Liu T (1984) Microlevel econometric analysis of retail closure. In: Pitfield D (ed) Discrete choice models in regional science. Pion, London
Maoh H (2005) Modeling firm demography in urban areas with an application to Hamilton, Ontario: towards an agent-based approach. PhD dissertation, McMaster University
Maoh H, Kanaroglou P (2007a) Geographic clustering of firms and urban form: a multivariate analysis. J Geogr Syst 9:29–52
Maoh H, Kanaroglou P (2007b) Business establishment mobility behavior in urban areas: a microanalytical model for the City of Hamilton in Ontario, Canada. J Geogr Syst 9:229–252
Maoh H, Kanaroglou P (2009) Intrametropolitan location of business establishments: microanalytical model for Hamilton, Ontario, Canada. Transport Res Rec 2133:33–45, Journal of Transportation Research Board
Mata J, Portugal P (1994) Life duration of new firms. J Ind Econ 42:227–246
Miller E, Hunt J, Abraham J, Salvini P (2004) Microsimulating urban systems. Comput Environ Urban Syst 28:9–44
Miron J (2002) Loschian spatial competition in an emerging retail industry. Geogr Anal 34:34–61
Moeckel R (2005) Microsimulation of firm location decision. In: Proceedings of 9th international conference on computers in urban planning and urban management, London
Schwartz M (2009) Beyond incubation: an analysis of firm survival and exit dynamics in the post graduation period. J Technol Transfer 34:403–421
Strotmann H (2007) Entrepreneurial survival. Small Bus Econ 28:87–104
Troske K (1996) The dynamic adjustment process of firm entry and exit in manufacturing and finance insurance and real estate. J Law Econ 39:705–735
Van-Wissen L (2000) A micro-simulation model of firms: applications of concepts of the demography of the firm. Pap Reg Sci 79:111–134
Vinodrai T (2001) A tale of three cities: the dynamics of manufacturing in Toronto, Montreal and Vancouver, 1976–1997. Research Paper Series No. 177, Analytical Studies Branch, Statistics Canada, Ottawa

Chapter 13
Choice Set Formation in Microscopic Firm Location Models

M. de Bok and F. Pagliara

Abstract This chapter presents a spatial firm demographic model (SFM) that simulates changes in states of individual firms and their location choice behaviour. Firm location choice in such disaggregate models is characterised by large numbers of alternatives and complex spatial interdependencies among them. This chapter deals with a particular issue of firm location choice: the choice set composition in a disaggregate spatial choice context. A choice model is presented with probabilistic choice sets assuming that choice alternatives that are dominated by others are not taken into consideration in the location decision. The estimated models have significant parameters for dominance, and they are implemented in the SFM model, to test to what extent the simulation results are improved.

13.1 Introduction

Choice behaviour can be considered as the process of selecting a certain alternative from a limited set of discrete opportunities in accordance with some decision rule (Golledge and Timmermans 1990; Thill 1992). Specifically in spatial detailed urban simulation models the formation of the choice set is challenging because of the large sets of possible alternatives. This makes it hard to assume beforehand that the individual is able to evaluate each and every one of them and then, make an educated decision. A portion of the universe is considered instead.

In the literature, three main strands addressing choice set formation for problems with a large number of alternatives can be identified: deterministic, probabilistic

M. de Bok (✉)
Significance, The Hague, The Netherlands
e-mail: debok@significance.nl

F. Pagliara
Department of Transportation Engineering, University of Naples Federico II, Naples, Italy
e-mail: fpagliar@unina.it

and sampling of alternatives. An overview of the different approaches of the first two strands is given in Thill (1992), who focused on destination choice and, in a more recent literature review, in Pagliara and Timmermans (2009), who focused on spatial contexts in general.

The models of the first strand are based on a deterministic specification of choice sets, where the choice sets are an exogenous input to the estimation step (Gautschi 1981; Weisbrod et al. 1984; Adler and Ben-Akiva 1976; Miller and O'Kelly 1983; Southworth 1981; Golledge and Timmermans 1990) and include also models of the time-geographic approach (Hägerstrand 1970; Landau et al. 1982; Thill and Horowitz 1997; Scott 2006). The second strand, which is often called the probabilistic approach, was founded by Manski (1977) and integrates the choice set formation step into the estimation procedure and jointly estimates the selection of a choice set and the choice of a particular alternative of this choice set. The third strand of choice set formation techniques are sampling of alternatives. This technique is commonly applied to avoid the computational burden involved in estimating choice models with large number of alternatives (Bierlaire et al. 2006).

Firm location choice is a typical example of choice behaviour in a spatial context. In particular in spatially disaggregated urban studies, the number of alternatives is large and complex interdependencies exist between clustered alternatives. However, most empirical studies (Anderstig and Mattsson 1991; Shukla and Waddell 1991; Waddell and Ulfarsson 2003; De Bok and Van Oort 2011; Maoh and Kanaroglou 2007) apply choice sets with random samples of alternatives in a multinomial logit (MNL) model, potentially violating the independence of irrelevant alternatives (IIA) assumption. The IIA assumption states that the probability ratio of a firm, selecting one alternative over the other, is independent over all other alternatives in the choice set. Some recent microscopic simulation studies try to improve the choice set formation in the context of firm location. Elgar (2011) applied a stratified sampling procedure to a priori defined search areas. Kim et al. (2008) is an example where count models are used to predict job locations in a microscopic urban simulation model. These count models are free from the IIA assumption underlying the MNL model, but they have less capacity to predict large count situations, often found in employment location. De Bok (2009) applied a probabilistic approach, the competing destinations model (Fortheringham 1983) to include spatial interdependencies between alternatives.

Furthermore it can be argued that in many choice contexts, some alternatives are not taken into account by the decision maker since they are "dominated" by other alternatives. In general, an alternative i is dominated by another alternative j if i is "worse" than j, with respect to one or more characteristics, without being better with respect to any characteristic. The concept of dominance among alternatives has been used within Random Utility (RU) theory only in destination choice modelling (Cascetta and Papola 2009) and in residential location choice (Cascetta et al. 2007). In this chapter the concept of dominance among alternatives is applied to the problem of choice set formation in the context of firm location.

This chapter first presents the microscopic Spatial Firm demographic Micro simulation (SFM) model that simulates firm location decisions and other transitions in the state of an individual firm (start-up, growth/decline and dissolution). Then the firm location model is presented that includes dominance on the choice set formation. The choice set parameters are estimated and presented. Next, these parameters are implemented in the SFM model and the simulation results are compared to the observed numbers of firms at neighbourhood level to test to what extent the concept of dominance improves the simulation results.

13.2 SFM Model

13.2.1 Firm Demography

The model presented in this chapter simulates the dynamics within a firm population in a disaggregate urban environment, with the objective to predict spatial economic development. This approach is founded on firm demographic micro simulation that has a long tradition started with Birch (1979). Many theories underlying the micro behaviour of firms have been described in the literature on demography of organisations (Carrol and Hannan 2000). An earlier empirical work on firm demographic simulation was presented by Van Wissen (2000), but at a regional scale. However, some recent international examples are available in which firm demographic micro simulation is applied in disaggregate urban simulation models: Moeckel (2007 and 2012), Maoh and Kanaroglou (2007 and 2012). The spatial dimension allows the inclusion of detailed location factors in the models for change in state (e.g. firm relocation or growth), but also introduces challenges for the model specification.

13.2.2 Structure of the Model

The SFM model describes the change in the state of individual firms that are beneath aggregate economic developments, similar to the SIMFIRMS model by Van Wissen (2000). These transitions are the result of a number of firm demographic events that are influenced by firm internal factors (e.g. size, age). In the SFM model, these events depend on location factors as well (available locations, accessibility, agglomeration). These events are probabilistic in nature: they can or may not occur.

The causal structure of the SFM model is outlined in Fig. 13.1. From the economic, planning and mobility scenario's at the top, the dependencies between elements are drawn. Bottom up, the approach simulates the firm population with distinctive sub-populations (industry sectors) and individual firms. The actual developments are simulated at the individual firm level. The firm demographic events that are simulated include firm formation, firm growth, firm migration and firm dissolution. These events are first of all influenced by the state of firms, defined

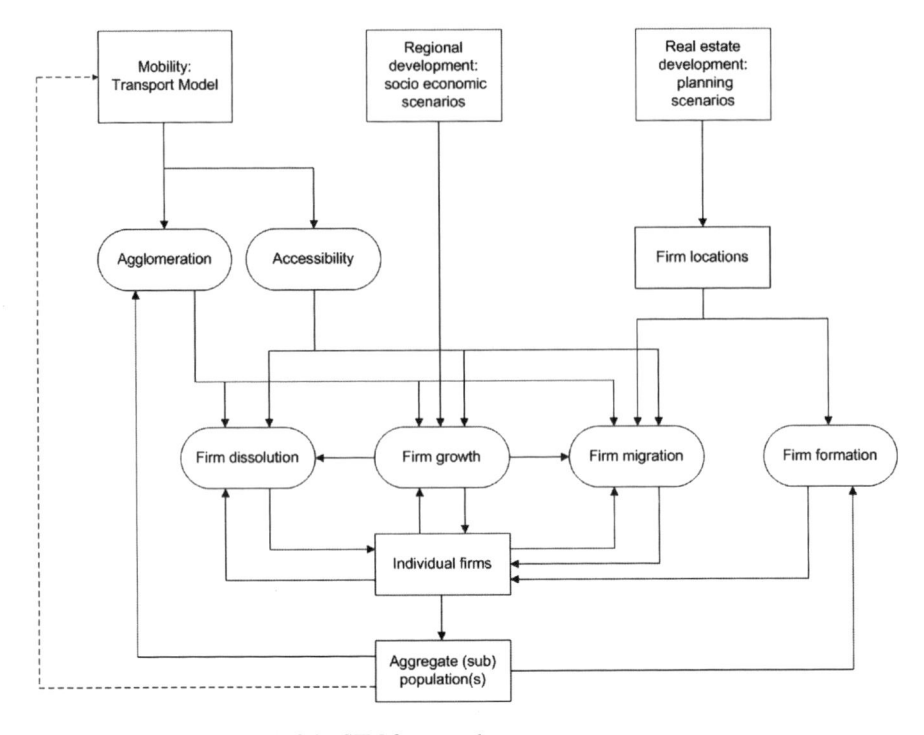

Fig. 13.1 Causal structure of the SFM framework

by attributes (e.g. age, size, location) that can change through firm demographic events: firm dissolution and firm migration can both be instigated by firm growth.

Accessibility and agglomeration are included as location factors in the firm demographic events. Accessibility is derived from a transport model. Agglomeration is derived from the (sub) firm population and travel times from the transport model.

The model represents the supply of industrial real estate at each firm location in order to account for constraints in location options explicitly. Each firm requires a corresponding firm location, in other words, an office related firm can only be located at a location that has the required amount of available office floor space. It is interesting to highlight that during its existence a firm might relocate to another location. Changes in the floor space supply, as a result of the construction of new industrial or office sites, are included through exogenous plans.

The regional economic development is conditional to the micro developments that are simulated in the SFM model. Thus, individual firm developments are influenced by structural developments in the industry sector or other macro influences that are outside the scope of this analysis. The implementation of exogenous scenarios for the regional economic development implies that the model cannot account for generative effects of transport infrastructure. This is assumed to be acceptable. First of all, these generative effects are much less significant compared to distributive effects (Rietveld 1994). Moreover, the scope of this urban simulation model is primary on (re-)distributive effects of infrastructure investments (allocation of employment growth and relocation of firms).

The SFM model distinguishes three geographical scale levels: the research area, zones, and locations. The research area comprises a region, province or state. Within this research area the microscopic developments in the firm population are simulated. Interaction with regions, outside this research area, takes place through exogenous scenarios. Zones are applied as an intermediate level to describe the characteristics of the urban environment around a location. Moreover, zones are required in order to enable interaction with the transport model. The finest geographical scale level that is applied is the six-digit zip code (comparable to building block). The simulation of individual firms and their detailed location enables the computation of detailed location attributes, such as proximity to train stations or highway onramps and the composition of the firm population in its surroundings.

13.2.3 Accessibility and Agglomeration in the SFM Model

The spatial environment of firm locations is an explicit and important dimension in the presented simulation approach. The quality of this spatial environment is described with accessibility and agglomeration attributes.

13.2.3.1 Proximity Measures

The transport based accessibility measures specifically are indicators of the transport infrastructure and the trips that can be made with it. The quality of the transport based accessibility is measured both with distance to infrastructure access points and logsum accessibility measures. The distance to train stations or highway onramps express a specific transport infrastructure quality that is easily interpreted. Previous empirical findings suggested that proximity measures to infrastructure access points are significant location factors in the location preference of firms (De Bok and Sanders 2005).

13.2.3.2 Urbanisation Economies

The logsum accessibility attributes and the travel time between zones in the study area are derived from the National Modelling System (NMS), the national transport model for the Netherlands (Hague Consulting Group 2000). The model is developed by the Transport Research Centre and it is applied in a back casting study that made the travel times and logsums available back to 1985. The NMS is based on disaggregate discrete choice models, and it provides the logsums for (non-home based) business trips and the (reflected) logsum for commuting trips. The logsum for business trips is assumed to be a representative measure for customer and supplier accessibility. This logsum is calculated as the sum of the trip utilities to all destination zones d for all person types p for all trips:

$$A_{om} = \log \sum_d \sum_p \exp(\mu_m \cdot V_{odpm}) \qquad (13.1)$$

with purpose $m =$ 'business trips', V_{odpm} as the expected utility for person type p to make a business trip from origin o to destination d and with the purpose specific scale parameter μ_m.

Labour market accessibility is derived from the utility of commuting trips in the transport model, from the perspective of the employer. For this reason, we analyse the commuting trips with a reflected logsum that measures the accessibility at the destination side of all commuting trips. The reflected logsum for commuting trips at firm location d, is specified as:

$$A_{dm} = \log \sum_o \sum_p \exp(\mu_m \cdot V_{odpm}) \qquad (13.2)$$

with trip purpose m 'commuting' and V_{odpm} as the expected utility for person type p to commute from origin o to destination d.

13.2.3.3 Localisation Economies

The congested travel time matrices are calculated in the peak hour, and they are used to determine a range band R_{jb} for each zone and to measure the composition of the firm population in this area. The composition is measured as the level of diversification or specialization within the range band from each location.

Specialization is measured as the concentration of an industry within a chosen travel range of 7.5 min that represents the local range band around the firm's location. For location j, the share of the employment in sector s in a range band R_{jb} from j is measured relative to the share of employment in that industry in the whole region. The production specialization index for location j and range band R_{jb} becomes:

$$PS_{jsb} = \dfrac{E_{sR_{jb}} \Big/ \sum_s E_{sR_{jb}}}{\sum_j E_{sR_{jb}} \Big/ \sum_j \sum_s E_{sR_{jb}}} \qquad (13.3)$$

with $E_{sR_{jb}}$ as the employment in industry within range band R_{jb}.

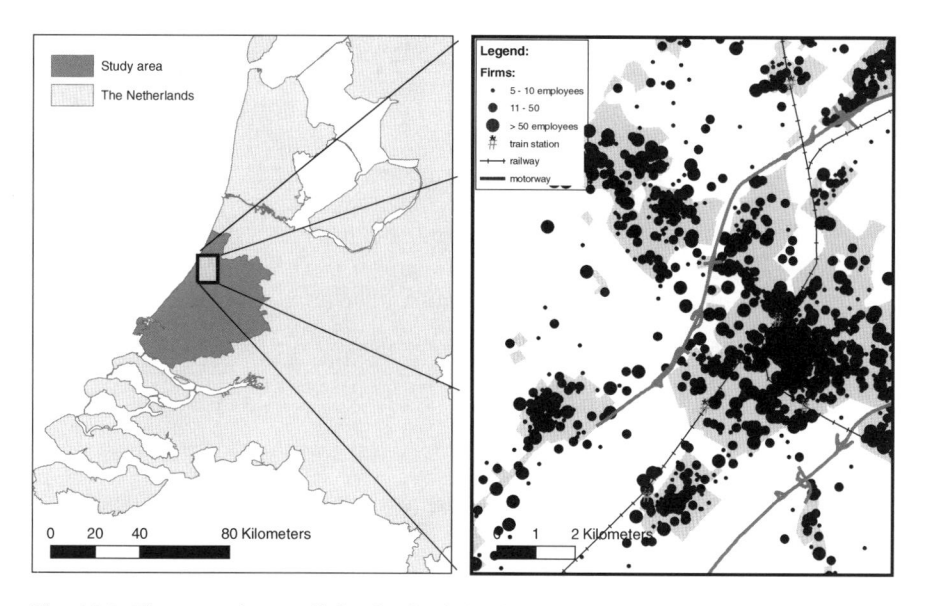

Fig. 13.2 The research area (*left*), detail of the firm data and their proximity to transport infrastructures (*right*)

Diversity of the firm population within a range band is measured with the productivity diversity index (Paci and Usai 1999). If S defines the number of industries and all industries are sorted in increasing order, then the production diversity index PD_{jb} for location j and range band R_{jb} is defined as:

$$PD_{jb} = \frac{1}{(S-1)E_{SR_{jb}}} \sum_{s=1}^{S-1} E_{sR_{jb}} \qquad (13.4)$$

with E_{sRjb} as the employment in the largest industry within range band R_{jb}.

13.2.4 Data Used for the SFM Model

The model has been estimated on an extensive longitudinal dataset of the firm population for the South Holland region, see Fig. 13.2, top left. This dataset is the so-called LISA business registration, a yearly updated database that includes the complete firm population. This data set is a common source for economic research in The Netherlands, for instance see Van Wissen (2000), Van Oort (2004) or De Bok (2009). For all firms in the population the dataset includes industry sector, size, the location (six digit zip code), the state in the previous year, and dummies for firm demographic events. For a graphical presentation of a fragment of the dataset see Fig. 13.2, top right. The firm population is segmented into 12 industry sectors. The model is estimated on period 1990–1996, and validated on the period 1996–2004 (De Bok 2009).

The attributes for diversification or specialisation in the direct surroundings of a location or in specific range band from that location, are computed from the travel times from the NMS and the location of all firms in the firm population.

In base year 1996 the study area hosted 95 thousand firms. The model was validated with the observed regional economic development in the study area between 1996 and 2004. The macro economic development is exogenous input that defines the size of the regional economy and in such forms constraints to the simulation and influences the outcomes at micro level. The supply of industrial real estate is a condition to the choice set formation when firms relocate. Changes in this industrial real estate supply are provided by synthetic real estate data for the corresponding period.

13.3 Location Choice Model with Probabilistic Choice Sets

The firm location choice model in this study is based on discrete choice models with probabilistic choice sets using the general form:

$$P_i = \frac{\exp(V_i) \cdot P(i \in C)}{\sum_j \exp(V_j) \cdot P(j \in C)} = \frac{\exp(V_i + \ln(P(i \in C)))}{\sum_j \exp(V_j + \ln(P(j \in C)))} \quad (13.5)$$

with the probability that an alternative i belongs to the decision-maker's choice set C, $P(i \in C)$, and the deterministic part of the location alternative i, V_i. An alternative's choice set membership is jointly estimated with its probability of being chosen within this set by adding the logarithm of $P(i \in C)$ in the utility function of alternative i. The theoretical proof is given in (Cascetta and Papola 2001). Different specifications of $P(i \in C)$ are introduced in the next section.

13.3.1 Choice Set Parameters

13.3.1.1 Centrality

The first proposition to measure competition between spatial clustered alternatives through a centrality measure was introduced by Fortheringham (1983). Based on the location of available alternatives, a centrality measure has been computed that measured the clustering of alternatives in each other's proximity. The closer two alternatives are located, the more likely they are to be substitutes to each other. This affects the choice probability of each individual alternative. The similarity between spatial alternatives is measured with a centrality measure c_i that is a proxy for the spatial cluster membership. The closer the alternatives are in space, the more likely they are to be substitutes for one another:

$$c_i = 1/(J-1) \sum_{j \neq i} w_j/d_{ij} \tag{13.6}$$

where J is the number of available firm locations, d_{ij} is the distance between alternative j and i and w_j as the size of alternative j. The size of an alternative is specified as the available (unoccupied) floor space or industrial area at a firm location. So, for each alternative that is selected in the consideration set, the centrality relative to all other available location alternatives is computed. It is important to stress that c_i measures the clustering of *available* locations. In the literature, centrality is often measured relative to current activities instead of available alternatives (Pelligrini and Fortheringham 2002). In those cases, centrality is similar to agglomeration. In this case, the model measures centrality relative to available firm locations. The influence of agglomeration economies is measured with the presented accessibility and agglomeration attributes.

The probability for choice set membership $p(i \in C)$ becomes:

$$P(i \in C) = (c_i)^\theta \tag{13.7}$$

Parameter θ is estimated. Values <0 indicate that alternatives that have many substitutes in proximity, have a high value for c_i, and have a smaller probability of being selected in a choice set.

13.3.1.2 Dominance

The dominance of location alternatives is determined at first a pairwise comparison of all alternatives in the universal choice set, see. The next step is to calculate the global degree of dominance. In the pairwise comparison different domination rules can be defined to determine if i is dominating j, given the available location attributes. The spatial dominance rule is defined considering the distance between two locations. A dominant alternative is an alternative close by relative to the other alternatives in the choice set. Specifically, for relocating firm n, a binary dominance variable $y^n{}_{ij}$ is defined equal to one if alternative i is dominated by j, if i is less close to n, compared to j:

$$y^n{}_{ij} = \begin{cases} 1 & \text{if } Dist_{ni} \geq Dist_{nj} \\ 0 & \text{otherwise.} \end{cases} \tag{13.8}$$

In a previous study it was shown that the specification of the dominance attribute is far from straightforward (De Bok and Pagliara 2011). Spatial dominance was a good predictor for choice set membership across all tested industry sectors.

Alternative dominance specifications that were tested proved to be less effective for improving the location choice models. The different specifications included dominance rules based on the accessibility of different alternatives (an alternative is dominated by another alternative if it has a worse proximity to train and highway onramp and lower logsum accessibility values) and the global dominance (defined on both the spatial location and accessibility).

The dominance degree of each alternative i is computed by summing the dominance of i compared to all alternatives j. This measure gives the number of alternatives within the feasible choice set of decision-maker n that dominate i. Thus, a low value for dominance means a high ranking:

$$dom^n{}_i = \sum_{j \in CS^n} y^n{}_{ij} \qquad (13.9)$$

The probability for choice set membership $P(i \in C)$ becomes:

$$P(i \in C) = (dom^n{}_i)^\theta \qquad (13.10)$$

Parameter θ is estimated. Values <0 indicate that alternatives that are dominated by other alternatives, and thus have a high value for $dom^n{}_i$, have a smaller probability of being selected in a choice set.

13.3.2 Estimation

The firm location models are estimated on a dataset with choice sets that are considered to be representative for the choice context of each observed firm relocation. The derivation of these choice sets is illustrated in Figure 13.3. First of all, the universal set of firm locations comprises of all firm sites (industrial-commercial- or office real estate) that exist in the research area. This set is reduced to a set of available locations by reducing the universal set with the locations occupied by the existing firm population. In case of a relocating firm, only those alternatives are assumed to be relevant that meet a minimal size constraint following from the size of the decision maker, the relocating firm.

This set of feasible alternatives is still a large set of alternatives that have to be brought down to a choice set with a limited but representative number of alternatives (20, including the chosen location). For estimation, choice sets were created by randomly sampling alternatives with equal probabilities. The measures for choice set membership, dominance and centrality (Eqs. 13.9 and 13.6), are computed for the selected alternatives and included in the estimation.

Fig. 13.3 Steps in the choice set formation

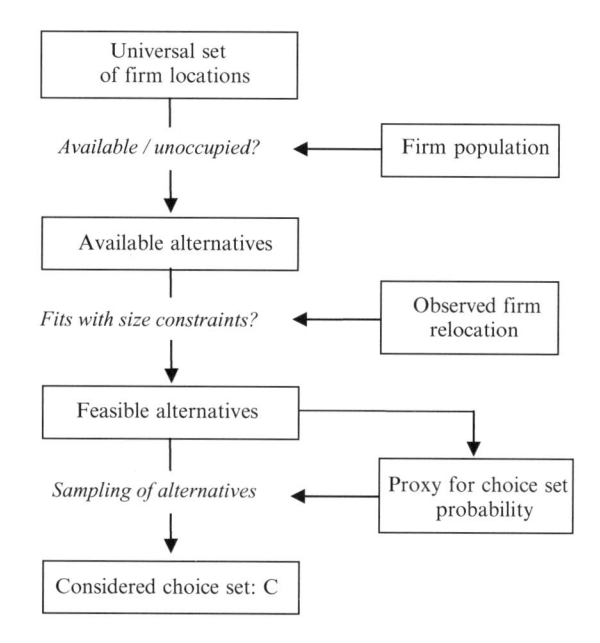

13.4 Estimation Results

Tables 13.1 and 13.2 present the estimated location choice models for the seven industry sectors for which we estimated models with dominance parameters. For each sector, two models are presented: the standard specification of the SFM model ('SFM') and the model including the dominance parameter ('SFM-D'). Since the models are used for simulation, the insignificant parameters were removed from the models.

The improved final log-likelihood of the models, with parameters for probabilistic choice sets, demonstrate that the model fit of location choice models is improved by the application of dominance.

Without exception the θ parameter for spatial dominance is significant and has a negative value. This means the less dominant an alternative is (high rank value) the lower the probability is part of the choice set. To translate this to the perspective of the decision maker, from the available alternatives, it is more likely that the decision maker (the firm) takes alternatives closer by into consideration. This is interpreted as an indication that alternatives that are more close to the original location are more likely to be considered and thus being part of the choice set. This is in line with the intervening opportunities concept formulated by Cascetta et al. (2007).

The spatial dominance attribute is derived from the distance to the original location, which is also an attribute in the deterministic utility. However, both measures represent a different aspect of the choice process, and judging by the stable signs for the estimated parameters the estimations are not affected by multicollinearity.

Table 13.1 Estimation of location choice models for manufacturing, construction and logistics

Sector	Manufacturing		Construction		Logistics	
Title	SFM	SFM-D	SFM	SFM-D	SFM	SFM-D
Observations	754	754	1,032	1,032	897	897
Final log (L)	−1352.6	−1333.5	−1673.6	−1664.5	−1749.5	−1712.9
D.O.F.	5	6	5	6	5	6
$Rho^2(c)$	0.401	0.41	0.459	0.462	0.349	0.363
Utility attributes:						
Distance to original loc.[$km^{1/2}$]	−1.687 (−29.3)	−0.9744 (−7.4)	−1.914 (−34.3)	−1.442 (−10.7)	−1.620 (−28.7)	−0.7699 (−7.3)
Infrastructure proximity						
α-location; near trainstation [−]						
β-location; near trainstation & HW onramp [−]			0.2508 (1.5)	0.2737 (1.6)	0.3114 (1.9)	0.3251 (1.8)
γ-location; near highway onramp [−]	0.2304 (2.4)	0.2452 (2.5)	0.2542 (3.0)	0.2587 (3.0)	0.2835 (3.5)	0.3262 (3.8)
Urbanisation economies						
Logsum business and commuting trips [−]						
Diversity attributes						
Diversity Rb < 7, 5 min. [−]	−0.5517 (−2.2)	−0.5060 (−1.9)				
Specialisation attributes						
Specialisation Rb < 7, 5 min. [−]	0.3779 (4.7)	0.3805 (4.7)	0.3928 (3.9)	0.3750 (3.7)	0.3500 (10.9)	0.3433 (10.1)
Parameters probabilistic choice sets:						
Theta; centrality [−]	−0.7575 (−2.7)	−0.6524 (−2.5)	−1.187 (−4.5)	−1.096 (−4.4)	−1.682 (−7.6)	−1.493 (−7.5)
Spatial dominance [−]		−0.4110 (−5.7)		−0.2559 (−3.8)		−0.4957 (−8.3)

13 Choice Set Formation in Microscopic Firm Location Models

Table 13.2 Estimation of location choice models for financial-, business services, government and general services

Sector	Financial services		Business services		Government		General Services	
Title	SFM	SFM-D	SFM	SFM-D	SFM	SFM-D	SFM	SFM-D
Observations	428	428	1992	1992	185	185	441	441
Final log (L)	−667.4	−647	−3583.2	−3509.8	−351.5	−349.3	−693	−675.6
D.O.F.	8	9	7	8	3	4	7	8
Rho^2(c)	0.479	0.495	0.4	0.412	0.366	0.37	0.475	0.489
Utility attributes:								
Distance to original loc. [km$^{1/2}$]	−2.261 (−20.5)	−1.152 (−7.1)	−1.886 (−44.0)	−1.026 (−13.7)	−1.899 (−13.1)	−1.385 (−4.8)	−2.229 (−22.6)	−1.204 (−7.1)
Infrastructure proximity								
α-location; near trainstation [−]	0.6198 (2.0)	0.6726 (1.9)	0.3758 (3.7)	0.3696 (3.5)			0.7096 (2.0)	0.7347 (2.0)
β-location; near trainstation & HW onramp [−]	0.6113 (2.6)	0.7113 (2.8)	0.1392 (2.2)	0.1401 (2.1)	0.7813 (2.8)	0.7951 (2.7)	0.6096 (2.4)	0.6008 (2.3)
γ-location; near highway onramp [−]	0.4074 (2.8)	0.4644 (3.0)	0.1773 (5.1)	0.1758 (5.0)	0.3603 (1.8)	0.3615 (1.8)	0.4393 (3.0)	0.4153 (2.8)
Urbanisation economies								
Logsum business and commuting trips [−]	0.2973 (3.6)	0.2964 (3.5)					0.2561 (3.0)	0.2449 (2.8)
Diversity attributes								
Diversity Rb < 7, 5 min. [−]	0.9742 (−2.5)	−0.8537 (−2.0)	−0.2936 (−1.8)	−0.2543 (−1.5)				
Specialisation attributes								
Specialisation Rb < 7, 5 min. [−]	0.3701 (4.1)	0.3395 (3.6)	0.5512 (8.2)	0.5880 (8.3)			−0.4815 (−3.3)	−0.4320 (−2.8)
Parameters probabilistic choice sets:								
Theta; centrality [−]	−2.119 (−5.7)	−1.995 (−5.8)	−1.265 (8.6)	−1.231 (−8.7)			−0.8816 (−2.7)	−0.7349 (−2.4)
Spatial dominance [−]		−0.5245 (−7.2)		−0.4517 (−12.3)		−0.2597 (−2.1)		−0.4885 (−5.9)

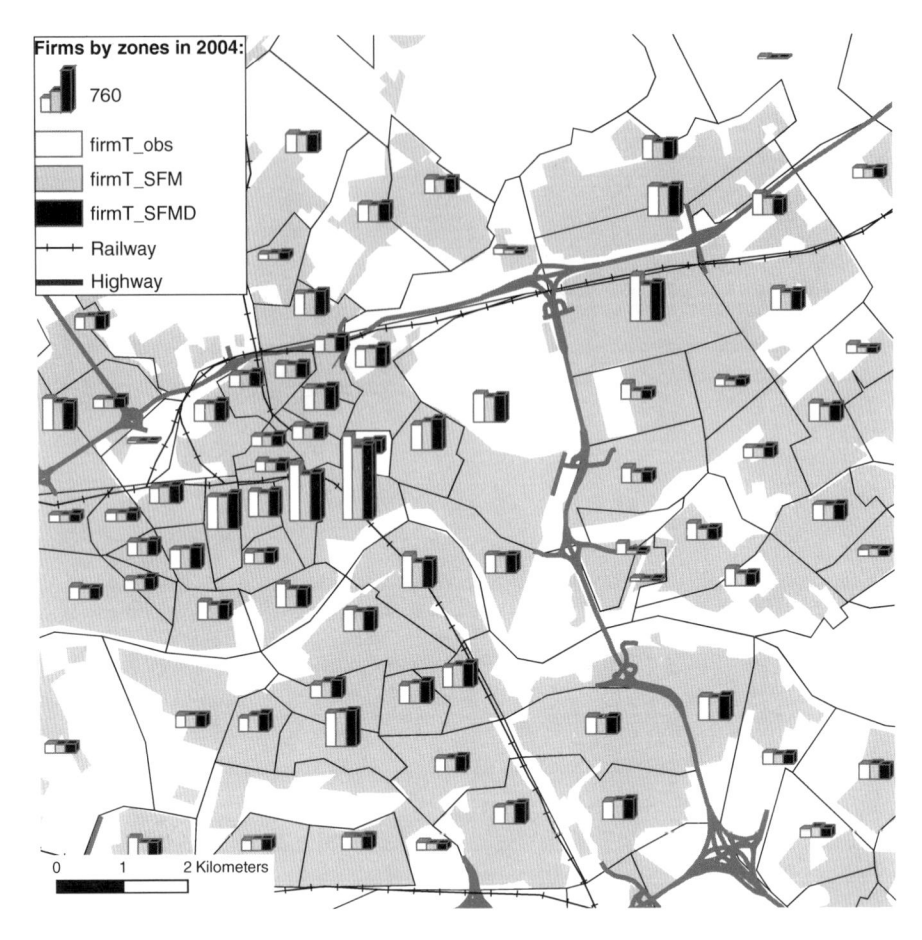

Fig. 13.4 Observed and simulated firms by zone for fragment of study area. Average over 10 replications with base model (SFM) and SFM model with dominance (SFMD)

13.5 Simulation Results

The estimated parameters are implemented in the SFM model to test if dominance can help to improve the simulation results. First of all, the evaluation of micro simulation models is not trivial. A simulation model should be validated relative to those measures of performance that will actually be used for decision making (Law and Kelton 1991). The scope of the presented micro simulation model is to make predictions of the transport demand and the demand for industrial real estate in a particular neighbourhood under different policy assumptions. As a consequence, the relevant level of aggregation for validation of the SFM-model seems to be an intermediate level. Therefore, to evaluate the goodness-of-fit of the simulations, the micro results from the SFM-zones are aggregated into zones.

Total number of firms by zone

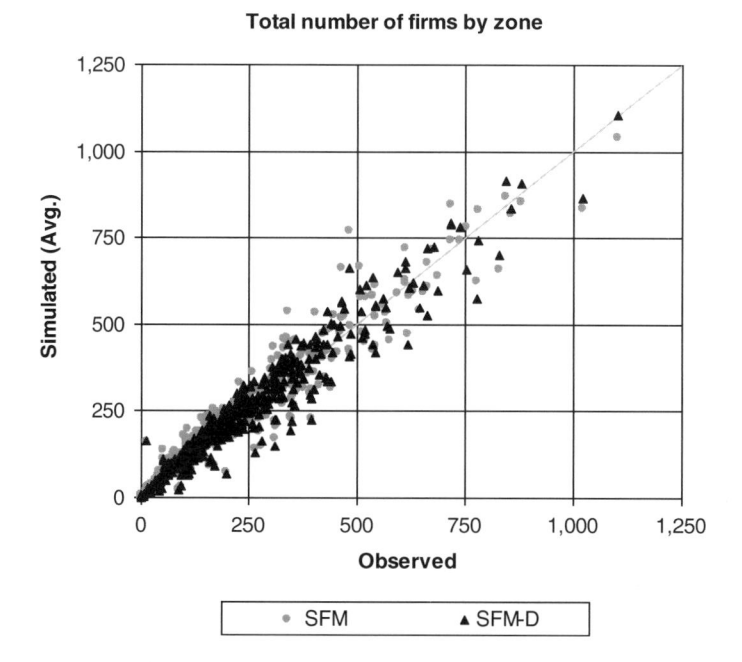

Fig. 13.5 Scatterplot of simulation results SFM model and SFM-D model

Figure 13.4 shows the total number of firms per zone over all industry sectors for a fragment of the study area (centre and eastern part of Rotterdam; one of the largest city in the study area). The bars show the observed number of firms and simulated averages over 10 replications with base model (SFM) and SFM model with dominance (SFMD). The grey and black bars show small differences resulting from the dominance in the choice sets. In some zones, the changes bring the simulation results (black bars) closer to the observed firm population (white bars), in other cases however the results from the simulations with dominance are less close to the observed population. This figure illustrates the results at zone level.

To illustrate the goodness of fit of the different model replications across the total study area, the simulated and observed number of firms is plotted in Fig. 13.5. Overall, the goodness of fit of both specifications seems quite similar; although there seem to be a few more 'outliers' in the normal SFM specification (grey dots that are further away from the diagonal) compared to the SFM-D specification (black triangles). This might be an indication that the models without dominance lead to less accurate predictions, which would result from the misspecification of the choice sets. From the figure only it is hard to evaluate if the SFM-D approach improves the simulation results.

For a more formal test of the goodness of fit of the models, the correlation between the observed and simulated average of the number of firms is presented in Table 13.3. Tests show improvement of goodness-of-fit (higher correlation) in all but one sectors. The largest increase was found for logistics (from .840 to .887). The

Table 13.3 Simulation results SFM model and model with dominance

	SFM		SFM-D	
	Correlation simulated means[a]	% within confidence interval[b]	Correlation simulated means[a]	% within confidence interval[b]
Manufacturing	.916[b]	83	.932[b]	82
Construction	.857[b]	71	.856[b]	73
Logistics	.840[b]	78	.887[b]	81
Finance	.873[b]	85	.877[b]	86
Businessservices	.878[b]	51	.878[b]	50
Government	.900[b]	79	.906[b]	81
Generalservice	.946[b]	86	.944[b]	84
N (number of zones)	433		433	

[a]The Pearson correlations is computed of observed and simulated zone averages
[b]The observed zone employment is compared to the simulated 90 % confidence intervals (CI)

improvement of the goodness of fit in these models is interpreted as (small) evidence that the choice behaviour of firms is improved by the inclusion of dominance in the choice set formation. In the tested specification, dominance means that alternatives that are closer to the existing firm location, are more likely to be known by the firm and more likely to be in the choice set of the firm. However, the improvement is rather small.

13.6 Conclusions

This chapter deals with the issue of firm location decisions in disaggregate urban simulation models. This choice context is characterized by large number of alternatives and complex spatial interdependencies among them, introducing challenges for the choice set composition. A location choice model is presented with probabilistic choice sets assuming that choice alternatives that are dominated by others are not taken into consideration in the location decision.

The estimations of the location choice models, with dominance parameter for choice set composition, have shown that the dominance attributes can improve the estimations of the choice sets. Estimations have demonstrated that the dominance attribute should be specified with attention, but the spatial dominance ('is an alternative close by relative to the other alternatives in the choice set') proved to be a good predictor for choice set membership across all tested industry sectors. This specification was implemented in the microscopic SFM model to test if the dominance attributes could lead to better predictions of firm location. The calculated correlations for the implemented approaches show that dominance can improve the simulation results, however to a small extent. In the tested specification dominance means that alternatives that are closer to the existing firm location are more dominant and more likely to be in the choice set of the firm.

The simulation results also show that the analysis of micro simulation models is not trivial due to the stochastic outcomes. This issue is also discussed by Moeckel in chapter (11). The stochastic choice processes, that are present in many micro simulation models, make it necessary to evaluate series of model replications and thus a distribution of outcomes. Statistics from these distributions can be used for evaluation (averages, standard deviations) but it is less trivial compared to models with deterministic outcomes. In this study, confidence intervals were used as a formal test to compare simulation results to observations. Furthermore it is emphasized that results of micro simulation are analysed at the aggregation level at which the simulation is expected to provide forecasts.

The absence of price in the location choice models can be viewed as a limitation of the presented analysis. Reliable and conclusive data on office or industrial real estate locations was not available for the study area so this could not be included in the models. As a consequence, the estimated models do not include trading behaviour between accessibility and price. As a consequence, there is no price adjustment mechanism if the supply of offices or other real estate changes. The real estate market is now simulated in a simple procedure where relocation events are simulated sequentially, firm by firm. For the supply of real estate this means that the principle of 'first-come-first-serve' applies. If prices could be included in the estimated models, it would be a sensible extension to include a price mechanism that simulates price adjustments on the real estate market reacting to changes in the supply/demand ratios.

References

Adler TJ, Ben-Akiva M (1976) Joint-choice model for frequency, destination and travel mode for shopping trips. Trans Res Rec 569:136–150

Anderstig C, Mattsson LG (1991) An integrated model of residential and employment location in a metropolitan region. Pap Reg Sci 70:167–184

Bierlaire M, Bolduc D, McFadden D (2006) The estimation of generalized extreme value models from choice-based examples. Report TRANSP-OR 060810, Transport and Mobility. Laboratory, School of Architecture, Civil and Environmental Engineering, Ecole Polytechnique Federale de Lausanne

Birch D (1979) The job generating process. Cambridge University Press, Cambridge

Carroll GR, Hannan MT (2000) The demography of corporations and industries. Princeton University Press, Princeton

Cascetta E, Papola A (2001) Random utility models with implicit availability/perception of choice alternatives for the simulation of travel demand. Transport Res C 9:249–263

Cascetta E, Papola A (2009) Dominance among alternatives in random utility models. Trans Res A Policy Pract 43:170–179

Cascetta E, Pagliara F, Axhausen KW (2007) The use of dominance variables in spatial choice set generation. In: Proceedings of the 11th World Conference on Transport Research, Berkeley, 24–28 June 2007

De Bok M (2009) Estimation and validation of a microscopic model for spatial economic effects of transport infrastructure. Trans Res A Policy Pract 43:44–59

De Bok M, Pagliara F (2011) How to obtain representative spatial choice sets? Dominance and centrality analysed for firm location choices. Paper presented at the CUPUM conference, Lake Alberta, 5–8 July 2011

De Bok M, van Oort F (2011) Agglomeration economies, accessibility and the spatial choice behaviour of relocating firms. J Trans Landuse 4(1):5–24

Elgar I (2011) Modelling office mobility and location decisions in a microsimulation environment. PhD Dissertation, University of Toronto (Canada), Canada. Retrieved 21 Feb2011, from Dissertations & Theses @ University of Toronto (Publication No. AAT NR39401)

Fortheringham AS (1983) A new set of spatial-interaction models: the theory of competing destinations. Environ Plann A 15:15–36

Gautschi DA (1981) Specification of patronage models for retail center choice. J Mark Res 18:162–174

Golledge RG, Timmermans H (1990) Application of behavioural research on spatial problems I: cognition. Prog Hum Geogr 14:57–99

Hägerstrand T (1970) What about people in regional science? Pap Reg Sci Assoc 24:7–21

Hague Consulting Group (2000) Het Landelijk Model Systeem versie 7.0 (in Dutch). Hague Consulting Group, The Hague

Hansen ER (1987) Industrial location choice in São Paulo, Brazil: a nested logit model. Reg Sci Urban Econ 17:89–108

Kim H, Waddell P, Shankar VN, Ulfarsson GF (2008) Modelling micro-spatial employment location patterns: a comparison of count and choice approaches. Geogr Anal 40:123–151

Landau U, Prashker JN, Alpern B (1982) Evaluation of activity constrained choice sets to shopping destination choice modelling. Trans Res A Policy Pract 16(3):199–207

Law AM, Kelton WD (1991) Simulation modelling and analysis. McGraw-Hill, New York

Manski C (1977) The structure of random utility models. Theory Decis 8:229–254

Maoh M, Kanaroglou P (2007) Business establishment mobility behaviour in urban areas: a microanalytical model for the City of Hamilton in Ontario, Canada. J Geogr Syst 9(3):229–252

Maoh H and Kanaroglou P (2012) Modelling firm failure: towards building a firmo-graphic microsimulation model. In: Pagliara F, de Bok M, Simmonds D and Wilson A (eds) Employment location in cities and regions – Models and applications. Advances in spatial sciences. Springer, Berlin, pp 243–262

Miller EJ, O'Kelly ME (1983) Estimating shopping destination models from travel diary data. Prof Geogr 35:440–449

Moeckel R (2007) Business location decisions and urban sprawl: a microsimulation of business relocation and firmography, vol 126, Blaue Reihe. Institut fur Raumplanung, Dortmund

Moekel R (2012) Firm location choice vs. job location choice in microscopic simulation models. In: Pagliara F, de Bok M, Simmonds D and Wilson A (eds) Employment location in cities and regions – Models and applications. Advances in spatial sciences. Springer, Berlin, pp 223–242

Paci R, Usai S (1999) Externalities, knowledge spillovers and the spatial distribution of innovation. GeoJournal 49:381–390

Pagliara F, Timmermans H (2009) Choice set generation in spatial contexts: a review. Trans Lett Int J Trans Res 1:181–196

Pelligrini PA, Fortheringham AS (2002) Modeling spatial choice: a review and synthesis in a migration context. Prog Hum Geogr 26:487–510

Rietveld P (1994) Spatial economic impacts of transport infrastructure supply. Transp Res Part A: Policy Pract 28:329–341

Scott DM (2006) Constrained destination choice set generation: a comparison of GIS-based approaches. In: Proceedings of the 85th annual meeting on Transportation Research Board, Washington, DC

Shukla V, Waddell P (1991) Firm location and land use in discrete urban space: a study of the spatial structure of Dallas-Fort worth. Reg Sci Urban Econ 21:225–253

Southworth F (1981) Calibration of multinomial logit models of mode and destination choice. Transport Res A 15:315–325

Thill JC (1992) Choice set formation for destination choice modeling. Prog Hum Geogr 16:361–382
Thill JC, Horowitz JL (1997) Travel time constraints on destination choice sets. Geogr Anal 29 (2):108–123
Van Oort FG (2004) Urban growth and innovation; spatially bounded externalities in the Netherlands. Ashgate, Aldershot
Van Wissen LJG (2000) A micro-simulation model of firms: applications of concepts of the demography of the firm. Pap Reg Sci 79:111–134
Waddell P, Ulfarsson GF (2003) Accessibility and agglomeration: discrete-choice models of employment location by industry sector. In: Conference proceedings of the 82nd annual meeting of the Transportation Research Board, Washington, DC
Weisbrod G, Parcells RJ, Kern C (1984) A disaggregate model for predicting shopping area market attraction. J Retail 60(1):65–83

Chapter 14
Employment Location Models: Conclusions

D. Simmonds and Michiel de Bok

Abstract The development of the theories underlying applied employment location models for urban simulation models is far from linear. This book presents an overview of contemporary studies, including continuing refinements to existing model packages, and new developments. We discuss the main drivers behind employment location, a generalization of approaches and relevant issues for the future development of these models.

14.1 Introduction

The focus of this book is the modelling of the location of economic activities, measured in terms of employment, in land-use and transportation systems. These measures are both important in themselves, and key inputs to transport models which represent the flows of persons and goods at intra-urban levels. Economic activities can be defined in terms of jobs or private-sector firms and public service organisations. Different levels of aggregation are used both in terms of organisational and geographical dimensions. In the case of firms and public organizations, a distinction can be made between the organizations themselves and corresponding establishments. For urban simulation models it is the location of establishments that is important. Comprehensive models are usually at coarser levels of aggregation and aggregate firms and organizations into employment sectors. Microscopic

---

D. Simmonds
David Simmonds Consultancy Ltd., Cambridge, UK

Honorary Professor, Heriot-Watt University, Edinburgh, Scotland
e-mail: david.simmonds@davidsimmonds.com

M. de Bok (✉)
Significance, The Hague, Zuid-Holland, Netherlands
e-mail: debok@significance.nl

approaches apply establishments as unit of analysis with individual characteristics such as age, size and sector of the establishment.

Over the past four decades, numerous operational modelling systems have been developed or and are applied in many urban regions in the US and Europe. However, the development of the theories underlying these models was far from linear. In the 1970s, Lee (1973) identified the seven sins of large scale models. In short: the models were accused of being too data hungry, non-transparent and in need of costly computer hardware. Despite the last argument faded out with advances in computer technology, many difficulties remained for developing large scale urban models (Timmermans 2003) and consequently integrated urban models have been on and off the research agenda over the years. The increase in computing power, data availability, and simulation and visualization tools did help to create more sophisticated models, and with increased transparency (Anderstig and Mattsson 1998). Thus, in the 1990s en early 2000s many urban simulation models became operational and were applied into policy studies. Some of these models are discussed in this book: IRPUD (Wegener 2012), MEPLAN (Jin and Echenique 2012), CREIM (Deal et al. 2012), DELTA (Simmonds and Feldman 2012), TIGRIS XL (De Graaff and Zondag 2012), rAps (Anderstig and Sundberg 2012).

This book has brought together a selection of studies on employment location models, including continuing refinements to existing model packages, and new developments. In this discussion we will highlight some issues that we find relevant and important in the development of employment location modelling. The purpose is not to give an overview of each individual contribution, but to try to make some general observations. For this it is necessary to make generalisations, which is always subject to dispute due to differences in interpretation. Therefore we emphasise that we do not strive to make rigid conclusions about the suitability or applicability of approaches, neither will we prescribe what is right or wrong. What we try to do is first to identify the main drivers behind employment location. Next we sketch a generalisation of the approaches that are applied in the employment location models that were discussed in these chapters. Finally we will highlight and discuss a few of the issues that we find relevant for the future evolution of employment location models.

14.2 What Are the Drivers Behind Employment Location?

The objective of the models presented in this book is to predict employment location. The techniques that are used or the geographic scale at which employment location is modelled vary, but a number of factors can be identified that influence employment location. The main, conventionally recognized ones are

- Demand (and selling prices)
- Intermediate inputs (and input costs)
- Capital (and capital costs)
- Labour (and labour costs)
- Space (and rents).

Capital is probably the factor most often omitted, because it is often assumed to be uniformly available to any one sector across the modelled city or region, and because the macroeconomic aspects (e.g. interest rates) are often built into a controlling scenario for the rates of overall economic growth. Beyond that, whilst the variables listed above are nearly always represented, the way they are represented and their exact functions in the model vary considerably. For example, in MEPLAN (Jin and Echenique 2012), "demand" is very much the dominant driver of the model: suppliers (firms) are hardly represented at all; they are mostly an implicit sets of actors which simply aggregate the costs of factors and inputs and pass them on to intermediate or final consumers, whose price-based choices between the zones from which they can obtain inputs then determine the distribution of production and of employment. In contrast, the more process oriented models such as DELTA (Simmonds and Feldman 2012) represent firms as "actors" in the process (albeit modelled in terms of the employment located, not in terms of firms or establishments per se), and whose choices are influenced by the costs of inputs and by accessibility to (current) demand. In the MEPLAN-like approaches, therefore, certain supplier choices are assumed always to be in equilibrium with the demand for the product or service in question, whereas in the DELTA-like approaches the supplier choices are not necessarily in equilibrium with demand, and some separate process of reconciliation is eventually required (later in the model sequence for one period of time, or later altogether) to ensure that sales by suppliers (and importers) are consistent with purchases by intermediate and final consumers. The questions of which choices are modelled as explicit "decisions" by actors, or categories of actors, and which are left as largely implicit "responses" is of course the subject of endless(and non-convergent) discussion between modellers.

Space is a conditional factor for employment location: sites or floorspace need to be available for employment to emerge. Existing models apply different methodologies to include available space in the model. On an increasing scale of complexity the following approaches can be distinguished: regions or zones have a stock of land available for employment location, availability is controlled by exogenous plans, or availability is modelled in a module that simulates developer's choices. The applicability of each approach depends on the scope and detail of the simulation model.

14.3 Overview of Employment Location Models

The drivers behind employment location, as listed above, are nearly always represented in the employment location models presented in the book. The differences between the models exist in the economic approaches behind them, and each model in the book is built on one, or a combination of some, of the economic approaches discussed in the Introduction (Wilson and Pagliara 2012):

- Input–output models;
- CGE (computable general equilibrium) models;
- Utility and/or profit maximising logit models;

- Microsimulation models;
- Cellular and/or agent-based models.

The employment location models presented in the book are all explicitly spatial but they vary in geographic dimension. Some models predict employment location at regional level while others contributions predict employment location at the intra urban level: at zones or micro level (parcel-, building block or grid level).

Table 14.1 presents an inventory of which techniques have been used in each chapter. The geographic detail (regional, zonal and micro models) is used as a dimension in the matrix on a scale from wider to finer geographic detail. Many systems are multi-level and apply different modules to simulate causalities of employment location at different scale levels. Sometimes, the higher level employment totals are an exogenous input, e.g. for a regional model this can be a national forecast.

Zonal models are without exception conditional on supply. This implies that the employment location probabilities take into account the availability of industrial locations (real estate or land). The need for reliable data or estimates of supply of locations for employment (or firms) is even more obvious for micro simulation models. This is also one of the challenges for micro simulation models: the availability of firm locations can be very different locally and show a large variability.

The overview illustrates the modular structure of most modelling systems, and which different approaches and coexist to work together to predict employment location. Anderstig and Sundberg (2012) illustrate how aggregate forecasts from a theoretically strong CGE model is combined with a input/output model for the sectoral and geographic disaggregation. This also fits in a trend in which models seem to become more comprehensive and interfaces are created between models for different domains (employment location, transport, land use). In Chap. 6, Deal et al. (2012) give an example of how an aggregate input/output employment location model, a transport model and a microscopic cellular land use model can be combined into an integrated system. In doing so, local externalities from traffic (congestion, noise, pollution) can be included in the simulation of land change, increasing the validity of the simulation results.

The trend of increasingly available research data also led to growing number of empirical studies into the development of micro simulation and agent based models. In an earlier overview of employment location models Wilson (2000) identified microsimulation as an important methodological development in this field. Many projects delivered useful empirical results, and some of these are presented in this book (Moeckel 2012; Maoh and Kanaroglou 2012 and De Bok and Pagliara 2012). In spite of the increased computing power and data availability, these models are not yet operational at a broader scale. Maybe some of the classical arguments, data hungriness and complexity, still hold against these models?

Suitability of a typical approach is largely dependent on the geographic level at which the causality behind employment location is modelled. In building and designing employment location models, many none trivial choices have to be made,

14 Employment Location Models: Conclusions

Table 14.1 Indicative overview of the geographic dimension of presented employment location models

Model (authors)	Geographic dimension: Regional	Zonal	Micro level
Chapter 2: IRPUD (Wegener)	Exogenous	Logit models	
Chapter 3: DELTA (Simmonds and Feldman)	Input/output model + logit models	Logit models	
Chapter 4: TIGRIS XL (de Graaff and Zondag)	Spatial interaction (spatial equilibrium)	Allocation models (rule based)	
Chapter 5: (Smith, Batty, Vargas-Ruiz)	Input/output model	Regression model	
Chapter 6: CREIM and LEAM (Deal, Kim and Hewings)	Input/output model (CREIM)		Cellular, agent based model (LEAM)
Chapter 7: MEPLAN (Jin and Echenique)	Recursive spatial equilibrium	Spatial social accounting matrix, logit models, floorspace clearing models	
Chapter 8: STRAGO/rAps (Anderstig and Sundberg)	CGE model	Input/output model	
Chapter 9: (Kim and Hewings)		Cellular land use spatial interaction	
Chapter 10: (Lambert and Birkin)		Agent based micro-simulation	
Chapter 11: ILUMASS (Moeckel)	Exogenous		Microsimulation, logit models
Chapter 12: Microsimulation SME's (Maoh and Kanaroglou)	Exogenous		Microsimulation, logit models
Chapter 13: SFM (de Bok and Pagliara)	Exogenous		Microsimulation, logit models

for which no general answers can be given. Choices depend on the scope and research questions that need to be analysed, and on the availability of resources. For example, research question on land use change, and competition between different forms of land use, require detailed cellular automata models such as presented by Kim and Hewings (2012). However, comprehensive simulation models typically apply zonal land markets, for instance IRPUD (Wegener 2012), DELTA (Simmonds and Feldman 2012) or TIGRIS XL (De Graaff and Zondag 2012).

14.4 Remaining Challenges and Issues

The inventory of model systems presented in the book illustrate that comprehensive modelling systems are generally built up from a combination of different approaches. The inventory is far from complete, but what kind of conclusions can we draw and to what extend are the classical arguments against large scale models (data hungriness, complexity, computing power) still valid?

There seems to be consensus about what typical employment models can be applied at regional level: input–output analysis and spatial interaction are approaches that have been widely applied in many systems that are used in practice. Input/output analysis is applied in one form or the other in DELTA (Simmonds and Feldman 2012, Smith et al. 2012), MEPLAN (Jin and Echenique 2012), CREIM (Deal et al. 2012) and rAps (Anderstig and Sundberg 2012). Simulation of economic growth remains an ongoing issue. CGE approaches can be effective but most of the models presented in the book rely on scenario input for the size of the economy and, in effect, only simulate redistribution of economic activities. For many policy studies this is fair enough but one can still argue that transport infrastructure investments can have an effect on the size of the economic system as a whole. Inclusion of productivity changes (as an agglomeration effect) is identified as a relevant area for improvement (Simmonds and Feldman 2012). The focus of urban modelling has for a decade or so been moving from dealing purely with questions about the distribution of economic activities to questions of how the spatial and transport arrangements of cities and regions can promote/ achieve additional growth, but much work still remains on this topic.

An important distinctive assumption behind models is equilibrium or dynamic modelling. Equilibrium models in which prices adjust to find the equilibrium (and nearly everything else adjusts in response to price) have advantages in interpretation in that the price variables effectively summarise everything else that is going in the model, particularly in terms of the balance between supply and demand wherever this can arise. This is undoubtedly helpful in the communication of what the model is doing, but at the same time there is a risk that if the equilibrium is more than very partial the model may ignore constraints on change (or on the rates of change) which are essential parts of the planning problem to be addressed by policy. Dynamic models in which prices are not necessarily at equilibrium and are only adjusted in a non-equilibrium fashion at pre-determined intervals may on the other hand not capture sufficiently the role that prices play in the economy. The balance between dynamics and equilibrium remains a matter of controversy between modellers (Simmonds et al. 2011).

Microsimulation was identified as an important methodological development in this field in an earlier overview of employment location models by Wilson (2000). Over the past decade microsimulation received increased attention in academic studies which led to successful applications of microsimulation for employment and/or firm location. In demographic simulation, microsimulation is applied successfully as well: Lambert and Birkin (2012) present a dynamic microsimulation model of the urban population that is operational for every city in Great Britain. Many challenges exist such as runtimes, stochastic outcomes, and data hungriness (for calibration/estimation and disaggregate data on supply of floorspace and industrial sites). Unpractical long runtimes that come with many microscopic simulation models can be overcome with increased computing power. Deal et al. (2012) illustrate how runtimes can be reduced from days to minutes by applying parallel processing, in their case using up to 128 processors. Many microscopic simulation procedures are very suitable to be broken up for parallel processing, though the exceptions – such as procedures which involve different actors competing for available units of space – can be a major constraint on parallelisation.

In his overview, Wilson correctly pointed out that microsimulation requires intelligent scanning of the outcomes of microsimulation models (Wilson 2000). Some of the contributions in the book illustrate the challenges that are involved in analysing the stochastic outcomes of these models effectively. Outcomes of microsimulation models are typically stochastic because of the Monte Carlo simulation that is involved to translate choice probabilities to discrete outcomes for individual agents. Deal et al. (2012) illustrate how from a set of possible land use maps that result from the stochastic microsimulation of landuse cells, one map is selected with typical land use changes, which is likely to be representative but remains arbitrary. De Bok and Pagliara (2012) use standard statistic measures from a set of outcomes from a simulation with a firm location model, such as averages and standard deviations. These are used to compute confidence intervals to validate the outcomes to observed employment developments.

The amount of data used for calibration of microscopic models is vast, and often requires access to business registers. However, Maoh and Kanaroglou (2012) and de Bok and Pagliara (2012) show that detailed data can be obtained that allow estimation procedures to calibrate the behavioural models, improving the empirical foundation of the models. This shows that the data hungriness problem is dissolving slowly with the rapidly growing number of digital databases. However, the fine urban scale makes the calibration also more challenging because of the spatial interdependencies between the alternatives. In some cases, when data are scarce, synthetic base data is derived from stochastic simulation. However, this introduces an interesting question on how to translate these stochastic outcomes to a base situation. Strictly speaking, in such cases, the model ought to be run from multiple bases.

Microscopic simulation remains data hungry, and a typical data issue is the availability of disaggregate data on available floorspace and industrial sites. In practice this can be solved by generating synthetic databases on supply of floorspace and industrial sites, but how reliable are such data? The simulation of changes behind urban land use and development of the supply of industrial real estate is also an important issue. Much work has been done in the IRPUD model (Wegener 2012) in which a real estate development module transfers land capacity (with planning policy controls) to employment locations.

It seems fair to ask if the additional data requirements and complexity of results from microsimulation models outweigh the advantages. This probably varies from the research and the "practical" policy-making point of view, where results may need to be defended in public hearings. From an academic point of view this is not a valid question, but a research challenge. There is an argument for saying that stochastic models make excellent research tools, but that the results of those models should be used to calibrate deterministic models which can more practically be used to produce results to advise decision-makers.

An important issue for debate is the question whether to simulate firm or employment location. Most conventional models simulate employment location, but some recent studies, presented in this book (Moeckel 2012; Maoh and Kanaroglou 2012; De Bok and Pagliara 2012), simulate firm location. Can it be said which approach is preferred: simulation of jobs or firms? If we are trying to model the decisions made by actors, from a behavioural perspective it is fair to conclude we should ideally be simulating firms. The problem then arises that firms are extraordinarily diverse, just in size, apart from any other consideration. Even if we focus on establishments (employment units in a single location, which may be part of a larger organization) there is still a range from single-worker establishments up to those employing thousands, which is far greater a range than we find in households. (There is also within this the complication that a firm can permanently split itself between multiple establishments, whilst a household that permanently splits itself between different dwellings is regarded as having split up into two or more households). There are broadly two responses to the heterogeneity of firms or establishments: one is to say that we have to microsimulate because any practical categorization will contain an unreasonably wide range of units in each category, the other is to say that it is simply infeasible (and also perhaps that given the divisibility of employment establishments it doesn't so much matter) and to work with models of sector employment (and possibly sector output) representing the collective decisions of all the firms in that sector. The heterogeneity in firm sizes introduces a larger variety (random noise) in stochastic simulation as is shown by Moeckel (2012) in a comparison between a firm based- and employment based simulation.

Urban economic theories can provide some interesting theories that can be relevant for the further development of employment location models. Most existing models simulate a static interaction between industries: service industries follow people and other industries. However, recent urban economic theories, that show that spatial externalities differ across sectors, and the lifecycle of firms (Duranton and Puga 2001; Rosenthal and Strange 2004). Concepts like 'nursery' cities from recent urban economic studies seem relevant for employment (or firm) location models. And what makes the modern industry seek urban locations? The knowledge intensive industries is an increasingly important sector in spatial planning and concepts like the 'creative class' by Florida (2002) can be relevant for employment location models.

In the end, the existence of employment location models is based on a need for policy analysis tools for policy questions. Now do the models described in the book provide relevant policy information? Most advancements that are presented, deal with improving the validity of the models applied, or to increase the functionalities of the model to better represent mechanisms that affect future employment location. Advancements are made across different lines: by applying more advanced approaches and data, by building more comprehensive models that incorporate more urban markets within the modelling systems. As a result, the validity of models can be improved or the models more suitable to evaluate a broader range of policy measures and scenario assumptions in the model, illustrating the practical value of the existing models. Many of the comprehensive models have a track record of policy studies: IRPUD (Wegener 2012), DELTA (Simmonds and Feldman 2012), TIGRIS XL (De Graaff and Zondag 2012), MEPLAN (Jin and Echenique 2012). Some of the current models are actively used for project appraisal of infrastructure investments. The list of models described in this book is not exhaustive, but it is our expectation that the list of model applications that are actively used to inform in project appraisal will continue to grow.

References

Anderstig C, Mattsson L (1998) Modelling land use and transport interaction: policy analysis using the IMREL model. In: Lundqvist L, Mattsson LG, Kim TJ (eds) Network infrastructure and the urban environment, Advances in spatial sciences. Springer, Berlin, pp 308–328

Anderstig C, Sundberg M (2012) Integrating SCGE and I-O in multiregional modelling. In: Pagliara F, de Bok M, Simmonds D, Wilson A (eds) Employment location in cities and regions – models and applications, Advances in spatial sciences. Springer, Berlin, pp 159–180

De Bok M, Pagliara F (2012) Choice set formation in microscopic firm location models. In: Pagliara F, de Bok M, Simmonds D, Wilson A (eds) Employment location in cities and regions – models and applications, Advances in spatial sciences. Springer, Berlin, pp 263–282

De Graaff T, Zondag B (2012) A population-employment interaction model as labour module in TIGRISXL. In: Pagliara F, de Bok M, Simmonds D, Wilson A (eds) Employment location in cities and regions – models and applications, Advances in spatial sciences. Springer, Berlin, pp 57–78

Deal B, Kim JH, Hewings GJD (2012) Complex urban systems integration: the LEAM experiences in coupling economic, land use, and transportation models in Chicago, IL. In: Pagliara F, de Bok M, Simmonds D, Wilson A (eds) Employment location in cities and regions – models and applications, Advances in spatial sciences. Springer, Berlin, pp 107–132

Duranton G, Puga D (2001) Nursery cities: urban diversity, process innovation, and the life cycle of products. Am Econ Rev 91:1454–1477

Florida R (2002) The rise of the creative class. Basic Books, New York

Jin Y, Echenique M (2012) Employment location modelling within an integrated land use and transport framework: taking cue from policy perspectives. In: Pagliara F, de Bok M, Simmonds D, Wilson A (eds) Employment location in cities and regions – models and applications, Advances in spatial sciences. Springer, Berlin, pp 133–158

Kim JH, Hewings GJD (2012) Interjurisdictional competition and land development: a micro-level analysis. In: Pagliara F, de Bok M, Simmonds D, Wilson A (eds) Employment location in cities and regions – models and applications, Advances in spatial sciences. Springer, Berlin, pp 181–199

Lambert P, Birkin M (2012) Occupation, education and social inequalities: a case study linking survey data sources to an urban microsimulation analysis. In: Pagliara F, de Bok M, Simmonds D, Wilson A (eds) Employment location in cities and regions – models and applications, Advances in spatial sciences. Springer, Berlin, pp 203–222

Lee D (1973) Requiem for large scale urban models. J Am Inst Plann 39:163–178

Maoh H, Kanaroglou P (2012) Modelling firm failure: towards building a firmo-graphic microsimulation model. In: Pagliara F, de Bok M, Simmonds D, Wilson A (eds) Employment location in cities and regions – models and applications, Advances in spatial sciences. Springer, Berlin, pp 243–262

Moeckel R (2012) Firm location choice vs. job location choice in microscopic simulation models. In: Pagliara F, de Bok M, Simmonds D, Wilson A (eds) Employment location in cities and regions – models and applications, Advances in spatial sciences. Springer, Berlin, pp 223–242

Rosenthal SS, Strange WC (2004) Evidence on the nature and sources of agglomeration economies. In: Henderson JV, Thisse JF (eds) Handbook of regional and urban economics, vol 4. Elsevier, Amsterdam, pp 2119–2167

Simmonds D, Feldman O (2012) Modelling the economic impacts of transport changes: experience and issues. In: Pagliara F, de Bok M, Simmonds D, Wilson A (eds) Employment location in cities and regions – models and applications, Advances in spatial sciences. Springer, Berlin, pp 33–56

Simmonds D, Waddell P, Wegener M (2011) Beyond equilibrium: advances in urban modelling. Paper presented at the CUPUM conference, Lake Alberta, 5–8 July 2011

Smith DA, Vargas-Ruiz C, Batty M (2012) Simulating the spatial distribution of employment in large cities: with applications to greater London. In: Pagliara F, de Bok M, Simmonds D, Wilson A (eds) Employment location in cities and regions – models and applications, Advances in spatial sciences. Springer, Berlin, pp 79–106

Timmermans H (2003) The saga of integrated land use-transport modeling: how many more dreams before we wake up? Conference keynote paper at the 10th international conference on travel behaviour research, Lucerne, 10–15 Aug 2003

Wegener M (2012) Employment and labour in urban markets in the IRPUD model. In: Pagliara F, de Bok M, Simmonds D, Wilson A (eds) Employment location in cities and regions – models and applications, Advances in spatial sciences. Springer, Berlin, pp 11–31

Wilson A (2000) Complex spatial systems: the modelling foundations of urban and regional analysis. Prentice-Hall/Pearson Education, Harlow

Wilson A, Pagliara F (2012) Employment location models: an overview. In: Pagliara F, de Bok M, Simmonds D, Wilson A (eds) Employment location in cities and regions – models and applications, Advances in spatial sciences. Springer, Berlin, pp 1–8

About the Editors

Dr. Francesca Pagliara is an Assistant Professor at the Department of Transportation Engineering of the University of Naples Federico II, Italy.

Dr. Michiel de Bok is a Senior Researcher at Significance, an independent research institute specialised in transport in Den Haag, The Netherlands.

Dr. David Simmonds is a Director of David Simmonds Consultancy Ltd. in Cambridge, United Kingdom; Honorary Professor, Heriot-Watt University, Edinburgh, Scotland.

Sir Alan Wilson is Professor at the Centre for Advanced Spatial Analysis of University College London, United Kingdom.

Printed by Printforce, the Netherlands